# Fundamentals of Microwave Transmission Lines

# WILEY SERIES IN MICROWAVE AND OPTICAL ENGINEERING

**KAI CHANG**, Editor
*Texas A&M University*

FIBER-OPTIC COMMUNICATION SYSTEMS • *Govind P. Agrawal*

COHERENT OPTICAL COMMUNICATION SYSTEMS • *Silvello Betti, Giancarlo De Marchis* and *Eugenio Iannone*

HIGH-FREQUENCY ELECTROMAGNETIC TECHNIQUES: RECENT ADVANCES AND APPLICATIONS • *Asoke K. Bhattacharyya*

COMPUTATIONAL METHODS FOR ELECTROMAGNETICS AND MICROWAVES • *Richard C. Booton, Jr*

MICROWAVE RING CIRCUITS AND ANTENNAS • *Kai Chang*

MICROWAVE SOLID-STATE CIRCUITS AND APPLICATIONS • *Kai Chang*

DIODE LASERS AND PHOTONIC INTEGRATED CIRCUITS • *Larry Coldren* and *Scott Corzine*

MULTICONDUCTOR TRANSMISSION-LINE STRUCTURES: MODAL ANALYSIS TECHNIQUES • *J. A. Brandão Faria*

PHASED ARRAY-BASED SYSTEMS AND APPLICATIONS • *Nick Fourikis*

FUNDAMENTALS OF MICROWAVE TRANSMISSION LINES • *Jon C. Freeman*

MICROSTRIP CIRCUITS • *Fred Gardiol*

HIGH-SPEED VLSI INTERCONNECTIONS: MODELING, ANALYSIS, AND SIMULATION • *A. K. Goel*

HIGH-FREQUENCY ANALOG INTEGRATED CIRCUIT DESIGN • *Ravender Goyal (ed.)*

OPTICAL COMPUTING: AN INTRODUCTION • *M. A. Karim* and *A. S. S. Awwal*

MICROWAVE DEVICES, CIRCUITS AND THEIR INTERACTION • *Charles A. Lee* and *G. Conrad Dalman*

ANTENNAS FOR RADAR AND COMMUNICATIONS: A POLARIMETRIC APPROACH • *Harold Mott*

INTEGRATED ACTIVE ANTENNAS AND SPATIAL POWER COMBINING • *Julio Navarro* and *Kai Chang*

FREQUENCY CONTROL OF SEMICONDUCTOR LASERS • *Motoichi Ohtsu (ed.)*

SOLAR CELLS AND THEIR APPLICATIONS • *Larry D. Partain (ed.)*

ANALYSIS OF MULTICONDUCTOR TRANSMISSION LINES • *Clayton R. Paul*

INTRODUCTION TO ELECTROMAGNETIC COMPATIBILITY • *Clayton R. Paul*

INTRODUCTION TO HIGH-SPEED ELECTRONICS AND OPTOELECTRONICS • *Leonard M. Riaziat*

NEW FRONTIERS IN MEDICAL DEVICE TECHNOLOGY • *Arye Rosen* and *Harel Rosen (eds.)*

FREQUENCY SELECTIVE SURFACE AND GRID ARRAY • *T. K. Wu (ed.)*

OPTICAL SIGNAL PROCESSING, COMPUTING AND NEURAL NETWORKS • *Francis T. S. Yu* and *Suganda Jutamulia*

# Fundamentals of Microwave Transmission Lines

JON C. FREEMAN
*NASA Lewis Research Center*

A WILEY-INTERSCIENCE PUBLICATION
**JOHN WILEY & SONS, INC.**
NEW YORK / CHICHESTER / BRISBANE / TORONTO / SINGAPORE

TK
7876
F74
1996

This text is printed on acid-free paper.

Copyright © 1996 by John Wiley & Sons, Inc.

All rights reserved. Published simultaneously in Canada.

Reproduction or translation of any part of this work beyond that permitted by Section 107 or 108 of the 1976 United States Copyright Act without the permission of the copyright owner is unlawful. Requests for permission or further information should be addressed to the Permissions Department, John Wiley & Sons, Inc., 605 Third Avenue, New York, NY 10158-0012.

*Library of Congress Cataloging in Publication Data:*

Freeman, Jon C.
  Fundamentals of microwave transmission lines / Jon C. Freeman.
     p.   cm.—(Wiley series in microwave and optical engineering)
  "A Wiley-Interscience publication."
  ISBN 0-471-13002-8 (cloth : alk. paper)
  1. Microwave transmission lines.  2. Electric circuit analysis.
I. Title.  II. Series.
TK7876.F734  1995
621.381'31—dc20
0471 13002 8                                              95-18094
                                                              CIP

Printed in the United States of America

10 9 8 7 6 5 4 3 2

*To My Family*
*Viola, Deborah, Michele, Maxwell, Kara, Kristen*

# Contents

| | |
|---|---|
| Preface | xi |

## 1  Introduction to Distributed Circuits — 1

| | |
|---|---|
| 1.1 Review of Maxwell's Equations | 1 |
| 1.2 Fields and Circuits | 6 |
| 1.3 Transmission Line Basics | 10 |
| 1.4 Transverse Electromagnetic Waves | 17 |
| 1.5 Equations for $\mathscr{V}$ and $\mathscr{I}$ for Lossless Lines | 27 |
| 1.6 Distributed Circuit Analysis | 36 |
| 1.7 Lossless Case Solution | 41 |
| 1.8 Lumped Equivalent Limitations | 45 |
| Supplementary Examples | 47 |
| Problems | 49 |
| References | 51 |
| Bibliography | 51 |

## 2  The Mathematics of Traveling Waves — 52

| | |
|---|---|
| 2.1 Physical Interpretations of Solutions | 52 |
| 2.2 Basic Notation and Methods | 62 |
| 2.3 Wave Launching by Generator | 64 |
| 2.4 Analysis at the Load Plane | 67 |
| 2.5 Reflection at the Source Plane | 70 |
| 2.6 Bounce Diagram | 71 |
| 2.7 Capacitive and Inductive Loads | 79 |
| 2.8 Laplace Transforms | 84 |
| Supplementary Examples | 92 |
| Problems | 127 |
| References | 131 |
| Bibliography | 131 |

## 3  Coupled Lines — 132

3.1  Equations for Coupled Lines — 132
3.2  Weak Coupling Analysis — 135
3.3  Lumped Approach to Coupled Lines — 143
3.4  General Solution for Coupled Lines — 148
3.5  Physical Interpretation of Even and Odd Modes — 150
Supplementary Examples — 154
Problems — 161
References — 162
Bibliography — 162

## 4  Time Domain Topics — 164

4.1  Time Domain Reflectometry — 164
4.2  Digital Applications — 169
4.3  Graphical Methods — 175
Problems — 185
References — 186
Bibliography — 186

## 5  Sinusoidal Steady State — 187

5.1  Sinusoidal Excitation — 187
5.2  Complex Representation (Phasors) — 193
5.3  General Solution in the Phasor Domain — 196
5.4  Properties of the General Solution — 199
5.5  Power Flow Along the Line — 202
5.6  Low Loss Conditions and Attenuation Formulas — 209
Supplementary Examples — 213
Problems — 244
References — 246

## 6  The Smith Chart — 247

6.1  Motivation — 247
6.2  Basis for the Smith Chart — 248
6.3  Basic Chart Calculations — 254
6.4  Properties of the Smith Chart — 260
6.5  The Admittance Chart — 265
6.6  Combined $Z$–$Y$ Chart — 270
6.7  Slotted Lines — 273
6.8  Lossy Lines and the Radial Scales — 280
Supplementary Examples — 291
Problems — 300
References — 302
Bibliography — 303

| | | |
|---|---|---|
| **7** | **Single Frequency Matching** | **304** |
| | 7.1 Basics, Stubs | 304 |
| | 7.2 Double-Stub Tuner | 321 |
| | 7.3 Quarter Wave Transformers | 327 |
| | 7.4 The $L$-Match | 349 |
| | Supplementary Examples | 358 |
| | Problems | 369 |
| | References | 371 |
| | Bibliography | 371 |
| **Index** | | **375** |

# Preface

This book is a self-study text and reference source for the solutions of microwave transmission line problems. It is assumed the reader has been introduced to Maxwell's equations, plane waves, the Poynting vector, and basic traveling wave concepts. This knowledge could come from a first course in electromagnetics, a sophomore physics course, or their equivalents. The self-study format used, is to first provide descriptive material to explain the underlying physics, then present the mathematics to handle the phenomena, and finally expand upon and clarify the topic, with completely solved example problems. A total of 126 examples are included toward that end. The examples range from simple exercises to descriptive design procedures, and have been tested in the classroom. Since transmission lines are at the core of distributed circuit analysis and microwave circuit design, the depth of coverage will help the reader in all subsequent electromagnetic courses and on-the-job endeavors. It will appeal especially to the engineer who did not study extensively in the electromagnetics area, but requires a sound understanding due to applications needed on the job. The examples serve to present preliminary information, for example the distinction between voltage and potential, or to expand upon a given topic. The end of an example is signalled with a solid triangle ▲. At the ends of most chapters a collection of informative supplementary examples is provided.

The book is suitable for the standard guided wave and transmission lines course, and can serve as background material for short courses in microwave circuit design. The format will free-up the instructor's time, as the basics are self-taught, so more time is available to cover CAD packages and other topics.

Chapter 1 reviews Maxwell's equations, then treats complex phasor-vectors, power and energy in the fields, and the loss tangent for microwave materials. The relationship between electromagnetic fields, lumped circuits, and distributed circuits is covered, then the concepts of phase are reviewed. Traveling waves on open two wire lines are used to demonstrate guided wave behavior, and characteristic impedance is developed from stored energy. The properties of TEM waves are presented; then the fields and the distributed elements, L & C, for coaxial lines are developed. The voltage and current wave variables $\mathscr{V}$, $\mathscr{I}$ and the differential equations for lines are developed from Faraday's law and the continuity equation. The line equations are developed again using the differential segment (the more standard approach), then the important and subtle assumptions inherent to distributed circuit analysis become clear. The solutions to the wave equation for the lossless case are discussed, and the limitations on the lumped equivalent circuit are addressed.

Chapter 2 explains the mathematics of traveling waves starting from the 1-dimensional wave equation. The notations and provisos of transmission line analysis are presented, then signal launching, reflection at load and source treated, and bounce diagrams introduced. Loads consisting of single energy storage elements and resistors are explained, then the Laplace transform is given. Chapter 3 is devoted to coupled lines. The treatment is applicable to crosstalk in high speed digital circuits, as well as all pulse applications of lines. The weak coupling approximations and their solutions are developed. A lumped approach follows immediately, then the general solutions are given with the introduction of the even/odd mode formalism. The physical interpretations for the even/odd mode decompositions end the chapter. Chapter 4 considers time domain reflectometer concepts for pedagogical reasons (the instrument has been superceded by others). Next the switching properties of high speed digital circuits are treated. Graphical solution methods for nonlinear terminations, (gates and diodes) are thoroughly covered.

Chapter 5 treats the sinusoidal steady state response, which in the past was the core of the courses/texts on the subject. The physics and mathematics of standing waves are treated in depth, then the complex representation is introduced. The general solution, and its properties are developed in detail, then power flow and lossy lines end the chapter. Chapter 6 is concerned with the Smith Chart; its construction, details, and properties are developed in depth. Its use is clarified by listing the five basic steps that are used repeatedly in all transmission line calculations. The slotted line is presented (again obsolete) for pedagogical reasons, and finally lossy lines are treated. Chapter 7 covers the single and double stub tuner matching concepts both analytically and on the Smith Chart. Single and cascaded quarter wave transformers are presented, the depth here illustrates the reference capability of the book. The last section treats the L-match procedure.

The objective of the book is to strive for clarity and rapid learning of material essential for distributed circuits, as well as prepare one for subsequent courses in electromagnetics.

*Cleveland, Ohio*                                                                            JON C. FREEMAN

# Fundamentals of Microwave Transmission Lines

CHAPTER ONE

# Introduction to Distributed Circuits

## 1.1 REVIEW OF MAXWELL'S EQUATIONS

The reader is assumed to have completed a one- or two-semester course in electromagnetics at the junior year in a typical electrical engineering program. We start our investigation by writing the field equations in a most basic form:

$$\nabla \times \bar{\mathcal{E}} = -\frac{\partial}{\partial t}\bar{\mathcal{B}} \tag{1.1}$$

$$\nabla \times \bar{\mathcal{B}} = \mu_0 \left( \varepsilon_0 \frac{\partial}{\partial t}\bar{\mathcal{E}} + \bar{\mathcal{J}}_f + \nabla \times \bar{m} + \frac{\partial}{\partial t}\bar{\mathcal{P}} \right) \tag{1.2}$$

$$\nabla \cdot \bar{\mathcal{E}} = \frac{1}{\varepsilon_0}\left( \rho_f - \nabla \cdot \bar{\mathcal{P}} \right) \tag{1.3}$$

$$\nabla \cdot \bar{\mathcal{B}} = 0 \tag{1.4}$$

where $\bar{\mathcal{E}}(\bar{r}, t)$ and $\bar{\mathcal{B}}(\bar{r}, t)$ are the electric field intensity and magnetic flux density. $\bar{\mathcal{J}}_f(\bar{r}, t), \bar{m}(\bar{r}, t)$, and $\bar{\mathcal{P}}(\bar{r}, t)$ are the current density due to free charges, the magnetization, and the electric polarization, respectively. The density of free electric charge is $\rho_f(\mathbf{r}, t)$. It is a net amount quantity, the net positive less the net negative charge at every point. The constants $\varepsilon_0, \mu_0$ are the permittivity and permeability of free space. The free charge density and current density are related by the continuity equation:

$$\nabla \cdot \bar{\mathcal{J}}_f = -\frac{\partial}{\partial t}\rho_f \tag{1.5}$$

which is the expression for the conservation of matter. The Lorentz force equation is

$$\bar{\mathscr{F}}(\bar{r}, t) = q[\bar{\mathscr{E}}(\bar{r}, t) + \bar{v}(\bar{r}, t) \times \bar{\mathscr{B}}(\bar{r}, t)] \tag{1.6}$$

where $\bar{v}(\bar{r}, t)$ is the velocity field through which the charge $q$ is moving. In other words, the velocity of $q$ at $(\bar{r}, t)$ is just $\bar{v}(\bar{r}, t)$. The force equation relates the basic mechanical terms, force and velocity, to the force fields $\bar{\mathscr{E}}(\bar{r}, t)$ and $\bar{\mathscr{B}}(\bar{r}, t)$.

To compress the first four equations, one introduces the vector fields $\bar{\mathscr{D}}(\bar{r}, t)$ and $\bar{\mathscr{H}}(\bar{r}, t)$ by

$$\bar{\mathscr{D}} = \varepsilon_0 \bar{\mathscr{E}} + \bar{\mathscr{P}} \tag{1.7}$$

$$\bar{\mathscr{H}} = \frac{1}{\mu_0} \bar{\mathscr{B}} - \bar{m} \tag{1.8}$$

and after inserting these into Eqs. (1.2) and (1.3), we find

$$\nabla \times \bar{\mathscr{E}} = -\frac{\partial}{\partial t} \bar{\mathscr{B}} \tag{1.9}$$

$$\nabla \times \bar{\mathscr{H}} = \bar{\mathscr{J}}_f + \frac{\partial}{\partial t} \bar{\mathscr{D}} \tag{1.10}$$

$$\nabla \cdot \bar{\mathscr{D}} = \rho_f \tag{1.11}$$

$$\nabla \cdot \bar{\mathscr{B}} = 0 \tag{1.12}$$

which are the more familiar forms for Maxwell's equations. To obtain a solution, the first step is to determine relationships between the vectors that are imposed by the medium under study. These relationships are called the constitutive equations and are often determined from experiments. They may be expressed as

$$\bar{\mathscr{B}} = f_1(\bar{\mathscr{E}}, \bar{\mathscr{H}}) \tag{1.13}$$

$$\bar{\mathscr{D}} = f_2(\bar{\mathscr{E}}, \bar{\mathscr{H}}) \tag{1.14}$$

$$\bar{\mathscr{J}}_f = f_3(\bar{\mathscr{E}}, \bar{\mathscr{H}}) \tag{1.15}$$

where the functions $f_1$, $f_2$, and $f_3$ can be extremely complicated. Complicated situations occur, for example, in the $\bar{\mathscr{B}} - \bar{\mathscr{H}}$ loop of iron.

If we assume the medium is linear, isotropic, homogeneous, lossless (the ideal simple medium), then

$$\bar{\mathscr{D}} = \varepsilon_0 \bar{\mathscr{E}} + \bar{\mathscr{P}} = \varepsilon_0 \bar{\mathscr{E}} + \chi_e \bar{\mathscr{E}} = \varepsilon_0 \left(1 + \frac{\chi_e}{\varepsilon_0}\right) \bar{\mathscr{E}}$$

$$= \varepsilon_0 \varepsilon_r \bar{\mathscr{E}} \tag{1.16}$$

$$\bar{\mathscr{H}} = \frac{1}{\mu_0} \bar{\mathscr{B}} - \bar{m} = \frac{1}{\mu_0} \bar{\mathscr{B}} - \chi_m \bar{\mathscr{H}}$$

or
$$\bar{\mathscr{B}} = \mu_0 \mu_r \bar{\mathscr{H}} \tag{1.17}$$
and
$$\bar{\mathscr{J}}_f = \sigma \bar{\mathscr{E}} \tag{1.18}$$

where $\varepsilon_r$ and $\mu_r$ are the relative permittivity and permeability, and $\sigma$ is the conductivity. These forms assume the polarization is linearly proportional to $\bar{\mathscr{E}}(\bar{r}, t)$ and the magnetization is related linearly to $\bar{\mathscr{H}}(\bar{r}, t)$. The final one is just Ohm's law. When we drop the explicit $(\bar{r}, t)$ dependence, the most common forms for Maxwell's equations are

$$\nabla \times \bar{\mathscr{E}} = -\mu_0 \mu_r \frac{\partial}{\partial t} \bar{\mathscr{H}} \tag{1.19}$$

$$\nabla \times \bar{\mathscr{H}} = \bar{\mathscr{J}}_f + \varepsilon_0 \varepsilon_r \frac{\partial}{\partial t} \bar{\mathscr{E}} \tag{1.20}$$

$$\nabla \cdot \bar{\mathscr{E}} = \frac{\rho_f}{\varepsilon_0 \varepsilon_r} \tag{1.21}$$

$$\nabla \cdot \bar{\mathscr{B}} = 0 \tag{1.22}$$

where

$$\bar{\mathscr{D}} = \varepsilon_0 \varepsilon_r \bar{\mathscr{E}}, \qquad \bar{\mathscr{B}} = \mu_0 \mu_r \bar{\mathscr{H}} \tag{1.23}$$

$$\bar{\mathscr{J}}_f = \sigma \bar{\mathscr{E}} \tag{1.24}$$

The above set relates the five field vectors $\bar{\mathscr{E}}, \bar{\mathscr{D}}, \bar{\mathscr{H}}, \bar{\mathscr{B}}$, and $\bar{\mathscr{J}}_f$ and the scalar $\rho_f$. The coupled equations are said to relate the four fields $\bar{\mathscr{E}}, \bar{\mathscr{D}}, \bar{\mathscr{H}}$, and $\bar{\mathscr{B}}$ to their sources $\bar{\mathscr{J}}_f$ and $\rho_f$. In most cases, we assume the sources of the fields are the free charges and currents. The charges that form the dipoles in matter that give rise to $\bar{\mathscr{P}}$ are bound to the material and, as such, are not treated as sources. Similarly, the microscopic currents that constitute $\bar{m}$ are bound in the magnetic materials.

Depending on the particular class of problem being treated, the solution processes tend to fall into definite categories. For the case of statics, electric and magnetic effects can be treated as independent (decoupled), since electrostatics can be considered a study with fixed assemblies of charges. Magnetostatics is the study of cases where steady currents and stationary magnets are present. In electrostatics, the vectors in question are $\bar{\mathscr{E}}, \bar{\mathscr{D}}$, and $\bar{\mathscr{P}}$ that relate to charges $\rho_f$. It turns out (by the nature of its definition) that $\bar{\mathscr{D}}$ is related to the free charge density $\rho_f$ by $\nabla \cdot \bar{\mathscr{D}} = \rho_f$, whereas $\bar{\mathscr{P}}$ is related to the dipoles in the media. Then $\bar{\mathscr{E}} = 1/\varepsilon_0 (\bar{\mathscr{D}} - \bar{\mathscr{P}})$ is the net "force" field in the general situation. In magnetostatics, the fields are $\bar{\mathscr{J}}_f, \bar{\mathscr{H}}, \bar{\mathscr{B}}$, and $\bar{m}$. The magnetic field intensity $\bar{\mathscr{H}}$ is defined

**4** INTRODUCTION TO DISTRIBUTED CIRCUITS

such that it is developed by currents in wires; that is: $\nabla \times \bar{\mathscr{H}} = \bar{\mathscr{J}}_f$. The atomic currents in magnetic materials generate $\bar{\mathscr{m}}$. An arbitrary magnet has an $\bar{\mathscr{m}}$ field. When a wire is wrapped about it, the current in that wire causes an $\bar{\mathscr{H}}$ field, which alters the existing $\bar{\mathscr{m}}$ field, which reacts back on the source that provides the current that sets up $\bar{\mathscr{H}}$. The net 'force' field is then $\bar{\mathscr{B}} = \mu_0(\bar{\mathscr{H}} + \bar{\mathscr{m}})$, which is the sum of macroscopic and microscopic current effects. In many cases, we treat free space conditions, then $\bar{\mathscr{P}}$ and $\bar{\mathscr{m}}$ are absent, and only $\bar{\mathscr{E}}$, $\bar{\mathscr{D}}$, $\bar{\mathscr{B}}$, and $\bar{\mathscr{H}}$ remain. Then $\bar{\mathscr{D}} = \varepsilon_0 \bar{\mathscr{E}}$ and $\bar{\mathscr{B}} = \mu_0 \bar{\mathscr{H}}$.

This section ends with a summary of power and energy of the fields. Since these are nonlinear quantities (products of field vectors), it is easiest to consider the sinusoidal steady state. This leads to the complex phasor-vector representation for the fields. In general, the $\bar{\mathscr{E}}(x, y, z, t)$ field can be expressed as

$$\bar{\mathscr{E}}(x,y,z,t) = E_{x_0}\cos(\omega t + \varphi_x)\bar{\mathbf{a}}_x + E_{y_0}\cos(\omega t + \varphi_y)\bar{\mathbf{a}}_y + E_{z_0}\cos(\omega t + \varphi_z)\bar{\mathbf{a}}_z \quad (1.25)$$

which we write as

$$\bar{\mathscr{E}}(x,y,z,t) = E_{x_0} Re\{e^{j\omega t}e^{j\varphi_x}\}\bar{\mathbf{a}}_x + E_{y_0} Re\{e^{j\omega t}e^{j\varphi_y}\}\bar{\mathbf{a}}_y + E_{z_0} Re\{e^{j\omega t}e^{j\varphi_z}\}\bar{\mathbf{a}}_z \quad (1.26)$$

$$= Re\{[E_{x_0}e^{j\varphi_x}\bar{\mathbf{a}}_x + E_{y_0}e^{j\varphi_y}\bar{\mathbf{a}}_y + E_{z_0}e^{j\varphi_z}\bar{\mathbf{a}}_z]e^{j\omega t}\}$$

$$= Re\{[E_x\bar{\mathbf{a}}_x + E_y\bar{\mathbf{a}}_y + E_z\bar{\mathbf{a}}_z]e^{j\omega t}\}$$

$$= Re\{\bar{\mathbf{E}}(x,y,z)e^{j\omega t}\} \quad (1.27)$$

Here, $\bar{\mathbf{E}}(x, y, z)$ is a vector with components that are complex numbers. For example, $E_x = E_{x_0}e^{j\varphi_x} = E_{x_r} + jE_{x_i}$. Thus, we have the correspondence:

$$\bar{\mathscr{E}}(x,y,z,t) = Re\{\bar{\mathbf{E}}(x,y,z)e^{j\omega t}\}$$

which is a generalization of phasors from circuit theory. With this spatial phasor representation for the fields, Maxwell's equations are

$$\nabla \times \bar{\mathbf{E}} = -j\omega\bar{\mathbf{B}} \quad (1.28)$$

$$\nabla \times \bar{\mathbf{H}} = \bar{\mathbf{J}}_f + j\omega\bar{\mathbf{D}} \quad (1.29)$$

$$\nabla \cdot \bar{\mathbf{D}} = \rho_f \quad (1.30)$$

$$\nabla \cdot \bar{\mathbf{B}} = 0 \quad (1.31)$$

$$\nabla \cdot \bar{\mathbf{J}}_f = -j\omega\rho_f \quad (1.32)$$

where all quantities are complex functions of $x$, $y$, $z$. We use boldface type for complex phasor quantities. For situations where energy loss occurs due to heat dissipation in the medium, the equations become

$$\nabla \times \bar{\mathbf{E}}(\bar{\mathbf{r}}) = -j\omega(\mu' - j\mu'')\bar{\mathbf{H}}(\bar{\mathbf{r}}) \tag{1.33}$$

$$\nabla \times \bar{\mathbf{H}}(\bar{\mathbf{r}}) = \sigma\bar{\mathbf{E}}(\bar{\mathbf{r}}) + j\omega(\varepsilon' - j\varepsilon'')\bar{\mathbf{E}}(\bar{\mathbf{r}}) \tag{1.34}$$

$$\nabla \cdot \bar{\mathbf{D}}(\bar{\mathbf{r}}) = \rho_f(\bar{\mathbf{r}}) \tag{1.35}$$

$$\nabla \cdot \bar{\mathbf{B}}(\bar{\mathbf{r}}) = 0 \tag{1.36}$$

where $\mu''$ and $\varepsilon''$ account for magnetization and polarization loss. For the general case, the time-averaged stored energy in the fields is

$$W_e = \tfrac{1}{4} Re \left\{ \int_V \bar{\mathbf{E}} \cdot \bar{\mathbf{D}}^* \, d\text{vol} \right\} \tag{1.37}$$

$$W_m = \tfrac{1}{4} Re \left\{ \int_V \bar{\mathbf{B}} \cdot \bar{\mathbf{H}}^* \, d\text{vol} \right\} \tag{1.38}$$

the subscripts denote electric and magnetic portions. The asterisk denotes complex conjugate. The power flow is given by the Poynting vector, and its time-averaged value is

$$\langle \bar{\mathbf{S}} \rangle = \tfrac{1}{2} Re \{ \bar{\mathbf{E}} \times \bar{\mathbf{H}}^* \} \tag{1.39}$$

***Example 1.1***  We discuss the loss tangent of a material and show the normal approximation for it at microwave frequencies.

Start with the last form for

$$\nabla \times \bar{\mathbf{H}}(\bar{\mathbf{r}}) = [\sigma + j\omega(\varepsilon' - j\varepsilon'')] \bar{\mathbf{E}}(\bar{\mathbf{r}})$$

Define the right side as $j\omega\hat{\varepsilon}\bar{\mathbf{E}}(\bar{\mathbf{r}})$. Then some manipulation shows

$$\hat{\varepsilon} = \varepsilon' - j\varepsilon'' - j\frac{\sigma}{\omega}$$

which is the "effective" dielectric coefficient. Define the loss tangent by

$$\tan \delta = \frac{\sigma/\omega + \varepsilon''}{\varepsilon'} = \frac{\sigma + \omega\varepsilon''}{\omega\varepsilon'}$$

and for many of the materials used in microwave circuits, we have $\sigma \ll \omega\varepsilon''$, and the above reduces to the standard approximation:

$$\tan \delta \doteq \frac{\varepsilon''}{\varepsilon'} \qquad \blacktriangle$$

# 6   INTRODUCTION TO DISTRIBUTED CIRCUITS

## 1.2  FIELDS AND CIRCUITS

The study of transmission lines (TLs) is called distributed circuit analysis, and it is intermediate between the low-frequency extreme of lumped circuits and the most general field equations. The relationship between electromagnetic fields, distributed circuit analysis, and lumped circuits is given in Figure 1.1. The two equations shown for distributed circuits will be developed later. The equations are written in phasor notation for convenience. Lumped circuit theory is associated with the following assumptions and approximations, which are not extremely precise in themselves, but rather a point of view. The physical size of the circuit is assumed to be much smaller than the wavelength of the signals that exist therein.

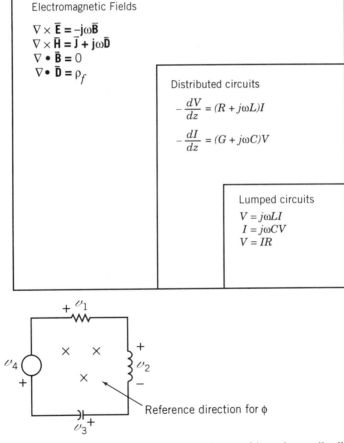

**FIGURE 1.1**  The relationship of field theory and two of its subsets: distributed and lumped circuit theory. The network indicates the reference direction for flux crossing the plane of the circuit, when used to develop Kirchoff's voltage law.

This implies no time delay between both voltages and currents at different parts of the network. In other words, the application of a voltage at one port is sensed immediately at any other port. For example, a circuit about 1 m on a side has a delay due to the finite speed of light of about 3.3 ns. If the signal of interest is at 1 MHz, this corresponds to about a thousandth of the period and is thus negligible. Also since the largest dimension of the circuit is much smaller than the wavelength, radiation is of little consequence. The stored energy in the region about an element is predominantly electric or magnetic, and it changes from one form being dominant to the other when the device goes through self-resonance. In other words, inductors can only store magnetic energy, whereas capacitors only store electric energy. The energy stored between currents and charges at different points in the circuit (stray inductance and capacitance) is assumed to be very small with respect to the energy in the truly lumped elements. The application of $\nabla \cdot \bar{J} = 0$ at nodes gives Kirchoff's current law:

$$\sum i_k = -\frac{dq}{dt}\bigg|_{\text{nodes}} = 0 \tag{1.40}$$

Application of $\nabla \times \bar{\mathscr{E}} = 0$ leads to Kirchoff's voltage law:

$$\sum_{i=1}^{4} v_i = -\frac{d\phi}{dt}\bigg|_{\substack{\text{through} \\ \text{the loop}}} = 0 \tag{1.41}$$

See the network in Figure 1.1 for the sign convention. The above conditions are generally valid up to about 100 MHz.

Figure 1.2 shows the conditions for the existence of a direct current developed by a chemical cell. The emf-producing "field" is shown confined to the cell, and it is derived by the chemical reaction taking place there. When the switch is closed, the reaction proceeds at a rate dependent on the current drawn. When the switch is opened, the reaction slows down and stops, and the cell remains at its "open" or standing "voltage." In this sense, current is just a by-product of the chemical process. Figure 1.2b depicts the development of an ac current by transformer action. Here, we use Faraday's law to describe the emf developed due to a time-varying magnetic flux cutting through the plane of the circuit:

$$\begin{aligned} \text{emf} &= -\frac{d\phi}{dt} = -\frac{d}{dt}\int \bar{\mathscr{B}} \cdot \overline{ds} = -\int \frac{\partial \bar{\mathscr{B}}}{\partial t} \cdot \overline{ds} \\ &= -\int \frac{\partial \bar{\mathscr{B}}_a}{\partial t} \cdot \overline{ds} - \int \frac{\partial \bar{\mathscr{B}}_I}{\partial t} \cdot \overline{ds} \end{aligned} \tag{1.42}$$

Here, the change from $d/dt$ to $\partial/\partial t$ assumes that the circuit is both rigid and stationary. The net flux $\phi$ is the sum of that applied by some external agent (the source loop) and the counterflux (or reaction flux) developed by the current induced $\phi_I$. The flux is applied normal to the plane of the loop and the gap at the

**8** INTRODUCTION TO DISTRIBUTED CIRCUITS

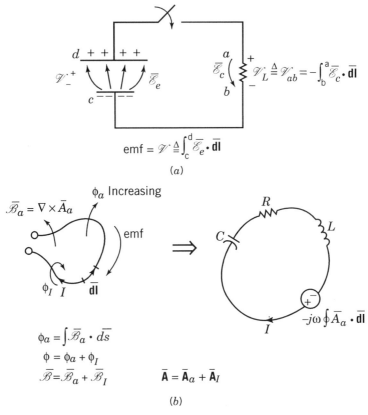

**FIGURE 1.2** (a) The emf produced by a chemical cell and its resulting dc response. (b) The lower part shows the transformation from a wire loop with a small gap and flux cutting its plane, to a lumped ac circuit. Here, the emf is that due to transformer action. The subscript $a$ means applied.

left is modeled as the capacitor $C$ in the lumped equivalent on the right. The lumped inductor is determined by $L = \Lambda/I$, $\Lambda = N\phi_I$, and the resistor is just the net resistance of the wire. The current-inducing property of the changing flux is modeled as the voltage source with value:

$$V = -j\omega \oint_l \bar{A}_a \cdot \overline{dl} \tag{1.43}$$

With some of the basics of lumped circuits in mind, we can state that the study of distributed circuits uses both field and circuit concepts interchangeably. Distributed analysis is the study of guided waves in circuits when the wavelength and the circuit dimensions are of the same order of magnitude. Different, and

FIELDS AND CIRCUITS 9

sometimes difficult to rigorously justify, assumptions are continually invoked while performing an analysis. This blending of both wave (field concept) and lumped ideas (currents, voltages) simultaneously is central to distributed circuit analysis. Referring again to Figure 1.1, we see that the distributed circuit equations are similar to the first two in the lumped section. Just drop the $-d/dz$ terms and the $R$ and $G$, and the distributed reduce to the lumped ones. It is not so obvious, but the distributed ones are the first two Maxwell equations for the special case of TEM waves on transmission lines. This correlation will be shown later.

This section ends with a discussion of phase. Fundamentally, the term "phase" refers to a particular instant in a sinusoidal oscillation with period $T$. It is necessary to define the starting instant (reference) of the oscillation, and generally, either a peak or a zero crossing is chosen. Then the phase of a particular instant is the difference in time units between the instant and the specified reference. Geometry is introduced into this concept of time intervals by the familiar rotating

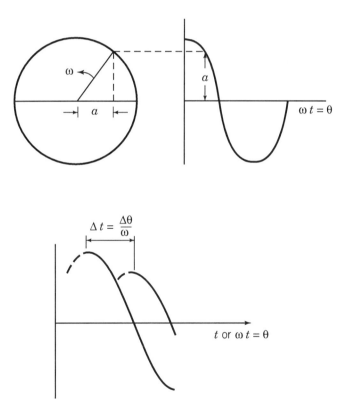

**FIGURE 1.3** The familiar rotating line segment; its horizontal projection describes the amplitude of a sinusoid. The difference in time between two peaks is referred to as a phase angle difference, phase shift, or phase delay between the two oscillations.

line segment in a plane; see Figure 1.3. The horizontal projection yields a cosine wave when the reference position is horizontal. The line rotates at angular velocity $\omega$ rad/s, and it completes one rotation (sweeps out $2\pi$ rad) per period $T$. Thus, the fundamental relationship is formed, $\omega T = 2\pi$. With this correspondence, angles and time intervals are related. Therefore, an angle corresponds to a phase, and we use the time interval ↔ angle relationship interchangeably. The relation is $\theta = \omega t$ and $\theta = 2\pi$ when $t = T$.

Quite often, the lack of synchronism between two oscillations is required (the time interval between the peaking is sought). This interval is called the relative phase (or phase angle) between the oscillations. Depending on which oscillation is chosen as the reference, the other one either leads or lags in phase. A sinusoidal electromagnetic traveling wave is composed of electric and magnetic fields that are sinusoidally distributed in space, which moves at its phase velocity $v_p$ (speed of light in free space). An observer sensing the wave moving past a particular plane will register a sinusoidal variation in time; the period will be the time for, say, successive peaks to pass the plane. Notice that observers at different planes will generally register a phase difference in their time signals. If they are separated by an integral number of wavelengths, they will register "in phase" signals. Therefore, in microwave applications the term "phase" can refer to the same wave at different spatial points, or two different waves at the same point, and any combination of these. The context of the discussion should make the distinction clear.

## 1.3 TRANSMISSION LINE BASICS

A transmission line (TL) guides electromagnetic energy from the source to the load in the form of a transverse electromagnetic (TEM) wave. Figure 1.4 depicts a segment of an open two-wire line propagating a sinusoidally distributed TEM wave. The principal quantities of interest are the electric and magnetic fields $\bar{\mathscr{E}}(x, y, z, t)$ and $\bar{\mathscr{H}}(x, y, z, t)$ (where script variables denote space and arbitrary time dependence). The TEM wave has the important property that $\mathscr{E}_z$ and $\mathscr{H}_z$ are both zero, which enables the development of unambiguous voltage and current wave variables that completely describe the details of the propagation. Figure 1.5 shows a section of coaxial line with both the transverse field components ($\mathscr{E}_r$ and $\mathscr{H}_\varphi$) and the associated voltage and current waves below. The wave energy is moving to the right, as can be seen by applying the right-hand rule with $\mathscr{E}_r$ and $\mathscr{H}_\varphi$. The dots for $\mathscr{H}_\varphi$ imply that the flux lines are coming out of the paper; the crosses mean they are into the paper. The waves for voltage and current shown below follow the convention that positive voltage occurs when the center conductor is positive with respect to the shield, and positive current flows to the right in the center conductor. Notice that at each plane, both $\mathscr{V}$ and $\mathscr{I}$ have the same sign; therefore, from a circuit sense, power is flowing to the right. That is, current is entering the positive node.

TRANSMISSION LINE BASICS 11

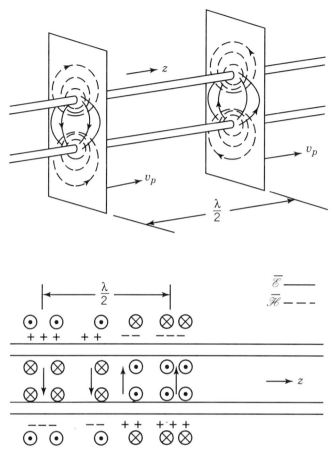

**FIGURE 1.4** The field distribution at a certain instant for a sinusoidally distributed TEM wave. The energy moves to the right, toward positive $z$. The wavefronts (any transverse plane) travel at the phase velocity $v_p$. The fields are shown in two planes spaced a half wavelength apart.

We will justify most of the following intuitive discussion later, but for now we seek an overall perspective of the ideas of distributed circuit analysis. Assume that the line is a two-conductor structure immersed in a uniform dielectric as shown in Figure 1.6. The load is shown as a block of microwave-absorbing material, and notice that it does not necessarily intercept all the field energy (shown by the field lines that extend beyond the boundaries of the load). In this situation, one would expect absorption, reflection, and radiation of some of the incident field energy. In TL analysis, radiation is neglected, so only absorption and reflection are permitted. The circuit diagrams are extensions of lumped circuit theory, as shown in Figure 1.6b. The heavy lines represent the TL where the wave nature of the fields must be considered, but the source and load are lumped elements.

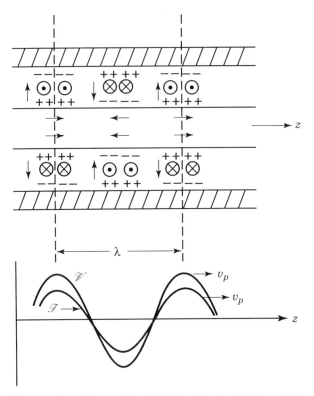

**FIGURE 1.5** Cross section of a coaxial cable indicating both field and circuit quantities. Here, the wave energy is moving to the right. For energy movement to the left, one would reverse the magnetic field lines and invert the current waveform. The direction of energy flow is always given by the direction of the Poynting vector.

Consider a pulse of electromagnetic energy launched onto a line (see Figure 1.7). The pulse has duration $T$ and consider the instant $t_1$, where $0 < t_1 < T$. During the interval $(0, T)$, the generator action can be thought of as leaking charge onto the line: positive on the upper wire and negative on the lower. The leading edges of the charge excesses move away from the launch plane at the speed $v_p$, called the phase velocity of the line. For the ideal line, this speed depends only on the $\mu$ and $\varepsilon$ of the surrounding medium, and not the shape of the conductors or their spacing; it is given by $v_p = 1/\sqrt{\mu\varepsilon}$. The generator maintains a constant voltage $\mathscr{V}_s$ at the input, and this value exists between conductors over the entire charged interval $(0 \leqslant z \leqslant v_p t_1)$. On the conductors, a surface current exists, and its value is the same at all points in the charged region. The current is directed to the right on the upper conductor (denoted by $\mathscr{I}^+$) and to the left on the lower one. The superscript means that the power is to the right (toward positive $z$). The currents on the two conductors are the same in magnitude, and

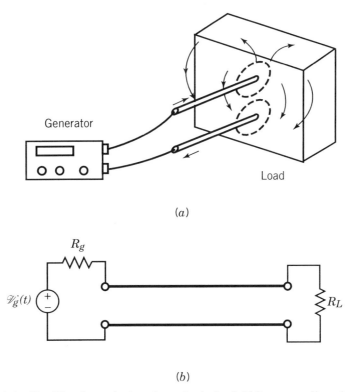

**FIGURE 1.6** The TL schematic that shows both the field lines as well as the surface currents on the condcutors. The heavy lines in (b) denote the TL, which means that the wave properties of the energy flow must be considered in that region. The source and load are treated as lumped elements, where ordinary circuit theory applies.

that on the lower one is directed to the left. This agrees with conventional notions; the direction of a current is in the direction of flow of positive charge, or opposite the direction of negative charge movement. Here, the upper conductor is positive with respect to the lower, and the positive current enters the positive node. Thus, power is moving to the right when viewed from a lumped circuit standpoint.

At the leading edge ($z_1 = v_p t_1$), the electric field is changing with time, since $\bar{\mathscr{E}}$ associated with the charge exists for $z < z_1$ and none exists for $z > z_1$ (we neglect fringing and transient precursors that actually exist). This edge may be thought of as constituting a displacement current $i_d$ that "returns" the current from the upper to the lower conductor, even though no such "return" concept is needed in distributed circuit analysis.

A main result of distributed circuit analysis is the modeling of the TL by a network consisting of distributed inductance $L$ and capacitance $C$. These are defined on a per unit (distributed) basis: H/m and F/m. Then the net inductance of

## 14　INTRODUCTION TO DISTRIBUTED CIRCUITS

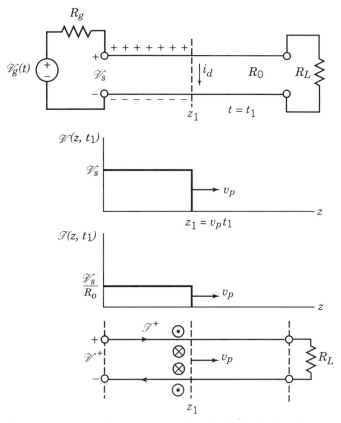

**FIGURE 1.7**  A generator applying a voltage $\mathscr{V}_s$ to the input of a line. The instant shown is $t_1$ seconds after the initiation of the voltage at the source end. The instantaneous voltage and current distributions on the line are shown below, along with the current directions in both conductors.

a segment of length $\Delta z$ is $L\Delta z$ H, with net capacity $C\Delta z$ F. It turns out that the wave speed is related to $L$ and $C$ by $v_p = 1/\sqrt{LC} = 1/\sqrt{\mu\varepsilon}$ so $LC = \mu\varepsilon$ in general. The current and voltage of the previously discussed pulse are related by $\mathscr{I}^+ = \mathscr{V}_s/R_0$ at any point in the charged interval. The quantity $R_0$ is called the characteristic resistance of the line; it is a function of the $\mu$ and $\varepsilon$ of the surrounding medium, as well as the line's geometry. A physical interpretation for $R_0$ can be developed by viewing the launching process as storing electromagnetic energy in the TL. At the instant $t_1$, the stored energy may be expressed as the sum of the electric ($W_e$) and magnetic ($W_m$) portions. We can use formulas from circuit theory while noting that at $t_1$, the line is charged up to the length $z_1 = v_p t_1$. Then the net charged capacitance and inductance are $Cz_1$ and $Lz_1$, respectively, we write

$$W_e = \tfrac{1}{2}(Cz_1)\mathscr{V}_s^2 \tag{1.44}$$

$$W_m = \tfrac{1}{2}(Lz_1)(\mathscr{I}^+)^2 \tag{1.45}$$

and after we use $\mathscr{I}^+ = \mathscr{V}_s/R_0$ and $R_0 \stackrel{\Delta}{=} \sqrt{L/C}$, the total energy is $W_{tot} = W_e + W_m = (Cz_1)\mathscr{V}_s^2$. Now the rate of change of stored energy at $t_1$ is

$$\frac{d}{dt}W_{tot} = \mathscr{V}_s^2 C \frac{dz_1}{dt} = \mathscr{V}_s^2 C v_p \tag{1.46}$$

and using $v_p = 1/\sqrt{LC}$, we find

$$\frac{dW_{tot}}{dt} = \frac{\mathscr{V}_s^2}{\sqrt{L/C}} \stackrel{\Delta}{=} \frac{\mathscr{V}_s^2}{R_0} \tag{1.47}$$

Since $dW_{tot}/dt$ is instantaneous power, the above expression looks like the power absorbed by a resistor $R_0$. Thus, the generator effectively "sees" a resistance $R_0$ at its terminals as it stores energy on the line. This characteristic resistance does not dissipate power into heat; here, the energy leaving the generator is stored in the line. The quantity $R_0$ is called either the characteristic or surge impedance of the line. The more general term impedance is used quite often, even though the impedance $Z_0 = R_0 + jX_0$ is normally mostly real $|X_0| \ll R_0$.

Figure 1.8 shows the pulse completely launched onto the line. The generator is off, and the pulse of energy spans the range $z_a$ to $z_b$, where $(z_b - z_a) = v_p T$. The upper sketch depicts the $\bar{\mathscr{E}}$ and $\bar{\mathscr{H}}$ flux lines. The right-hand rule for $\bar{\mathscr{H}}$ and $\mathscr{I}$ applies well as $\bar{\mathscr{P}} = \bar{\mathscr{E}} \times \bar{\mathscr{H}}$ pointing in the direction of energy flow; $\bar{\mathscr{P}}$ is the Poynting vector. The lower sketch indicates the charge motion and associated voltage and current. The electrostatic relationships

$$\mathscr{V}_{21} = -\int_1^2 \bar{\mathscr{E}} \cdot \overline{dl} \tag{1.48}$$

$$\mathscr{I} = \oint \bar{\mathscr{H}} \cdot \overline{dl} \tag{1.49}$$

will be shown to hold later. The "charges" in both conductors move to the right (toward positive z) at speed $v_p$, so the currents are oppositely directed. By 'charges', we mean the deviation from total neutrality at a particular point. The charges don't move at the phase velocity $v_p$, just the energy of the pulse.

Consider the superposition of two pulses moving in opposite directions on the line as shown in Figure 1.9. The electric fields are directed downward so the net field during overlap is the sum of both. The magnetic fields are in opposition, as

**16** INTRODUCTION TO DISTRIBUTED CIRCUITS

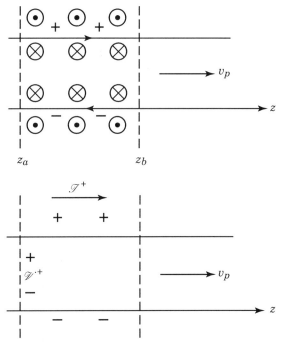

**FIGURE 1.8** A sketch of a moving pulse of energy on an ideal TL. The upper part shows the electric and magnetic field lines along with the excess and deficiency of negative charge and associated currents. The lower part shows the normal voltage and current wave variables along with the induced charge on both conductors. The superscript + means the energy is moving to the right (toward positive $z$).

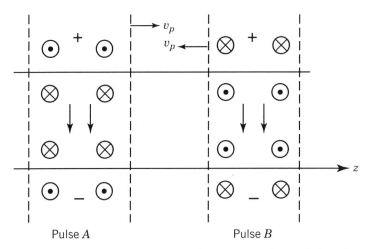

**FIGURE 1.9** Two pulses moving in opposite directions at an instant before they overlap. Pulse $A$ moves to the right and pulse $B$ to the left.

they wrap about the same conductor in opposite directions. So during overlap, the net current is the difference of the currents.

Finally, the process of the pulse impinging on the load is covered. In general, some of the wave energy is absorbed, with the remainder reflected back toward the generator. Suppose that the physical load shown in Figure 1.6 is a carbon or iron powder-filled foam block. Such materials absorb nearly all the wave energy and are said to "match" the line, since no reflected energy is returned to the generator. Thus, the load is effectively a lumped resistor of value $R_0 \, \Omega$. In practice, loads of this type have tapers that end in a sharp point that is directed toward the generator end. The details of the taper are determined by trial and error in the laboratory.

## 1.4 TRANSVERSE ELECTROMAGNETIC WAVES

The transverse electromagnetic (TEM) wave is the primary mode of wave propagation on a transmission line, so we investigate some of its properties. The TL structure is assumed to be uniform in the $z$ direction and composed of perfect conductors that support surface charge and current. An ideal dielectric surrounds the conductors, and all wave fields exist in this material. Under these assumptions, the wave equation developed from Maxwell's equations yields TEM solutions for the structure. In the space about the conductors, the equation for the $\bar{\mathscr{E}}$ field is, in general,

$$\nabla^2 \bar{\mathscr{E}} - \mu\varepsilon \frac{\partial^2 \bar{\mathscr{E}}}{\partial t^2} = 0 \qquad (1.50)$$

and a solution of the form

$$\bar{\mathscr{E}}(x, y, z, t) = \bar{e}(x, y) f\left(t \pm \frac{z}{v_p}\right) \qquad (1.51)$$

is possible since the equation is separable. Both $\bar{\mathscr{E}}$ and $\bar{\mathscr{H}}$ are constrained to the $xy$ plane (recall that $\mathscr{E}_z = \mathscr{H}_z = 0$, for the TEM wave), and it turns out (see Example 1.2) that the field pattern may be expressed as

$$\bar{e}(x, y) = -\nabla \phi(x, y) \qquad (1.52)$$

where $\phi(x, y)$ is a scalar potential function that satisfies Laplace's equation $\nabla^2 \phi(x, y) = 0$. The function $f(t \pm z/v_p)$ serves to move the $\bar{e}(x, y)$ pattern at velocity $v_p$ along the structure. The negative sign case moves $\bar{e}(x, y)$ toward positive $z$, whereas the positive one moves it toward negative $z$.

The transverse pattern $\bar{e}(x, y)$ has the properties

$$\nabla_t \times \bar{e}(x, y) = \nabla_t \cdot \bar{e}(x, y) = 0 \qquad (1.53)$$

$$\nabla_t = \frac{\partial}{\partial x} \bar{a}_x + \frac{\partial}{\partial y} \bar{a}_y$$

which makes it identical to a static field in a charge-free region. This fact permits the use of static calculations (when appropriate) in the analysis of high-frequency behavior. The voltage wave associated with the traveling electric field wave is denoted by

$$\mathscr{V}(x, y, z, t) = \phi(x, y) f\left(t \pm \frac{z}{v_p}\right) \tag{1.54}$$

and we define

$$\mathscr{V}_{21}(z, t) = -\int_1^2 \bar{\mathscr{E}} \cdot \overline{dl} \tag{1.55}$$

where the path is in the $xy$ plane. Figure 1.10 summarizes some of the above points. The power flow is along $+z$, as may be seen by the Poynting vector $\bar{\mathscr{S}} = \bar{\mathscr{E}} \times \bar{\mathscr{H}}$. From the definitions of $\mathscr{V}$ and $\mathscr{I}$ Eqs. (1.55) and (1.49), we observe that they are both positive in the figure. Notice that the product $\mathscr{V}\mathscr{I}$ denotes power flow to the right, since the current in the upper conductor is entering the positive node. This figure summarizes the fact that the fields $\{\bar{\mathscr{E}}_t, \bar{\mathscr{H}}_t\}$ (the

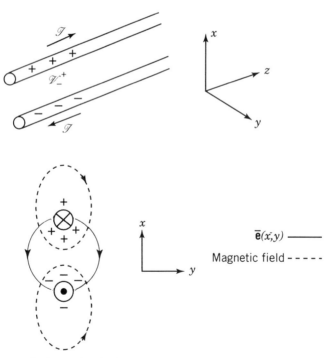

**FIGURE 1.10** The lower sketch shows the electric and magnetic field lines at an instant, whereas the upper one shows the corresponding voltage and current waves. The field vectors $\bar{\mathscr{E}}_t$ and $\bar{\mathscr{H}}_t$ are related to the voltage and current variables by Eqs. (1.48) and (1.49).

subscript $t$ means transverse components only) may be represented by the scalar set $\{\mathscr{V}, \mathscr{I}\}$ consisting of voltage and current waves. The information needed for the representation are the values of $L$ and $C$ for the line.

We can summarize the steps needed to apply distributed circuit analysis to TEM waves on transmission lines as follows. First, find the expression for $\phi(x, y)$ once the geometry is known. This function is found using analytic techniques from statics. Next determine $C$ from $\phi(x, y)$, then $L$ follows from the fact that $LC = \mu\varepsilon$. Consequently, one has the wave speed $v_p = 1/\sqrt{LC}$ and the characteristic impedance $R_0 = \sqrt{L/C} = (\varepsilon/C)\sqrt{\mu/\varepsilon}$, where the latter form shows the dependence on geometry via $C$. The term $\sqrt{\mu/\varepsilon}$ is called the intrinsic wave impedance of the dielectric. It turns out that an $N$ conductor structure can support $(N-1)$ different TEM modes [different $\bar{e}(x, y)$ patterns]. This result is useful in coupled lines, which are covered later.

***Example 1.2*** Verify Eqs. (1.52) and (1.53).

Write the del operator as

$$\nabla = \frac{\partial}{\partial x}\bar{a}_x + \frac{\partial}{\partial y}\bar{a}_y + \frac{\partial}{\partial z}\bar{a}_z = \nabla_t + \nabla_z$$

where

$$\nabla_z = \frac{\partial}{\partial z}\bar{a}_z$$

Write the fields as $\bar{\mathscr{E}} = \bar{\mathscr{E}}_t + \bar{\mathscr{E}}_z$, where $\bar{\mathscr{E}}_t = \mathscr{E}_x\bar{a}_x + \mathscr{E}_y\bar{a}_y$ and $\bar{\mathscr{E}}_z = \mathscr{E}_z\bar{a}_z$. Then Maxwell's first equation is

$$\nabla \times \bar{\mathscr{E}} = -\frac{\partial}{\partial t}\mu\bar{\mathscr{H}}$$

or

$$(\nabla_t + \nabla_z) \times (\bar{\mathscr{E}}_t + \bar{\mathscr{E}}_z) = -\mu\frac{\partial}{\partial t}(\bar{\mathscr{H}}_t + \bar{\mathscr{H}}_z)$$

Expand

$$\nabla_t \times \bar{\mathscr{E}}_t + \nabla_z \times \bar{\mathscr{E}}_t + \nabla_t \times \bar{\mathscr{E}}_z + \nabla_z \times \bar{\mathscr{E}}_z = -\mu\frac{\partial}{\partial t}\bar{\mathscr{H}}_t - \mu\frac{\partial}{\partial t}\bar{\mathscr{H}}_z$$

Now use the fact that $\bar{\mathscr{E}}_z = \bar{\mathscr{H}}_z = 0$ defines a TEM wave, so we have

$$\nabla_t \times \bar{\mathscr{E}}_t + \nabla_z \times \bar{\mathscr{E}}_t = -\mu\frac{\partial}{\partial t}\bar{\mathscr{H}}_t$$

**20** INTRODUCTION TO DISTRIBUTED CIRCUITS

Observe that $\nabla_t \times \bar{\mathscr{E}}_t$ is oriented along $\bar{\mathbf{a}}_z$ and the right side has no $\bar{\mathbf{a}}_z$ component. Thus,

$$\nabla_t \times \bar{\mathscr{E}}_t = 0 \tag{1}$$

Now $\bar{\mathscr{E}}_t = \bar{\mathbf{e}}(x, y) f(t \pm z/v_p)$, and since the $\nabla_t$ operator does nothing to $f$ [since $f$ is only a function of $(z, t)$] and

$$\nabla_t = \frac{\partial}{\partial x} \bar{\mathbf{a}}_x + \frac{\partial}{\partial y} \bar{\mathbf{a}}_y$$

we have

$$\nabla_t \times \bar{\mathscr{E}}_t = f\left(t \pm \frac{z}{v_p}\right) \nabla_t \times \bar{\mathbf{e}}(x, y) = 0$$

Hence for nontrivial solutions,

$$\nabla_t \times \bar{\mathbf{e}}(x, y) = 0 \tag{2}$$

which is Eq. (1.53). From the identity $\nabla \times \nabla A = 0$ for a scalar function $A$ and from observing Eq. (2), we can write $\bar{\mathbf{e}}(x, y)$ as the gradient of some scalar. Let

$$\bar{\mathbf{e}}(x, y) = -\nabla \phi(x, y) \tag{3}$$

where the negative sign is by convention. The total del symbol is permissible since $\phi(x, y)$ does not depend on $z$; notice that this is Eq. (1.52). From Maxwell's third equation,

$$\nabla \cdot \varepsilon \bar{\mathscr{E}}_t = \rho$$

we are in a perfect charge-free dielectric so $\rho = 0$, and since $\bar{\mathscr{E}}_t$ has no $z$ component, we have $\nabla_t \cdot \bar{\mathscr{E}}_t = 0$ and again $f$ can be divided out. Thus,

$$\nabla_t \cdot \bar{\mathbf{e}}(x, y) = 0 \tag{4}$$

which is Eq. (1.53). ▲

***Example 1.3*** Find the expression for $\bar{\mathscr{E}}_t$ and $\bar{\mathscr{H}}_t$ for a coaxial cable.
From Eqs. (3) and (4) of the previous example, we have

$$\nabla_t \cdot \bar{\mathbf{e}}(x, y) = \nabla \cdot \bar{\mathbf{e}}(x, y) = -\nabla \cdot \nabla \phi(x, y) = 0$$

or

$$\nabla^2 \phi(x, y) = 0 \tag{1}$$

## TRANSVERSE ELECTROMAGNETIC WAVES

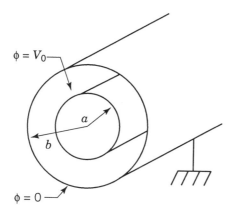

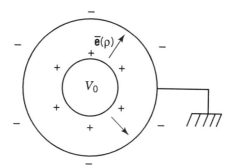

**FIGURE 1.11**

Use cylindrical coordinates $(\rho, \varphi, z)$ and assume that the boundary conditions (bc) are

$$\phi(\rho = a) = V_0$$
$$\phi(\rho = b) = 0$$

See Figure 1.11. Then

$$\nabla^2 \phi(\rho, \varphi) = \frac{1}{\rho} \frac{\partial}{\partial \rho}\left(\rho \frac{\partial \phi}{\partial \rho}\right) + \frac{1}{\rho^2} \frac{\partial^2 \phi}{\partial \varphi^2} = 0$$

and from symmetry, we observe that there should be no $\varphi$ variation. Thus,

$$\nabla^2 \phi(\rho) = \frac{1}{\rho} \frac{\partial}{\partial \rho}\left(\rho \frac{\partial \phi}{\partial \rho}\right) = 0 \qquad (2)$$

The solution to Eq. (2) is

$$\phi(\rho) = C_1 \ln \rho + C_2 \qquad (3)$$

To determine the constants $C_1$ and $C_2$, we apply the bc:

$$\phi(\rho = a) = V_0 = C_1 \ln a + C_2$$
$$\phi(\rho = b) = 0 = C_1 \ln b + C_2$$

Solving for $C_1$ and $C_2$ permits us to write Eq. (3) as

$$\phi(\rho) = \frac{V_0}{\ln(a/b)} \ln(\rho/b) \qquad (4)$$

Then

$$\bar{e}(\rho) = -\nabla\phi = -\frac{V_0}{\ln(a/b)} \frac{1}{\rho} \bar{a}_\rho \qquad (5)$$

so $\bar{e}(\rho)$ is along $\bar{a}_\rho$ since $\ln(a/b) < 0$. Now we can write

$$\bar{\mathscr{E}}_t(\rho, \varphi, z, t) = -\frac{V_0}{\ln(a/b)} \frac{1}{\rho} \left[ f\left(t \pm \frac{z}{v_p}\right) \right] \bar{a}_\rho \qquad (6)$$

where $f(t \pm z/v_p)$ is some function as yet to be determined. In general, it is defined by the generator. It serves to translate the $\bar{e}(\rho)$ spatial pattern in Eq. (5) at the speed $v_p$ down the cable.

To find $\bar{\mathscr{H}}_t(\rho, \varphi, z, t)$, we can use magnetostatics. From any introductory text, we find the static transverse pattern to be

$$\bar{h}(\rho) = \frac{I}{2\pi\rho} \bar{a}_\varphi, \qquad a < \rho < b \qquad (7)$$

where $I$ is the current on the center conductor. Then

$$\bar{\mathscr{H}}_t(\rho, \varphi, z, t) = \frac{I}{2\pi\rho} f\left(t \pm \frac{z}{v_p}\right) \bar{a}_\varphi \qquad (8)$$

▲

***Example 1.4*** Find $L$, $C$, and $Z_0$ for a coaxial cable.

Here, we write the electric and magnetic fields by assuming first a linear charge density $\rho_l$ on the conductors, then a uniform current $I$. The inductance found is the external value, that is, only that associated with the flux between the cylinders, and not including that which exists in them. The $\bar{\mathscr{E}}$ field is

$$\bar{\mathscr{E}} = \frac{\rho_l}{2\pi\varepsilon\rho} \bar{a}_\rho$$

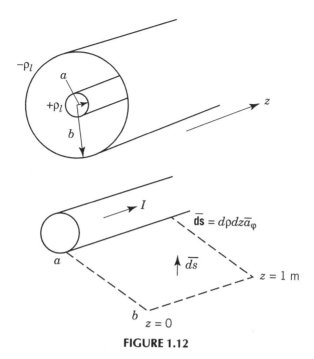

**FIGURE 1.12**

The potential $\phi$ is

$$\phi_{ba} = -\int_a^b \bar{\mathscr{E}} \cdot \overline{dl} = -\int_a^b \frac{\rho_l}{2\pi\varepsilon\rho} \bar{a}_\rho \cdot d\rho\bar{a}_\rho$$

$$= -\frac{\rho_l}{2\pi\varepsilon} \ln\left(\frac{b}{a}\right)$$

See Figure 1.12. The capacity is defined as the charge on one conductor to the potential between them:

$$C = \frac{Q}{\phi_{ba}} = \frac{-\rho_l(1\,\text{m})(2\pi\varepsilon)}{-\rho_l \ln(b/a)} = \frac{2\pi\varepsilon}{\ln(b/a)} \quad \text{F/m}$$

The magnetic field is

$$\bar{\mathscr{H}} = \frac{I}{2\pi\rho} \bar{a}_\varphi$$

and the flux density is

$$\bar{\mathscr{B}} = \mu\bar{\mathscr{H}} = \frac{\mu I}{2\pi\rho} \bar{a}_\varphi$$

The net flux linking the cylinders is (See Figure 1.12)

$$\psi_m = \int \overline{\mathcal{B}} \cdot \overline{ds} = \frac{\mu I}{2\pi} \int_0^1 \int_a^b \frac{1}{\rho} d\rho\, dz$$

$$= \frac{\mu I}{2\pi} \ln\left(\frac{b}{a}\right)$$

The inductance is the flux divided by the current on one conductor:

$$L \triangleq \frac{\psi_m}{I} = \frac{\mu}{2\pi} \ln\left(\frac{b}{a}\right)$$

The characteristic impedance is (by its definition)

$$Z_0 = \sqrt{\frac{L}{C}} = \frac{\sqrt{\mu/\varepsilon}}{2\pi} \ln\left(\frac{b}{a}\right)$$

Notice that the product $LC = \mu\varepsilon$ as stated earlier. ▲

***Example 1.5*** Find the characteristic impedance of the open two-wire line. First, obtain an approximate expression by assuming that the charge distribution is uniform over the periphery of the conductors. Then compare this result with the exact expression.

Assume that the cylinders have radius $b$ and are separated a distance $d$ as shown in Figure 1.13. The left conductor has a total charge of $-q$ per unit length. If we focus on the line of centers, the electric field there is just (directed from right to left)

$$\mathscr{E} = \frac{q}{2\pi\varepsilon r} - \frac{-q}{2\pi\varepsilon(d-r)} \qquad (1)$$

By symmetry, the plane between the wires is a constant potential surface, and we define it to be our zero reference. Then if each cylinder is either $V_0$ above or below zero, the integral of $\mathscr{E}$ from $b$ to $(d-b)$ will yield the value $2V_0$, which is the total potential between the wires. Then

$$2V_0 = \int_b^{d-b} \frac{1}{\varepsilon}\left[\frac{q}{2\pi r} + \frac{q}{2\pi(d-r)}\right] dr = \frac{q}{\pi\varepsilon} \ln\left(\frac{d-b}{b}\right)$$

or

$$V_0 = \frac{q}{2\pi\varepsilon} \ln\left(\frac{d-b}{b}\right) \qquad (2)$$

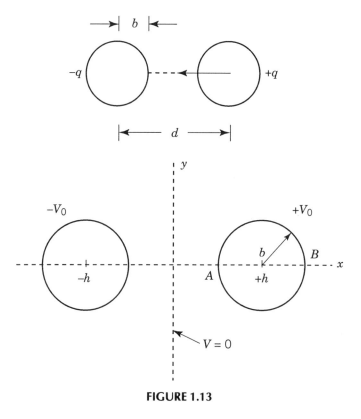

**FIGURE 1.13**

The capacitance is

$$C = \frac{q}{2V_0} = \frac{\pi\varepsilon}{\ln\left(\frac{d-b}{b}\right)} \tag{3}$$

For the inductance calculation, the flux density along the centerline is

$$\mathscr{B} = \mu\mathscr{H} = \frac{\mu I}{2\pi r} + \frac{\mu I}{2\pi(d-r)} = \frac{\mu I}{2\pi}\left(\frac{1}{r} + \frac{1}{d-r}\right)$$

Then the net flux between the wires is

$$\psi_m = \frac{\mu I}{2\pi}\int_b^{d-b}\left(\frac{1}{r} + \frac{1}{d-r}\right)dr = \frac{\mu I}{\pi}\ln\left(\frac{d-b}{b}\right)$$

# 26 INTRODUCTION TO DISTRIBUTED CIRCUITS

The inductance is therefore

$$L = \frac{\psi_m}{I} = \frac{\mu}{\pi} \ln\left(\frac{d-b}{b}\right) \tag{4}$$

$$Z_0 = \sqrt{\frac{L}{C}} = \frac{\sqrt{\mu/\varepsilon}}{\pi} \ln\left(\frac{d-b}{b}\right) \doteq \frac{1}{\pi}\sqrt{\frac{\mu}{\varepsilon}} \ln\left(\frac{d}{b}\right), \qquad d \gg b \tag{5}$$

The exact expression for $\phi(x, y)$ will not be developed here, but it can be found in many textbooks, for example, Refs. 1, 2. For this case, we use rectangular coordinates; see Figure 1.13 for details. Define $a = \sqrt{h^2 - b^2}$. Then

$$V = \frac{q}{4\pi\varepsilon} \ln\left[\frac{(x+a)^2 + y^2}{(x-a)^2 + y^2}\right] \tag{6}$$

or

$$V = \frac{V_0/2}{\ln\left(\frac{h+\sqrt{h^2-b^2}}{b}\right)} \ln\left[\frac{(x+a)^2 + y^2}{(x-a)^2 + y^2}\right]$$

Thus

$$q = 4\pi\varepsilon \, \frac{V_0/2}{\ln\left(\frac{h+a}{b}\right)} \tag{7}$$

The capacity is

$$C = \frac{q}{2V_0} = \frac{\pi\varepsilon}{\ln\left(\frac{h+\sqrt{h^2-b^2}}{b}\right)} = \frac{\pi\varepsilon}{\cosh^{-1}(h/b)} \tag{8}$$

where we have used the identity

$$\cosh^{-1} x = \ln(x + \sqrt{x^2 - 1}), \qquad x \geq 1$$

We can omit the inductance calculation by using $LC = \mu\varepsilon$ so

$$Z_0 = \sqrt{\frac{L}{C}} = \frac{\sqrt{\mu\varepsilon}}{C} \tag{9}$$

Thus,

$$Z_0 = \frac{1}{\pi}\sqrt{\frac{\mu}{\varepsilon}} \cosh^{-1}\left(\frac{h}{b}\right) \tag{10}$$

To compare Eqs. (5) and (10), we list $\mathbf{Z}_0$ values along with the ratio of the charge density at points $A$ and $B$ in the figure. To obtain the charge density, we proceed as follows. The $\bar{\mathscr{E}}$ field is found from Eq. (8):

$$\bar{\mathscr{E}} = -\nabla V$$

$$\bar{\mathscr{E}}(x,y) = \frac{q}{\pi\varepsilon}\left\{\frac{a(x^2 - y^2 - a^2)\bar{a}_x + 2axy\bar{a}_y}{[(x+a)^2 + y^2][(x-a)^2 + y^2]}\right\} \quad (11)$$

Then the charge density is $\rho_s = \varepsilon\bar{\mathscr{E}}\cdot\bar{n}$, where $\bar{n}$ is the unit normal. The points $A$ and $B$ correspond to $x = (h-b)$ and $x = (h+b)$, respectively. Then

$$\rho_s(A) = \frac{q}{2\pi}\frac{a}{b(h-b)}$$

$$\rho_s(B) = \frac{q}{2\pi}\frac{a}{b(h+b)}$$

so we have the simple relationship

$$\frac{\rho_s(A)}{\rho_s(B)} = \frac{h+b}{h-b}$$

The table below compares the two approaches:

| $h/b$ | $\mathbf{Z}_0$ (approx.) | $\mathbf{Z}_0$ (exact) | $\frac{\rho_s(A)}{\rho_s(B)}$ (approx.) | (exact) |
|---|---|---|---|---|
| 1.5 | 83.18 | 115.49 | 1 | 5 |
| 6.3 | 294.12 | 303.28 | 1 | 1.38 |
| 10.0 | 353.3 | 359.19 | 1 | 1.22 |

The comparison shows that the error in $\mathbf{Z}_0$ is 28% and 1.6% for $h/b = 1.5$ and 10, respectively. The main point of this example was to give a feel for the difference in complexity between an easily found approximate result vs. the exact one. In actual practice, the exact solutions to practical structures are often nearly impossible to determine, so many approximate solutions can be found in the literature. ▲

## 1.5 EQUATIONS FOR $\mathscr{V}$ AND $\mathscr{I}$ FOR LOSSLESS LINES

The development of the partial differential equations for $\mathscr{V}$ and $\mathscr{I}$ for lossless lines follows directly from Maxwell's equations. Before we show this, a review of voltage, potential difference, and emf is helpful.

**Example 1.6** Discuss the distinctions between the terms voltage, potential difference, and emf. See refs. 2, 3, and 4.

In general, voltage is defined as a line integral of the electric field. For the time-varying case, we have

$$\mathscr{V}_{21}(t) = -\int_1^2 \bar{\mathscr{E}}(\bar{r}, t) \cdot \overline{dl}$$

$$= -\int_1^2 -\nabla\Phi(\bar{r}, t) \cdot \overline{dl} - \int_1^2 -\frac{\partial}{\partial t}\bar{A}(\bar{r}, t) \cdot \overline{dl} \quad (1)$$

where the field is expressed in terms of the scalar and vector potentials $\Phi(\bar{r}, t)$ and $\bar{A}(\bar{r}, t)$. The first term may be written as

$$\int_1^2 \frac{\partial \Phi}{\partial l} dl = \Phi_2(t) - \Phi_1(t)$$

Thus,

$$\mathscr{V}_{21}(t) = \Phi_2(t) - \Phi_1(t) + \int_1^2 \frac{\partial}{\partial t}\bar{A}(\bar{r}, t) \cdot \overline{dl} \quad (2)$$

which, if we consider the static case,

$$\mathscr{V}_{21} = \mathscr{V}_2 - \mathscr{V}_1 = \Phi_2 - \Phi_1$$

Thus, the voltage difference $\mathscr{V}_{21}$ is the same as the potential difference $\Phi_2 - \Phi_1$ in the static case. In the general case, the integral in Eq. (2) causes the voltage and potential differences to be different. Also, $\mathscr{V}_{21}(t)$ is not unique, since the value of the integral is path-dependent.

The definition of potential difference is generally reserved for electrostatics, and it is

$$\Phi_{21} = \Phi_2 - \Phi_1 = -\int_1^2 \bar{\mathscr{E}}_c \cdot \overline{dl} \quad (3)$$

where $\bar{\mathscr{E}}_c$ means a field developed by a static distribution of charge. Here, $\Phi_{21}$ is the work done per unit positive charge as it is moved from point 1 to point 2. Another subscript convention exists, and it is $\Phi_{12} = \Phi_2 - \Phi_1$. Although both terms have the same physical meaning and numerical value, confusion exists if one also uses the conventional idea that reversing the order of subscripts changes the sign of the quantity. When the initial point (point 1) is chosen at infinity and the potential there defined to be zero, then

$$\Phi(r_2) = -\int_\infty^{r_2} \bar{\mathscr{E}}_c \cdot \overline{dl} \quad (4)$$

is known as the absolute potential. In electrostatics, the field is directed from positive to negative charge, and it is conservative:

$$\oint \bar{\mathscr{E}}_c \cdot \overline{\mathrm{dl}} = 0$$

This allows us to describe it as the gradient of a scalar potential:

$$\bar{\mathscr{E}}_c = -\nabla\Phi(r) \tag{5}$$

Now consider electromotive forces (emfs) in the static case. An emf is represented by the normal source element of circuit theory, and it creates a current in a dc circuit. Recall that the field due to just static charges cannot create a steady current in a dc circuit, since the only energy available is that required to set up the charges in the particular configuration at hand. An emf, on the other hand, has an internal energy conversion mechanism that creates electric energy. Examples are chemical storage batteries, solar cells, and thermocouples. Consider the dc battery of circuit theory. The chemical reaction inside the battery causes charge separation as shown in the uppermost sketch in Figure 1.14. It is customary to define a field $\bar{\mathscr{E}}_e$ inside the battery that apparently moves the charges onto the plates. This field is called an emf-producing one. If we use the basic notion that

$$\bar{\mathscr{E}} = \frac{\bar{\mathscr{F}}}{q}$$

apparently $\bar{\mathscr{E}}_e$ must be oriented as shown in the figure. Notice that it is in the opposite direction of the electrostatic field set up by the charges on the plates ($\bar{\mathscr{E}}_c$ is not shown in this sketch). For clarity in this problem, we will sketch an emf-producing field with dotted field lines. The emf developed by the device is defined as (if a 3-V battery is used)

$$\mathrm{emf} \triangleq \mathscr{V}_{AB} = \int_B^A \bar{\mathscr{E}}_e(\bar{r}) \cdot \overline{\mathrm{dl}} = +3 \, \mathrm{V} \tag{6}$$

Notice that emf is the work per unit charge moving from $B$ and $A$ and is measured in volts. Thus, the term electromotive force is a misnomer (its units are not those of forces), so sometimes the term electromotance is used. Apparently, $\bar{\mathscr{E}}_e$ can only exist in the paste (or we define it to exist only there, which seems reasonable since the actual force that separates the charges is confined to the paste), and thus it is not a conservative field. From the figure, we observe that the integral along path 1 is zero, whereas that over path 2 is $+3$ V.

Equation (6) can be written as

$$\mathscr{V}_{AB} = \int_B^A \bar{\mathscr{E}}_e(\bar{r}) \cdot \overline{\mathrm{dl}} = +3 \, \mathrm{V}$$

# 30 INTRODUCTION TO DISTRIBUTED CIRCUITS

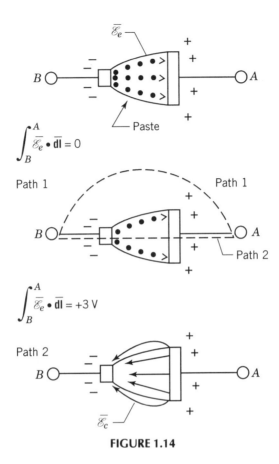

**FIGURE 1.14**

where $\mathscr{V}_{AB}$ is defined as the *voltage rise* from B to A. Now sketching only the electrostatic field lines (neglect the $\bar{\mathscr{E}}_e$ lines) and using the definition in Eq. (3), we have

$$\Phi_{AB} = -\int_B^A \bar{\mathscr{E}}_c(\bar{r}) \cdot \overline{dl} = +3 \text{ V}$$

Thus, $\Phi_{AB} = \mathscr{V}_{AB}$ for an open-circuited battery. Note that care should be exercised, since the definitions are rather similar, except for the signs, and the two fields are oppositely directed.

The battery produces current in a dc loop since the chemical reaction can replenish the charges on the plates as fast as they leak away and move through the loop. When current is indeed present in the loop, the potential and emf across the battery are not equal, as a net field must exist in the battery to transport charge through it. The difference is ascribed to the internal resistance of the source. Thus, the battery produces the voltage rise in the loop, and the internal resistance and

EQUATIONS FOR $\mathscr{V}$ AND $\mathscr{I}$ FOR LOSSLESS LINES    31

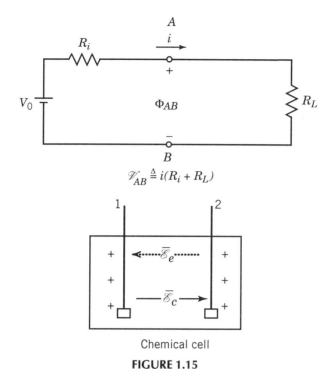

**FIGURE 1.15**

external resistance produce the voltage drops. See the top sketch in Fig. 1.15. For the cell shown in Figure 1.15, the voltage $\mathscr{V}_{12}$ is given by

$$\mathscr{V}_{12} = \int_2^1 \overline{\mathscr{E}}_e \cdot \overline{\mathbf{dl}} = -\underbrace{\int_2^1 \overline{\mathscr{E}}_c \cdot \overline{\mathbf{dl}}}_{\text{inside or outside the battery}} = \mathscr{V}_1 - \mathscr{V}_2$$

and we note that the net field $\overline{\mathscr{E}}_{tot}$ inside the battery is zero. The emf is zero outside the electrolyte, and since its field is not conservative, it cannot be expressed as the gradient of some scalar field.

Now consider a static emf caused by motion. The situation is shown below in Figure 1.16, which is the basic "flux cutting" or motional emf. For constant $\overline{\mathscr{B}}$ and $\bar{v}$, $\overline{\mathscr{E}}_e$ is constant. To summarize the static case: When work is done by movement in any static electric field, regardless of the source of the field, one defines the work per unit charge as the voltage:

$$\mathscr{V}_{AB} = -\int_B^A \overline{\mathscr{E}}_c \cdot \overline{\mathbf{dl}} + \int_B^A \overline{\mathscr{E}}_e \cdot \overline{\mathbf{dl}} \qquad (7)$$

**32** INTRODUCTION TO DISTRIBUTED CIRCUITS

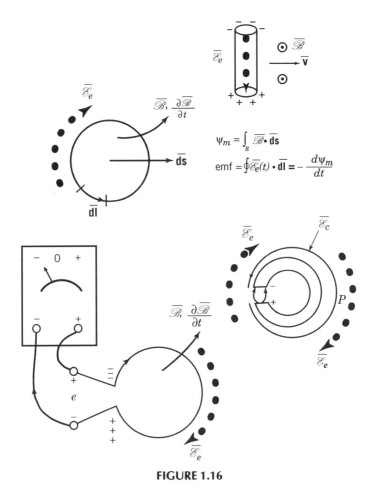

**FIGURE 1.16**

Here, we have chosen to break up a general static field into that produced by charges placed about the region $\bar{\mathscr{E}}_c$ and those produced by sources of electromotance $\bar{\mathscr{E}}_e$.

For the time-varying case, the emf is that derived by the changing magnetic field as expressed by Faraday's law. We assume that the loop is stationary. Then the emf developed is called a transformer type of electromotance; see Figure 1.16. Notice that $\bar{\mathscr{E}}_e(t)$ and $\overline{dl}$ are oppositely directed. The counterclockwise movement of $\overline{dl}$ and the orientation of $\overline{ds}$ is a consequence of the basic definitions of line integrals, etc., from calculus. The minus sign is by convention, which we will show forces agreement with our definitions of polarity and the direction of current developed by a source. For right-handed coordinate systems, we recall that the rules for $\overline{dl}$ and $\overline{ds}$ are as follows: Walk around the contour in the counterclockwise sense (so the area is over your left shoulder); $\overline{ds}$ points upward from the floor. The clockwise direction of $\bar{\mathscr{E}}_e$ is found by experiment and has no bearing on the line integral rules. The direction of $\bar{\mathscr{E}}_e$ is easily found by applying Lenz's law; the induced current in a loop is in a direction to always resist the change in flux through that loop. $\bar{\mathscr{E}}_e$ is in the direction of the induced current. It is important to note that

$$\psi_m = \psi_m(\text{applied}) + \psi_m(\text{induced}) \tag{8}$$

where the first term is that applied and would be the net flux if the induced current were prohibited from flowing. The second term is due to the induced current. In many cases when a loop is broken, so the "emf" can be brought out, one assumes that the capacity of the gap is zero, so no induced current is possible. Just a dc open is not sufficient to eliminate the induced current. Displacement current exists across the gap, which completes the loop for the conduction current in the metal. Notice that, unlike $\bar{\mathscr{E}}_c$ lines, which start and end on positive and negative charge, $\bar{\mathscr{E}}_e$ lines form closed loops. Often, a conducting loop is placed in the $\bar{\mathscr{E}}_e$ field and a small gap is made so that the 'emf' can be sensed with a meter; see Figure 1.16. The reference sign convention for $e$ is shown as well as the instantaneous charge distribution at the gap, in addition to the current in the loop. Observe that $e$ is negative at the instant shown, and the connected voltmeter reads a negative number. At the instant shown, $\bar{\mathscr{E}}_c$ is codirectional with $\bar{\mathscr{E}}_e$ in the gap region. However, at point $P$, they are contradirectional. Thus, in general, $\bar{\mathscr{E}}_c$ and $\bar{\mathscr{E}}_e$ can have any relative orientation.

There are several ways voltage may be expressed in the general time-varying case. If we start from the lead developed in statics, we have

$$\Phi_{AB} = -\int_B^A \bar{\mathscr{E}}_c(r) \cdot \overline{\mathbf{dl}} = -\int_B^A \frac{\bar{\mathscr{F}}_c}{q} \cdot \overline{\mathbf{dl}} \tag{9}$$

where we have used the fact that $\bar{\mathscr{E}}_c = \bar{\mathscr{F}}_c/q$. In the general case, the net force is

$$\bar{\mathscr{F}} = q[\bar{\mathscr{E}}(\bar{\mathbf{r}},t) + \bar{\mathbf{v}}(\bar{\mathbf{r}},t) \times \bar{\mathscr{B}}(\bar{\mathbf{r}},t)]$$

then $\bar{\mathscr{F}}/q$, when inserted into Eq. (9), yields

$$\mathscr{V}_{AB} = -\int_B^A \bar{\mathscr{E}}(\bar{\mathbf{r}},t) \cdot \overline{\mathbf{dl}} - \int_B^A \bar{\mathbf{v}} \times \bar{\mathscr{B}} \cdot \overline{\mathbf{dl}}$$

The second term on the rhs can be the Hall voltage in the case of semiconductors. In most cases, the first expression given for $\mathscr{V}$ (at the beginning of this example) is the one chosen. ▲

**Example 1.7** Show that the voltage defined for a TEM mode in a two-conductor system is unique.

Start from the general expression for the total electric field in terms of the scalar and vector potentials:

$$\bar{\mathscr{E}}(\bar{\mathbf{r}},t) = -\nabla\Phi(\bar{\mathbf{r}},t) - \frac{\partial}{\partial t}\bar{\mathbf{A}}(\bar{\mathbf{r}},t)$$

**34** INTRODUCTION TO DISTRIBUTED CIRCUITS

Define $\mathscr{V}_{21}$ by

$$\mathscr{V}_{21}(t) = -\int_1^2 \bar{\mathscr{E}} \cdot \overline{\mathbf{dl}} = \int_1^2 \nabla \Phi(\bar{\mathbf{r}}, t) \cdot \overline{\mathbf{dl}} + \int_1^2 \frac{\partial}{\partial t} \bar{\mathbf{A}}(\bar{\mathbf{r}}, t) \cdot \overline{\mathbf{dl}}$$

$$= \int_1^2 \frac{\partial \Phi}{\partial l} \, dl + \frac{d}{dt} \int_1^2 \bar{\mathbf{A}} \cdot \overline{\mathbf{dl}}$$

$$= \Phi_2(t) - \Phi_1(t) + \frac{d}{dt} \int_1^2 \bar{\mathbf{A}}(\bar{\mathbf{r}}, t) \cdot \overline{\mathbf{dl}}$$

For the voltage to be uniquely derivable from $\Phi$, we want the last term to vanish. For our uniform line, all surface currents are $z$-directed: $\bar{\mathbf{K}}_s = |K_s|\bar{\mathbf{a}}_z$. From the defining relationship between $\bar{\mathbf{K}}_s$ and $\bar{\mathbf{A}}$, $\bar{\mathbf{K}}_s$ and $\bar{\mathbf{A}}$ are always oriented in the same direction. Thus, $\bar{\mathbf{A}} = A\bar{\mathbf{a}}_z$ and since our integration is confined to the $xy$ plane, we have $\overline{\mathbf{dl}} = dx\bar{\mathbf{a}}_x + dy\bar{\mathbf{a}}_y$. Then $\bar{\mathbf{A}}(\bar{\mathbf{r}}, t) \cdot \overline{\mathbf{dl}} = 0$. Thus, the last term vanishes as it must if $\mathscr{V}_{21}$ is to be unique. ▲

The equation for $\mathscr{V}$ is developed from Faraday's law:

$$\oint \bar{\mathscr{E}} \cdot \overline{\mathbf{dl}} = -\frac{d}{dt} \int_s \bar{\mathscr{B}} \cdot \overline{\mathbf{ds}} = -\frac{d\psi_m}{dt} \quad (1.56)$$

where $\psi_m$ is the magnetic flux enclosed by the contour defined on the left side. The left side is (see (Figure 1.17a) evaluated about the rectangle 1234 as

$$\oint \bar{\mathscr{E}} \cdot \overline{\mathbf{dl}} = \int_1^2 + \int_2^3 + \int_3^4 + \int_4^1 = \int_2^3 + \int_4^1$$

To contributions on the paths 1–2 and 3–4 are zero for either of two reasons: Either $\bar{\mathscr{E}}_{\tan}$ is zero on a perfect conductor, or $\bar{\mathscr{E}}_t \cdot \overline{\mathbf{dl}} = 0$ as $\bar{\mathscr{E}}_t$ is transverse (confined to the $xy$ plane). The charge and voltage distributions are shown for clarity; then from the definition [Eq. 1.48], we have

$$\mathscr{V}_{32}(z + \Delta z, t) = -\int_2^3 \bar{\mathscr{E}} \cdot \overline{\mathbf{dl}}$$

$$\mathscr{V}_{14}(z, t) = -\int_4^1 \bar{\mathscr{E}} \cdot \overline{\mathbf{dl}}$$

so Eq. (1.56) becomes

$$-\mathscr{V}_{32}(z + \Delta z, t) - \mathscr{V}_{14}(z, t) = -\frac{d}{dt} \psi_m$$

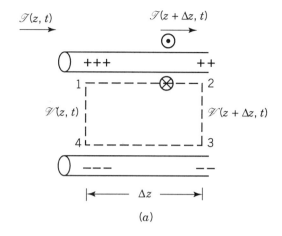

**FIGURE 1.17** The application of Eq. (1.56) about the rectangle 1234 (a); and the application of the continuity expression part (b). The structure is assumed to be propagating a TEM wave moving to the right.

but $\mathscr{V}_{23} = -\mathscr{V}_{32}$, and $\psi_m \triangleq L\Delta z \mathscr{I}(z + \Delta z/2, t)$, so

$$\mathscr{V}_{23}(z + \Delta z, t) - \mathscr{V}_{14}(z, t) = -L\Delta z \frac{\partial}{\partial t} \mathscr{I}\left(z + \frac{\Delta z}{2}, t\right)$$

Recall that $L$ is a distributed quantity (H/m); thus, $L\Delta z$ is the net inductance associated with $\psi_m$ in the rectangle 1234. We drop the subscripts by henceforth defining the reference polarity of $\mathscr{V}$ to be + to − with the upper conductor positive. Now divide by $\Delta z$ and take the limit:

$$\lim_{\Delta z \to 0} \frac{\mathscr{V}(z + \Delta z, t) - \mathscr{V}(z, t)}{\Delta z} = \lim_{\Delta z \to 0} \left[ -L \frac{\partial}{\partial t} \mathscr{I}\left(z + \frac{\Delta z}{2}, t\right) \right]$$

or

$$\frac{\partial \mathscr{V}}{\partial z} = -L \frac{\partial \mathscr{I}}{\partial t} \qquad (1.57)$$

## 36  INTRODUCTION TO DISTRIBUTED CIRCUITS

The equation for $\mathscr{I}$ is derived from the continuity expression:

$$\oint \bar{\mathscr{J}} \cdot \overline{\mathbf{ds}} = -\frac{\partial Q}{\partial t} \tag{1.58}$$

Apply this to the upper conductor by using a small cylinder as the surface of integration. (See Figure 1.17b). At plane 1–4, we have

$$\int_s \bar{\mathscr{J}} \cdot \overline{\mathbf{ds}}_1 = -\mathscr{I}(z,t)$$

The negative sign occurs since $\bar{\mathscr{J}}$ and $\overline{\mathbf{ds}}_1$ are oppositely directed. At plane 2–3, we find

$$\int_s \bar{\mathscr{J}} \cdot \overline{\mathbf{ds}}_2 = \mathscr{I}(z+\Delta z, t)$$

There is no contribution from the sidewall, since $\bar{\mathscr{J}}$ is z-directed. From the definition of $Q$, $C$, and $\mathscr{V}$ about the rectangle, we have

$$Q \triangleq C\Delta z \mathscr{V}\left(z+\frac{\Delta z}{2}, t\right)$$

Then Eq. (1.58) becomes

$$\mathscr{I}(z+\Delta z, t) - \mathscr{I}(z,t) = -\frac{\partial}{\partial t}\left[C\Delta z \mathscr{V}\left(z+\frac{\Delta z}{2}, t\right)\right]$$

Again divide by $\Delta z$ and pass to the limit:

$$\frac{\partial \mathscr{I}(z,t)}{\partial z} = -C\frac{\partial \mathscr{V}(z,t)}{\partial t} \tag{1.59}$$

Therefore, the set of field vectors $\{\bar{\mathscr{E}}_t, \bar{\mathscr{H}}_t\}$ can be studied using the set of scalars $\{\mathscr{V}, \mathscr{I}\}$ once $L$ and $C$ of the line are specified.

## 1.6  DISTRIBUTED CIRCUIT ANALYSIS

The previous section developed the differential equations for the voltage and current waves on a TL if we assume lossless conditions and a pure TEM mode propagating. The developments started with Faraday's law and the continuity equation. The electric and magnetic fields, in that case, were the fundamental quantities. Distributed circuit analysis shows how the TL equations may be developed using Kirchoff's laws of lumped circuit theory, with the necessary provisos outlined. Then losses can be included in the differential equations for $\mathscr{V}$ and $\mathscr{I}$ with minimal effort. The benefit of this method is that it bypasses the use of the field equations. Its utility is evident when treating coupled lines, which are covered later.

Now assume losses in both the conductors and the surrounding dielectric. Conductor loss is due to the wave energy penetrating the metal (skin effect), whereas polarization and leakage currents constitute the dielectric loss mechanisms. Wave damping due to polarization current is dominant in plasticlike dielectrics (e.g., Teflon), whereas actual conductive current losses occur when the conductors are separated by high-resistivity semiconductors. The net effect of losses is to alter the mode of propagation, and a pure TEM mode no longer exists. The quasi-TEM conditions give electric and magnetic field components in directions other than the transverse plane.

Figure 1.18 shows a small section of TL where series and shunt loss elements have been added to the principal $L$ and $C$ distributed portions. The inductance results from the flux lines between the conductors (lossless case condition), as well as that residing in the skin depths. The traveling electric and magnetic fields are accompanied by surface currents in the conductors in addition to the displacement current between them. Recall that displacement current is just $\partial \mathscr{D}/\partial t$. The

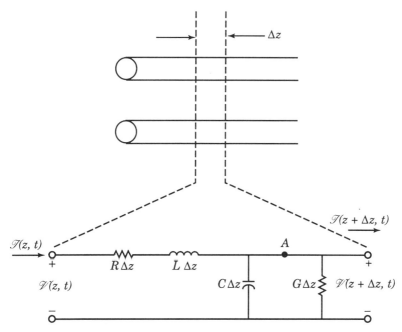

**FIGURE 1.18** A differential length of line is assumed to possess distributed series inductance $L$ and resistance $R$, as well as shunt capacitance $C$ and conductance $G$. Kirchoff's laws are assumed to hold for the small length.

**38** INTRODUCTION TO DISTRIBUTED CIRCUITS

differential length is much smaller than the wavelength of the traveling wave field, and under that condition Kirchoff's laws of lumped circuit theory are assumed to hold. Start by writing a node equation at $A$:

$$\mathscr{I}(z+\Delta z, t) - \mathscr{I}(z,t) = -G\Delta z \mathscr{V}(z+\Delta z, t) - C\Delta z \frac{\partial}{\partial t}\mathscr{V}(z+\Delta z, t)$$

Divide by $\Delta z$ and take the limit:

$$\frac{\partial \mathscr{I}(z,t)}{\partial z} = -C\frac{\partial \mathscr{V}(z,t)}{\partial t} - G\mathscr{V}(z,t) \qquad (1.60)$$

which is the extension of Eq. (1.59) for the lossy case. Next, consider a loop equation. The notions of voltage, emf, and potential considered in Example (1.6) are useful here. In lumped circuit theory, no time-varying flux is assumed to be threading the plane of the loop when a loop equation is written. If the condition occurs, the effect of the emf is modeled by a voltage source included in the loop. For this case, the magnetic flux between the conductors is removed and its effect incorporated in the voltages at both ends of the section. Simultaneously, the voltages at the ends are effectively line integrals of the electric field of the traveling wave. Therefore, the voltages at the ends serve to indicate the change in electric field with position along the line, as well as absorbing the needed emf due to removal of magnetic flux between the conductors. The loop equation is

$$\mathscr{V}(z+\Delta z, t) = \mathscr{V}(z,t) - \mathscr{I}(z,t)R\Delta z - L\Delta z \frac{\partial}{\partial t}\mathscr{I}(z,t)$$

Notice that we have assumed voltage drops across the series elements. Passing to the limit results in

$$\frac{\partial \mathscr{V}(z,t)}{\partial z} = -L\frac{\partial \mathscr{I}(z,t)}{\partial t} - R\mathscr{I}(z,t) \qquad (1.61)$$

When the distributed losses are eliminated, we see the above equations reduce to the lossless case. In real transmission lines, the distributed elements $L, C, R, G$ are weak functions of frequency, which is often ignored.

**Example 1.8** Find the complete expression for the time-varying fields for a coaxial cable. From the field relationships, develop the scalar equations for $\mathscr{V}(z,t)$ and $\mathscr{I}(z,t)$. Start the analysis with Maxwell's equations.

In the region between the conductors ($a < \rho < b$) (see Example 1.3), we apply the field equations:

$$\nabla \times \bar{\mathscr{E}} = -\mu \frac{\partial \bar{\mathscr{H}}}{\partial t} \qquad (1a)$$

$$\nabla \times \bar{\mathscr{H}} = \varepsilon \frac{\partial \bar{\mathscr{E}}}{\partial t} \qquad (1b)$$

## DISTRIBUTED CIRCUIT ANALYSIS

$$\nabla \cdot \varepsilon \bar{\mathscr{E}} = 0 \tag{1c}$$

$$\nabla \cdot \mu \bar{\mathscr{H}} = 0 \tag{1d}$$

The dielectric is perfect, so no conduction current appears in Eq. (1b), and no free charge occurs in Eq. (1c). We want the TEM solution, so set $\mathscr{E}_z = \mathscr{H}_z = 0$. We know that the field components of the static solution are the same as those in the time-varying TEM case. Thus, $\mathscr{E}_\varphi = \mathscr{H}_\rho = 0$. Then the nonzero components are $\mathscr{E}_\rho$ and $\mathscr{H}_\varphi$. Expanding the above equations with this information yields.

$$\nabla \times \bar{\mathscr{E}} = \frac{\partial \mathscr{E}_\rho}{\partial z} \bar{\mathbf{a}}_\varphi$$

Then Eq. (1a) becomes

$$\frac{\partial \mathscr{E}_\rho}{\partial z} = -\mu \frac{\partial \mathscr{H}_\varphi}{\partial t} \tag{2a}$$

In a similar fashion, Eq. (1b) yields

$$-\frac{\partial \mathscr{H}_\varphi}{\partial z} = \varepsilon \frac{\partial \mathscr{E}_\rho}{\partial t} \tag{2b}$$

Instead of solving the previous two equations, we can write the solutions down from the results of Example (1.3):

$$\mathscr{E}_\rho = \mathscr{E}_t(\rho, \varphi, z, t) = -\frac{V_0}{\ln(a/b)} \frac{1}{\rho} \left[ f\left(t \pm \frac{z}{v}\right) \right] \tag{3a}$$

$$\mathscr{H}_\varphi = \mathscr{H}_t(\rho, \varphi, z, t) = \frac{I}{2\pi\rho} \left[ f\left(t \pm \frac{z}{v}\right) \right] \tag{3b}$$

for simplicity, we assume that just forward waves (traveling toward positive $z$ exist. Then $f(t - z/v)$ applies. We find the relationship between the arbitrary amplitudes $V_0$ and $I$ by using Eqs. (2a) and (2b). The first implies

$$V_0 \frac{\partial f}{\partial z} = \frac{\mu I}{2\pi} \ln\left(\frac{a}{b}\right) \frac{\partial f}{\partial t}$$

and the second requires

$$\frac{I}{2\pi} \frac{\partial f}{\partial z} = \frac{\varepsilon V_0}{\ln(a/b)} \frac{\partial f}{\partial t}$$

**40** INTRODUCTION TO DISTRIBUTED CIRCUITS

Equating $\partial f/\partial z$ yields

$$\frac{V_0}{I} = \sqrt{\frac{\mu}{\varepsilon}} \frac{1}{2\pi} \ln\left(\frac{a}{b}\right) = Z_0 \tag{3c}$$

where the last result is obtained from Example (1.4). We now express Eqs. (3a) and (3b) in final form:

$$\mathscr{E}_\rho = \frac{Z_0 I}{\ln(a/b)} \frac{1}{\rho} f\left(t - \frac{z}{v}\right) \tag{3d}$$

$$\mathscr{H}_\varphi = \frac{I}{2\pi\rho} f\left(t - \frac{z}{v}\right) \tag{3e}$$

which are the complete expressions for the coaxial fields. The form of the function $f$ depends on the particular generator exciting the line. In the remaining parts of this example, we will use the results of Example (1.4).

To develop the scalar equations for $\mathscr{V}(z,t)$ and $\mathscr{I}(z,t)$, we note

$$\mathscr{V}_{ab} = -\int_b^a \vec{\mathscr{E}} \cdot \vec{dl} = -\int_b^a \mathscr{E}_\rho \, d\rho = Z_0 If$$

and using Eq. (3c) reduces this to [use Eq. (3a)]

$$\mathscr{V}_{ab} = \ln\left(\frac{b}{a}\right) \rho \mathscr{E}_\rho \tag{4}$$

Then

$$\frac{\partial \mathscr{V}_{ab}}{\partial z} = \ln\left(\frac{b}{a}\right) \rho \frac{\partial \mathscr{E}_\rho}{\partial z} = \ln\left(\frac{b}{a}\right) \rho \left(-\frac{\partial}{\partial t} \mathscr{H}_\varphi\right)$$

$$= \frac{-\mu}{2\pi} \ln\left(\frac{b}{a}\right) \frac{\partial}{\partial t} (If) \tag{5}$$

Defining $If(t - z/v) \triangleq \mathscr{I}(z,t)$, we have

$$\frac{\partial \mathscr{V}_{ab}}{\partial z} = -L \frac{\partial \mathscr{I}(z,t)}{\partial t} \tag{6}$$

which is our first desired result. The other expression follows from Eqs. (2b), (3d), and (4). We find

$$\frac{\partial \mathscr{H}_\varphi}{\partial z} = \frac{\partial}{\partial z}\left(\frac{\mathscr{I}}{2\pi\rho}\right) = -\varepsilon \frac{\partial}{\partial t} \mathscr{E}_\rho$$

$$= -\varepsilon \frac{\partial}{\partial t}\left[\frac{1}{\rho} \frac{1}{\ln(b/a)} \mathscr{V}_{ab}\right]$$

then

$$\frac{\partial \mathscr{I}}{\partial z} = -\frac{2\pi\varepsilon}{\ln(b/a)}\frac{\partial \mathscr{V}_{ab}}{\partial t}$$

$$= -C\frac{\partial \mathscr{V}_{ab}}{\partial t} \tag{7}$$

which completes the problem. ▲

## 1.7  LOSSLESS CASE SOLUTION

The lossless case equations ($R = G = 0$) for the traveling waves of $\mathscr{V}$ and $\mathscr{I}$ are

$$\frac{\partial \mathscr{V}(z,t)}{\partial z} = -L\frac{\partial \mathscr{I}(z,t)}{\partial t} \tag{1.62}$$

$$\frac{\partial \mathscr{I}(z,t)}{\partial z} = -C\frac{\partial \mathscr{V}(z,t)}{\partial t} \tag{1.63}$$

and the solution to them proceeds as follows. Take $\partial/\partial z$ of the first and use the second while noting $\partial^2/\partial z \partial t = \partial^2/\partial t \partial z$ for $\mathscr{I}$, since all functions are assumed differentiable to at least second order. We obtain

$$\frac{\partial^2 \mathscr{V}}{\partial z^2} = LC\frac{\partial^2 \mathscr{V}}{\partial t^2} \tag{1.64}$$

which is the one-dimensional wave equation. The solutions are of the form:

$$\mathscr{V}^+(z,t) = f\left(t - \frac{z}{v}\right) \tag{1.65a}$$

$$\mathscr{V}^-(z,t) = f\left(t + \frac{z}{v}\right) \tag{1.65b}$$

where the superscripts $+$ and $-$ denote wave propagation (energy flow) in the positive $z$ direction (forward wave) and negative $z$ direction, respectively. A few specific examples of permissible functions are

$$f_1(z,t) = \cos\left[\omega\left(t - \frac{z}{v}\right)\right]$$

$$f_2(z,t) = C_2 \cdot \left(t - \frac{z}{v}\right)^2, \quad C_2 \text{ a constant}$$

**42** INTRODUCTION TO DISTRIBUTED CIRCUITS

To show these forms satisfy Eq. (1.64), we first define auxiliary variables $u$ and $w$:

$$u = t - \frac{z}{v} \tag{1.66a}$$

$$w = t + \frac{z}{v} \tag{1.66b}$$

Then

$$f(z,t) = f\left(t - \frac{z}{v}\right) = f(u)$$

so

$$\mathscr{V}^+(z,t) = \mathscr{V}^+(u)$$

Start by forming $\partial \mathscr{V}^+(z,t)/\partial z$ and $\partial \mathscr{V}^+(z,t)/\partial t$:

$$\frac{\partial \mathscr{V}^+(z,t)}{\partial z} = \frac{d\mathscr{V}^+(u)}{du}\frac{\partial u}{\partial z}$$

$$\frac{\partial \mathscr{V}^+(z,t)}{\partial t} = \frac{d\mathscr{V}^+(u)}{du}\frac{\partial u}{\partial t}$$

but

$$\frac{\partial u}{\partial z} = -\frac{1}{v}, \qquad \frac{\partial u}{\partial t} = 1$$

so

$$\frac{\partial \mathscr{V}^+(z,t)}{\partial z} = \frac{d\mathscr{V}^+(u)}{du}\left(-\frac{1}{v}\right) \tag{1.66c}$$

$$\frac{\partial \mathscr{V}^+(z,t)}{\partial t} = \frac{d\mathscr{V}^+(u)}{du}(1) \tag{1.66d}$$

Take second partial derivatives:

$$\frac{\partial^2 \mathscr{V}^+}{\partial z^2} = \frac{\partial}{\partial z}\left[-\frac{1}{v}\frac{d\mathscr{V}^+(u)}{du}\right] = -\frac{1}{v}\frac{d^2\mathscr{V}^+(u)}{du^2}\frac{\partial u}{\partial z}$$

$$= \frac{1}{v^2}\frac{d^2\mathscr{V}^+(u)}{du^2} \tag{1.67}$$

$$\frac{\partial^2 \mathscr{V}^+(z,t)}{\partial t^2} = \frac{d^2\mathscr{V}^+(u)}{du^2}\frac{\partial u}{\partial t} = \frac{d^2\mathscr{V}^+(u)}{du^2} \tag{1.68}$$

Equate the term $d^2 \mathcal{V}^+(u)/du^2$ in Eqs. (1.67) and (1.68) to find

$$v^2 \frac{\partial^2 \mathcal{V}^+(z,t)}{\partial z^2} = \frac{\partial^2 \mathcal{V}^+(z,t)}{\partial t^2}$$

which satisfies Eq. (1.64) when $v = 1/\sqrt{LC}$. Since $v$ is the velocity, it is always considered a positive number. Similar steps will show that $\mathcal{V}^-(z,t)$ also satisfies Eq. (1.64), where $v$ is the same. The equation for the current is

$$\frac{\partial^2 \mathcal{I}(z,t)}{\partial z^2} = LC \frac{\partial^2 \mathcal{I}(z,t)}{\partial t^2} \tag{1.69}$$

which is identical in form with Eq. (1.64), so the solutions are of the same type. Since $\mathcal{V}$ and $\mathcal{I}$ are coupled, it is important to ascertain the relationships between them. First, consider waves traveling to the right. Then

$$\mathcal{V} = \mathcal{V}^+\left(t - \frac{z}{v}\right), \qquad \mathcal{I} = \mathcal{I}^+\left(t - \frac{z}{v}\right)$$

Assume that $\mathcal{I}^+$ and $\mathcal{V}^+$ are related by $\mathcal{I}^+ = \mathcal{V}^+/k$, where $k$ is a constant to be determined. Then using Eqs. (1.62) and (1.63), we note:

$$\frac{\partial \mathcal{V}^+}{\partial z} = -\frac{L}{k} \frac{\partial \mathcal{V}^+}{\partial t} \tag{1.62a}$$

$$\frac{1}{k} \frac{\partial \mathcal{V}^+}{\partial z} = -C \frac{\partial \mathcal{V}^+}{\partial t} \tag{1.62b}$$

which means $L = Ck^2$, or $k = \sqrt{L/C}$. The positive square root is appropriate since Eq. (1.66c) must hold. Now use Eqs. (1.66c) and (1.66d):

$$-\frac{1}{v} \frac{d\mathcal{V}^+}{du} = -\frac{L}{k} \frac{d\mathcal{V}^+}{du}$$

or $k = vL = \sqrt{L/C}$, or $k \triangleq R_0$. Thus,

$$\mathcal{I}^+(z,t) = \frac{1}{R_0} \mathcal{V}^+(z,t) \tag{1.70}$$

Next, for waves traveling to the left, we have

$$\mathcal{V} = \mathcal{V}^-\left(t + \frac{z}{v}\right), \qquad \mathcal{I} = \mathcal{I}^-\left(t + \frac{z}{v}\right)$$

## 44 INTRODUCTION TO DISTRIBUTED CIRCUITS

and assume that

$$\mathscr{I}^- = \frac{1}{a}\mathscr{V}^-$$

In a similar fashion, we find $a^2 = L/C$, but we must choose the negative square root for the following reason. First,

$$\frac{\partial \mathscr{V}^-}{\partial z} = \frac{d\mathscr{V}^-(w)}{dw}\frac{\partial w}{\partial z} = \frac{1}{v}\frac{d\mathscr{V}^-}{dw}$$

$$\frac{\partial \mathscr{V}^-}{\partial t} = \frac{d\mathscr{V}^-(w)}{dw}\frac{\partial w}{\partial t} = \frac{d\mathscr{V}^-}{dw}$$

Then Eqs. (1.62) and (1.63) become

$$\frac{1}{v}\frac{d\mathscr{V}^-}{dw} = -\frac{L}{a}\frac{\partial \mathscr{V}^-}{\partial t} = -\frac{L}{a}\frac{d\mathscr{V}^-}{dw}$$

or

$$a = -vL = -\sqrt{L/C} = -R_0$$

so

$$\mathscr{I}^-(z,t) = -\frac{1}{R_0}\mathscr{V}^-(z,t) \tag{1.71}$$

The negative sign is important and needs explanation. Note for forward waves (energy moving to the right):

$$\mathscr{I}^+ = \frac{1}{R_0}\mathscr{V}^+$$

whereas for reverse waves (energy moving to the left):

$$\mathscr{I}^- = -\frac{1}{R_0}\mathscr{V}^-$$

In the development of the coupled partial differential equations, the reference polarity for both $\mathscr{V}^+$ and $\mathscr{V}^-$ was the same (upper positive with respect to the lower). The reference for both $\mathscr{I}^+$ and $\mathscr{I}^-$ was to the right. Assume that a wave is moving to the left, and without loss of generality assume that $\mathscr{V}^-$ is a positive number. Equation (1.71) indicates that the current $\mathscr{I}^-$ (directed to the right) associated with this energy movement is negative. Thus, $-\mathscr{I}^-$ (a positive number) is directed to the left. Then physically, the positive current $-\mathscr{I}^-$ is entering the positive node (at any transverse plane), and from the laws of lumped circuit theory, power is being transported to the left of said plane.

In summary, our fundamental equations are

$$\mathcal{V}(z,t) = \mathcal{V}^+\left(t - \frac{z}{v}\right) + \mathcal{V}^-\left(t + \frac{z}{v}\right) \tag{1.72a}$$

$$\mathcal{I}(z,t) = \frac{1}{R_0}\left[\mathcal{V}^+\left(t - \frac{z}{v}\right) - \mathcal{V}^-\left(t + \frac{z}{v}\right)\right] \tag{1.72b}$$

where the superscripts $^+$ and $^-$ mean propagation to the right and left, respectively. Each quantity $\mathcal{V}^+, \mathcal{V}^-, \mathcal{I}^+, \mathcal{I}^-$ can have positive or negative values, so long as they are consistent with both equations. The equations are general in that waves of different shapes [different functional expressions of the basic argument $(t \pm z/v)$] and amplitudes are permitted to move in both directions simultaneously. Physically, this can happen if generators at both ends of the line exist, and each is delivering a different signal onto the line.

## 1.8 LUMPED EQUIVALENT LIMITATIONS

If one scans 20 or more texts on electromagnetics, Ref. 5, for example, he or she finds many variations on the basic theme discussed in the previous sections. One major variation is the choice of signs in the coupled equations for $\mathcal{V}$ and $\mathcal{I}$:

$$\frac{\partial \mathcal{V}}{\partial z} = \pm L \frac{\partial \mathcal{I}}{\partial t} \tag{1.73a}$$

$$\frac{\partial \mathcal{I}}{\partial z} = \pm C \frac{\partial \mathcal{V}}{\partial t} \tag{1.73b}$$

A majority choose the negative sign; however, it makes no difference, since either set of signs leads to the wave equation:

$$\frac{\partial^2 \mathcal{V}}{\partial z^2} = LC \frac{\partial^2 \mathcal{V}}{\partial t^2} \tag{1.74}$$

which is the actual equation to be solved. The change in voltage $\partial \mathcal{V}/\partial z$ is best interpreted as a change with respect to distance of the instaneous charge distribution on the conductors with the corresponding change of the electric field between them. The apparent voltage drops across the series-lumped elements yield the same final equations. The choice of the negative sign is therefore more natural since "voltage drops" are more familiar. One conceptual problem with the lumped model occurs for the lossless case; there are no z components of electric field, but the "drop" across the inductor must be derived from such a component. Therefore, the inductance should be defined in such a way that the stored energy per unit length associated with it is equal to the stored energy in the magnetic field per unit length. Although the lumped, or distributed, circuit approach is not always easy to justify, it is our workhorse. Predictions from it agree with experiments, which constitutes the justification for its use.

***Example 1.9*** Discuss the definitions of the L, C, R, and G elements with respect to the actual fields about the transmission line.

A satisfactory way to define the line elements L, C, R, and G is to relate the stored energy and dissipated power of them with corresponding expressions for the fields. for the distributed inductor, we have

$$W_m = \tfrac{1}{4} II^* L$$

which is the stored energy per unit length (from lumped circuit theory). The corresponding energy in the magnetic field is

$$W_m = \frac{\mu}{4} \int \mathbf{H} \cdot \bar{\mathbf{H}}^* \, ds$$

where $\bar{\mathbf{H}}(\bar{r})$ is the phasor representation (phasors always in boldface type) for the $\mathcal{H}(\bar{r}, t)$ field. Here, we should use the lossfree field solution. Then equating the stored energy yields

$$L = \frac{\mu}{II^*} \int \bar{\mathbf{H}} \cdot \bar{\mathbf{H}}^* \, ds$$

For the capacitor,

$$W_e = \tfrac{1}{4} VV^* C = \varepsilon' \int \bar{\mathbf{E}} \cdot \bar{\mathbf{E}}^* \, ds$$

or

$$C = \frac{\varepsilon'}{VV^*} \int \bar{\mathbf{E}} \cdot \bar{\mathbf{E}}^* \, ds$$

where $\varepsilon = \varepsilon' - j\varepsilon''$, and the $\varepsilon''$ term represents the loss of the dielectric. For the resistor and conductance, we equate dissipated power:

$$R = \frac{R_m}{II^*} \int_{s_1 + s_2} \bar{\mathbf{H}} \cdot \bar{\mathbf{H}}^* \, dl$$

where $dl$ is a contour enclosing the surfaces $s_1$ and $s_2$ of the two conductors in a transverse plane. $R_m$ is the surface resistivity:

$$G = \frac{\omega \varepsilon''}{VV^*} \int \bar{\mathbf{E}} \cdot \bar{\mathbf{E}}^* \, ds$$

Note that

$$\frac{G}{C} = \frac{\omega \varepsilon''}{\varepsilon'} = \omega \tan \delta \quad \text{[see Example (1.1)]}$$

which permits the determination of $G$ from $C$ and the measured values for $\varepsilon'$ and $\varepsilon''$. ▲

## SUPPLEMENTARY EXAMPLES

**1.1** To develop an expression using $\mathscr{V}$ and $\mathscr{I}$ that is similar to the Poynting expression for the field vectors, start with the TL equations that include losses and multiply by $\mathscr{V}$ and $\mathscr{I}$:

$$\mathscr{V}\frac{\partial \mathscr{I}}{\partial z} = \mathscr{V}\left(-C\frac{\partial \mathscr{V}}{\partial t} - G\mathscr{V}\right)$$

$$\mathscr{I}\frac{\partial \mathscr{V}}{\partial z} = \mathscr{I}\left(-L\frac{\partial \mathscr{I}}{\partial t} - \mathscr{I}R\right)$$

add these:

$$\mathscr{V}\frac{\partial \mathscr{I}}{\partial z} + \mathscr{I}\frac{\partial \mathscr{V}}{\partial z} = \frac{\partial}{\partial z}(\mathscr{V}\mathscr{I}) = -C\mathscr{V}\frac{\partial \mathscr{V}}{\partial t} - L\mathscr{I}\frac{\partial \mathscr{I}}{\partial t} - G\mathscr{V}^2 - \mathscr{I}^2 R$$

and use the fact that

$$\frac{\partial}{\partial t}(\tfrac{1}{2}C\mathscr{V}^2) = C\mathscr{V}\frac{\partial \mathscr{V}}{\partial t}$$

Then

$$\frac{\partial}{\partial z}(\mathscr{V}\mathscr{I}) = -\frac{\partial}{\partial t}(\tfrac{1}{2}C\mathscr{V}^2 + \tfrac{1}{2}L\mathscr{I}^2) - G\mathscr{V}^2 - \mathscr{I}^2 R$$

which is the desired expression. It may be interpreted like the Poynting equation in that the left side represents the rate of change of power with position, and the right is the change in stored energy in the lumped $C$ and $L$ elements, along with the losses in $R$ and $G$.

**1.2** We have treated $L$ and $C$ as primary variables and developed $v_p$ and $Z_0$ from them. In practice, however, one measures $\lambda$ and $Z_0$ and then calculates $L$ and $C$ if desired. For example, the wavelength can often be measured with a probe at a given frequency $f$. Then the velocity is $v_p = f\lambda$. The characteristic impedance can be measured in several ways, then $C$ and $L$ follow from

$$C = \frac{1}{v_p Z_0}$$

$$L = \frac{Z_0}{v_p}$$

Suppose that a coaxial cable has $R_0 = 50\,\Omega$ and $v_p = 2 \times 10^8$ m/s. Then

**48** INTRODUCTION TO DISTRIBUTED CIRCUITS

$L$ and $C$ are (assume lossless conditions)

$$C = \frac{1}{v_p R_0} = 100 \, \text{pF/m}$$

$$L = CZ_0^2 = CR_0^2 = 0.25 \, \mu\text{H/m}$$

**1.3** We have stated that the TL equations require the principal mode of propagation to exist, which is the TEM mode. It exists from dc to the cutoff frequency of the first "higher-order" mode. The higher-order mode is the $TE_{11}$ mode, and it is avoided by restricting the frequencies used in the cable to below the cutoff value. The cutoff wavelength for this mode in an air-filled line is approximately (see sketch below for dimensions)

$$\lambda_c \doteq \frac{\pi}{2}(d + D)$$

Then that for a line filled with material of dielectric constant $\varepsilon_r$ is

$$\lambda_c' = \sqrt{\varepsilon_r} \lambda_c$$

The cutoff frequency is then (with $c$ the speed of light in air)

$$f_c' = \frac{c}{\lambda_c' \sqrt{\varepsilon_r}} = \frac{2c}{\pi \sqrt{\varepsilon_r}(d + D)}$$

which is accurate to about 8%. The sketch below gives a crude display of the field pattern:

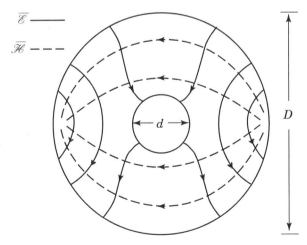

**FIGURE 1.19**

**1.4** Why is 50 Ω used as the "standard" reference impedance? It can be shown that different impedance levels of a coaxial line optimize various quantities, such as minimum loss, or maximum power-carrying capability (see Ref. 6). Depending on the assumed constraints, different $Z_0$ values are optimum. For example, the minimum loss under the constraint that no $TE_{11}$ mode (the "higher-order" mode) occurs, is when $Z_0 = 92.6\,\Omega$. However, for a fixed outer diameter, the minimum loss occurs for $Z_0 = 77\,\Omega$. Similarly, the maximum power-handling capacity under the above constraints is 44.4 and 30 Ω, respectively. For a fixed outer diameter, maximum field strength before breakdown in air occurs for $Z_0 = 60\,\Omega$. Finally for resonant lines, the highest quality factor $Q$ occurs for $Z_0 = 133\,\Omega$. The value of 50 is a reasonable compromise of all the above and has been chosen as the standard.

**1.5** We find the internal inductance of a nonmagnetic wire of radius $a$ using stored energy concepts. Assume that the wire carries a uniformly distributed current $I$. Then at some radius $r$, $0 < r < a$ the current enclosed is

$$I_{enclosed} = \oint \mathcal{H} \cdot \overline{dl} = I \frac{\pi r^2}{\pi a^2}$$

Also, the magnetic field is just

$$\oint \mathcal{H} \cdot dl = \mathcal{H}_\varphi (2\pi r) = I$$

$$\therefore \quad \mathcal{H}_\varphi = \frac{I}{2\pi a^2} r$$

The stored energy in the wire is

$$W_m = \frac{1}{2} \int \mu_0 \mathcal{H}^2 \, dvol = \frac{1}{2} \int \mu_0 \left( \frac{I^2}{4\pi^2 a^4} r^2 \right) dvol$$

$$= \frac{\mu_0 I^2}{8\pi^2 a^4} \int_0^a \int_0^{2\pi} r^2 \, d\varphi \, dr = \frac{\mu_0 I^2}{16\pi} J$$

Equating this to the stored energy in a lumped inductor gives

$$L = \frac{2W_m}{I^2} = \frac{\mu_0}{8\pi} \text{ H/m}$$

which is the desired result.

## PROBLEMS

**1.1** Find the time average of the squared magnitude of the electric field in Eq. (1.25). Show that it may be expressed as $\overline{E} \cdot \overline{E}^* / 2$.

**1.2** Find the phasor notation for the following time-harmonic vector:

$$\vec{\mathscr{E}}(t) = (3\cos\omega t + 5\sin\omega t)\bar{a}_x + 8(\cos\omega t - 2\sin\omega t)\bar{a}_z$$

**1.3** The complex Poynting vector is defined as $\bar{S} = \bar{E} \times \bar{H}^*/2$, and its time average was given in Eq. (1.39). Using the identity

$$Re\{\bar{A}\} = \tfrac{1}{2}(\bar{A} + \bar{A}^*)$$

for a complex vector, show

$$Re\{\bar{E}\} \times Re\{\bar{H}\} = \tfrac{1}{2}Re\{\bar{E} \times \bar{H}^* + \bar{E} \times \bar{H}\}$$

Then show that the instantaneous Poynting vector may be expressed as

$$\mathscr{S}(\bar{r}, t) = \tfrac{1}{2}Re\{2\bar{S} + \bar{E} \times \bar{H}e^{j2\omega t}\}$$

**1.4** Develop Eq. (1.40) starting with $\nabla \cdot \bar{J} = 0$.

**1.5** Develop Eq. (1.41) starting with $\nabla \times \vec{\mathscr{E}} = 0$.

**1.6** Derive Eq. (1.50), the wave equation for $\vec{\mathscr{E}}$. Start with Maxwell's equations and state all assumptions.

**1.7** Find the expression for the time-averaged power flow in a coaxial cable. Use the following expression:

$$P_{avg} = \int \bar{S} \cdot \bar{ds} = \frac{1}{2}\int_{\text{surf}} Re\{\bar{E}_t \times \bar{H}_t^*\} \cdot \bar{a}_z r\, dr\, d\theta$$

**1.8** A ferrite is an anisotropic material formed by mixing and sintering a metal oxide with ferric oxide, $Fe_2O_3$. The constitutive relation between $\bar{B}$ and $\bar{H}$ shows that, in general, they are not collinear. The resistivity is high ($\rho > 10^{12}\,\Omega\cdot\text{cm}$), so the loss at microwave frequencies is low. The dielectric constants are between 5 and 20. The material is used with a static magnetic field $H_0\bar{a}_z$, and under this condition the constitutive relation is

$$\bar{B} = \vec{\mu}\bar{H}$$

where $\vec{\mu}$ is

$$\vec{\mu} = \mu_0 \begin{bmatrix} \mu & -j\kappa & 0 \\ j\kappa & \mu & 0 \\ 0 & 0 & 1 \end{bmatrix}, \quad \begin{aligned} \mu &= 1 - \frac{f_c f_m}{f^2 - f_c^2} \\ \kappa &= \frac{f f_m}{f^2 - f_c^2} \\ \mu_0 &= 4\pi \times 10^{-7}\,\text{H/m} \end{aligned}$$

where $f_c$ = cyclotron or gyromagnetic frequency = $2.8 \times 10^6 \, H_0$ and $H_0$ is in Oersteds (Oe). 1 Oe = $10^3/4\pi$ A/m. $f_m$ = magnetization frequency = $2.8 \times 10^6 \, M_S$ (in G). 1 G = $10^{-4}$ Wb/m². And $M_S$ is the saturation magnetization. Let $M_S = 500$ G, $H_0 = 400$ Oe, and assume that $f = 4$ GHz. Calculate the elements of $\vec{\mu}$ and use

$$\begin{bmatrix} B_x \\ B_y \\ B_z + B_0 \end{bmatrix} = \mu_0 \begin{bmatrix} \mu & -j\kappa & 0 \\ j\kappa & \mu & 0 \\ 0 & 0 & 1 \end{bmatrix} \begin{bmatrix} H_x \\ H_y \\ H_z + H_0 \end{bmatrix}$$

where $B_x$, $B_y$, $B_z$, $H_x$, $H_y$, and $H_z$ are small rf field components, such that $H_z \ll H_0$. Let $H_x = 20$ Oe, and $H_y = H_z = 0$ at a given instant. Then find the direction of $\bar{\mathbf{B}}$ at that instant. What is the angle between $\bar{\mathbf{B}}$ and $\bar{\mathbf{H}}$?

**1.9** Surface current is defined by $\bar{\mathbf{K}} = \bar{\mathbf{n}} \times \bar{\mathbf{H}}_a$, where $\bar{\mathbf{H}}_a$ is the tangential magnetic field. For the TEM mode in a coax, find the expression for $\bar{\mathbf{K}}$.

**1.10** For a teflon-filled semirigid coaxial cable with $Z_0 = 50 \, \Omega$, the inner diameter of the outer conductor is 0.090 in. What is the useful upper frequency limit for this cable? Here, $\varepsilon_r = 2.0$.

## REFERENCES

1. Zahn, Markus, *Electromagnetic Field Theory. A Problem Solving Approach*, John Wiley & Sons, New York, 1979, p. 94.
2. Hayt, William H., *Engineering Electromagnetics*, 4th ed., McGraw-Hill, New York, 1981.
3. Walsh, John B., *Electromagnetic Theory and Engineering Applications*, Ronald Press, New York, 1960, Chap. 8 and 9.
4. Haus, Herman A. and Melcher, James R., "Electric and Magnetic Fields," *Proc. IEEE*, 59, June 1971, pp. 887–894.
5. Durney, Carl H. and Johnson, Curtis C., *Introduction to Modern Electromagnetics*, McGraw-Hill, New York, 1969.
6. Ragan, George L., *Microwave Transmission Circuits*, MIT Radiation Laboratory Series, Boston Technical Publishers, Cambridge, MA, 1964, Chap. 4.

## BIBLIOGRAPHY

Reitz, J. R., and Milford, F. J., *Foundations of Electromagnetic Theory*, Addison-Wesley, Massachusetts, Reading, 1960.

Harrington, R. F., *Time-Harmonic Electromagnetic Fields*, McGraw-Hill, New York, 1961.

# CHAPTER TWO

# The Mathematics of Traveling Waves

## 2.1 PHYSICAL INTERPRETATIONS OF SOLUTIONS

The solutions to the one-dimensional wave equation were shown to be of the form:

$$\mathscr{V}^+ = f\left(t - \frac{z}{v}\right) \tag{2.1}$$

$$\mathscr{V}^- = f\left(t + \frac{z}{v}\right)$$

or of the form:

$$\mathscr{V}^+ = f(z - vt) \tag{2.2}$$

$$\mathscr{V}^- = f(z + vt)$$

The first forms express $f$ in terms of an argument that is the difference (or sum) of two "times," $t$ and $z/v$, whereas the latter have arguments that are the sum or difference of two "distances," $z$ and $vt$. For simplicity, we consider just forward waves ($\mathscr{V}^+$).

Recall the basic identity $\sin(\omega t) = \cos(\omega t - \pi/2)$, which shows that a sine wave is a cosine wave pushed $\pi/2$ radians to the right. This identity is a special case of the general rule that states a cosine wave pushed $\theta$ radians to the right is expressed as $\cos(\omega t - \theta)$. If it is pushed $\varphi$ radians to the left, its representation is $\cos(\omega t + \varphi)$. For a general function $f(t)$, a push to the right by $t_0$ means replacing $t$ by $t - t_0$ everywhere $t$ exists in $f$. The equation for the pushed curve is $f(t - t_0)$. Physically, $t_0$ is the time interval that $f(t - t_0)$ is "delayed" from $f(t)$. In an analogous manner, $g(z - z_0)$ is $g(z)$ pushed $z_0$ units to the right; see Figure 2.1c. The push rule

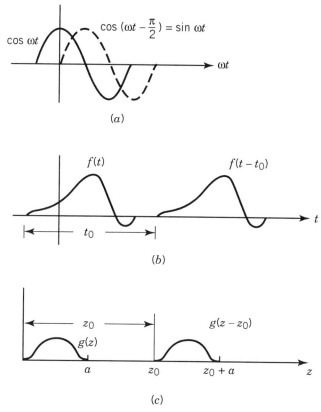

**FIGURE 2.1** (a) A cosine wave pushed $\pi/2$ to the right. (b) An arbitrary function delayed $t_0$ seconds. (c) An arbitrary function pushed $z_0$ units to the right.

was demonstrated for functions of a single variable. Our solution functions for lines are, however, functions of both $z$ and $t$. In this case, the push rule can be expanded on somewhat, with more involved connotations.

Since our functions $f(z, t)$ are developed by generators at either or both ends of a TL, they are assumed to be single-valued and well behaved. Specific expressions for $f(z, t)$ are

$$f_1(z, t) = \cos\left[\omega\left(t - \frac{z}{v}\right)\right] = \cos\left(\omega t - \frac{\omega}{v}z\right)$$

$$f_2(z, t) = (at - bz)^2, \quad a, b \text{ constants}$$

$$f_3(z, t) = AU\left(t - \frac{z}{v}\right)$$

The last function is a step of amplitude $A$ occurring at $t = z/v$. A nonpermissible function is

$$f_4(z, t) = z^2 - 3t$$

which can be verified by trying it in the wave equation. For forward waves, the variables $z$ and $t$ must be in the combination:

$$u = (at - bz)$$

where $a$ and $b$ must have the same sign. Then any well-behaved function $f(u)$ satisfies the wave equation. If $a = 1$, $b = 1/v$. Then

$$u = \left(t - \frac{z}{v}\right)$$

and the argument $u$ is the difference of "times." When $b = -1$ and $a = -v$,

$$u = (z - vt) \qquad (2.3)$$

and $u$ is the difference of "distances." Consider the function

$$f(z, t) = 2(z - vt)^2 = 2u^2 \qquad (2.3a)$$

which is a parabola and is sketched at two instants, $t = 0$ and $t = t_1$, in Figure 2.2. We may write $f$ as

$$f(z, t) = 2(z - z_0)^2 \qquad (2.4)$$

where $z_0$ is an implied function of time: $z_0 = vt$. This form is recognized as the parabola $2z^2$ pushed $z_0$ units to the right. Since $z_0$ changes linearly with time, the above represents the curve $2z^2$ sliding to the right (shape unchanged) at speed $v$.

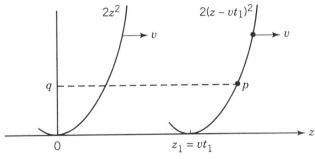

**FIGURE 2.2** The function $2u^2$ $(u = z - vt)$, at $t = 0$ and $t = t_1$. The point $p$ has ordinate $q$.

PHYSICAL INTERPRETATIONS OF SOLUTIONS    55

Now focus on the point $p$ that has ordinate $q$. As the parabola moves, the $z$ coordinate of $p$ changes, but its ordinate remains the same, at the value $q$. Let $z_p(t)$ be the coordinate of $p$. Now write $f(z,t)$ as

$$f(z,t) = f(u) = 2u^2$$

and let $u_p = z_p(t) - vt$ represent the value of $u$ for the point $p$. Since $f(u_p) = q$ for any time, then $u_p$ must be constant with time, or

$$\frac{du_p}{dt} = 0 = \frac{dz_p(t)}{dt} - v$$

or

$$\frac{dz_p(t)}{dt} = v \qquad (2.5)$$

so $p$ moves at the speed $v$. Often, one finds the above argument stated as "to a moving observer $u$ is constant, thus"

$$\frac{du}{dt} = 0 = \frac{dz}{dt} - v$$

or

$$\frac{dz}{dt} = v$$

This can be misinterpreted if one does not realize the '$z$' above is that for a particular point on the curve, that is, $z_p$. For one could ask, "But $z$ and $t$ are independent variables here, so what does $dz/dt$ mean"?

Notice that we have not labeled the vertical axes in Figures 2.1 and 2.2, since different functions have been sketched with respect to the abcissa, that is, $g(z)$ and $g(z - z_0)$. So, in general, the vertical axis in this chapter is not labeled. To sum up this section, consider a TL of infinite length with a signal applied at $z = 0$. Assume that $f(t)$ is nonzero only over the interval $(0, 1)$, and it is repeated by the generator every $T$ seconds. To an observer stationed at $z = 0$ (input plane of the line), the complete signal as a function of time would be

$$F(t) = f(t) + f(t - T) + f(t - 2T) + \cdots + f(t - nT) + \cdots$$

Now, consider an observer placed a distance $z_0$ down the line. The first signal does not arrive until $t_0 = z_0/v$ seconds; then the next one follows $T$ seconds later. The complete signal at $z_0$ would be

$$G(t) = f(t - t_0) + f(t - t_0 - T) + f(t - t_0 - 2T) + \cdots + f(t - t_0 - nT)$$

$$= f\left(t - \frac{z_0}{v}\right) + f\left(t - \frac{z_0}{v} - T\right) + f\left(t - \frac{z_0}{v} - 2T\right) + \cdots$$

**56**   THE MATHEMATICS OF TRAVELING WAVES

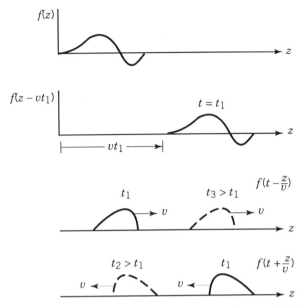

**FIGURE 2.3**  Examples of notation and sketching of waveforms with time delays and spatial shifts.

Then let the observation point $z_0$ become any point on the line $z$. At such a point,

$$H(t) = f\left(t - \frac{z}{v}\right) + f\left(t - \frac{z}{v} - T\right) + f\left(t - \frac{z}{v} - 2T\right) + \cdots \tag{2.6}$$

Figure 2.3 shows functions sliding both to the right and left, along with the normal notation. A concrete example follows.

***Example 2.1***  Assume that the excitation of a TL at $z = 0$ (its input) is (forward wave only)

$$f^+(0, t) = 2E_0 U(t) - E_0 U(t - T) - E_0 U(t - 2T) \tag{2.7}$$

as sketched in Figure 2.4. The signal travels at speed $v$ along positive $z$, and at the observation point $z_0$, all portions of $f$ are delayed by $t_0 = z_0/v$ seconds. Then at $z_0$, we have

$$f^+(z_0, t) = f^+(0, t - t_0) = 2E_0 U\left(t - \frac{z_0}{v}\right) - E_0 U\left(t - \frac{z_0}{v} - T\right)$$

$$- E_0 U\left(t - \frac{z_0}{v} - 2T\right) \tag{2.8}$$

PHYSICAL INTERPRETATIONS OF SOLUTIONS   57

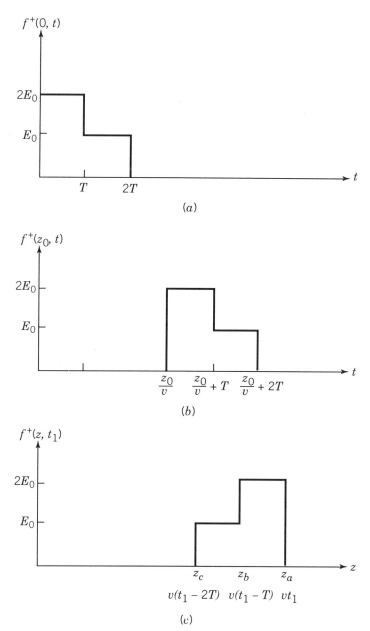

**FIGURE 2.4** (a) $f^+(0,t)$, the signal at the input to a TL. (b) The signal observed at a point $z_0$ down the line. (c) A snapshot at $t = t_1$, showing the distribution of the signal energy with $z$.

## 58  THE MATHEMATICS OF TRAVELING WAVES

In Equation 2.8, the expression has jumps at $z_0/v$, $z_0/v + T$, and $z_0/v + 2T$ [recall that $U(x)$ jumps when $x = 0$]. This function is also shown below. Therefore, for any arbitrary point $z$,

$$f^+(z,t) = 2E_0 U\left(t - \frac{z}{v}\right) - E_0 U\left(t - \frac{z}{v} - T\right) - E_0 U\left(t - \frac{z}{v} - 2T\right) \quad (2.9)$$

Now, consider the signal on the line at the particular instant $t_1$. This snapshot gives the distribution of the signal as a function of $z$:

$$f^+(z,t_1) = 2E_0 U\left(t_1 - \frac{z}{v}\right) - E_0 U\left(t_1 - \frac{z}{v} - T\right) - E_0 U\left(t_1 - \frac{z}{v} - 2T\right) \quad (2.10)$$

The jump points as well as the height of the jumps are

$$z_a = vt_1, \qquad 2E_0$$
$$z_b = v(t_1 - T), \qquad -E_0$$
$$z_c = v(t_1 - 2T), \qquad -E_0$$

which is sketched in Figure 2.4c. Note that $z_c < z_b < z_a$ and the pulse when plotted wrt $z$ is "spun around" from its representation wrt $t$ (at $z = 0$). Physically, this makes sense as the larger value ($2E_0$) should be further down the line than ($E_0$), since the value ($2E_0$) left the source earlier. ▲

*Example 2.2*  Consider the waveform in Figure 2.5 that is the staircase used in the previous example. Assume that now the TL extends to the left of the origin, so the signal propagates to the left (toward increasing negative $z$). Develop sketches similar to those of the example.

The upper sketch (a) is $f(0,t)$ and the signal is thereafter denoted as $f^-(z,t) = f(t + z/v)$, so

$$f^-(0,t) = 2E_0 U(t) - E_0 U(t - T) - E_0 U(t - 2T)$$

and

$$f^-(z,t) = 2E_0 U\left(t + \frac{z}{v}\right) - E_0 U\left(t - T + \frac{z}{v}\right) - E_0 U\left(t - 2T + \frac{z}{v}\right) \quad (1)$$

At some point $z_0 (z_0 < 0)$, we have the middle sketch (b). Now, fix $t$ at $t_1$. Then the expression in Eq. (1) jumps at $z_a = -vt_1$, $z_b = -v(t - T)$, and $z_c = -v(t - 2T)$. Notice $|z_c| < |z_b| < |z_a|$ as shown in the lower sketch (c). We observe that the waveform when sketched vs. $z$ has the same orientation as when sketched vs. $t$. This is unlike that for the forward-traveling wave in the previous example; it is not

PHYSICAL INTERPRETATIONS OF SOLUTIONS 59

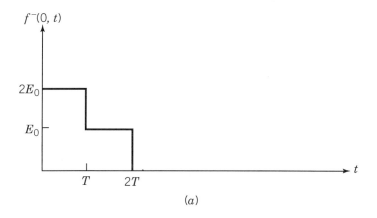

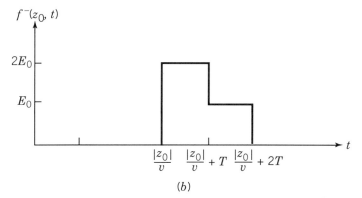

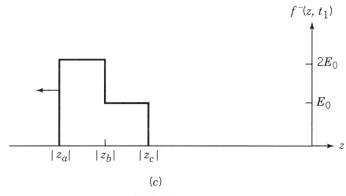

**FIGURE 2.5**

60   THE MATHEMATICS OF TRAVELING WAVES

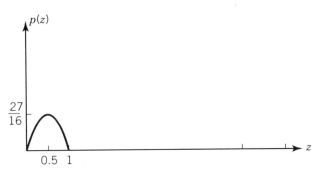

**FIGURE 2.6**

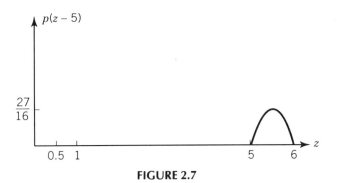

**FIGURE 2.7**

"spun around." This makes sense as the $2E_0$ value should be further from the origin, since that value left the source first.  ▲

***Example 2.3*** Sketch the function. See Figure 2.6.

$$p(z) = \frac{27}{4} z(1-z)[U(z) - U(z-1)]$$

Sketch $p(z - 5)$ and express it analytically (See Figure 2.7):

$$p(z-5) = \frac{27}{4}(z-5)[1 - (z-5)][U(z-5) - U(z-6)]$$

That is, replace $z$ by $(z - 5)$ everywhere in $p(z)$.  ▲

***Example 2.4*** An alternate way to look at the push rule is as follows. For the sketch below, the circles are described as

Circle A
$$x^2 + y^2 = r^2$$

# PHYSICAL INTERPRETATIONS OF SOLUTIONS

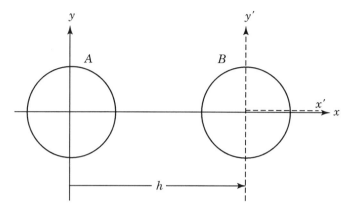

**FIGURE 2.8**

Circle B

$$x'^2 + y'^2 = r^2$$

where circle B has been described by the $(x', y')$ coordinates. To express the circle in terms of $x$ and $y$, we use the transformation between systems. Visually, we note that

$$x = x' + h$$
$$y = y'$$

so circle B becomes

$$(x - h)^2 + y^2 = r^2$$

which we observe is circle A pushed $h$ units to the right. If $h \to h(t) = v_0 t$, then the circle slides to the right at speed $v_0$. See Figure 2.8. ▲

**Example 2.5** Consider a cosinusoidal traveling wave:

$$y(z, t) = A \cos(\omega t - kz)$$
$$= A \cos\left[\omega\left(t - \frac{k}{\omega}z\right)\right] \quad (1)$$

which from our previous results shows the waveform to be moving at velocity $v = \omega/k$. To show this in another manner, examine $y$ at two instants, $t_1$ and $t_1 + \Delta t$. We flag a particular point at $t_1$ with ordinate $\bar{y}$ and follow it over the interval $\Delta t$. At $t_1$, its $z$ coordinate is $z_1$, and at $t_1 + \Delta t$, the coordinate is $z_2$. Write

## 62 THE MATHEMATICS OF TRAVELING WAVES

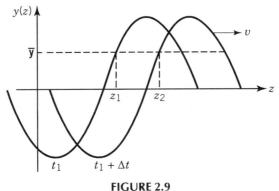

**FIGURE 2.9**

Eq. (1) at these instants:

$$\bar{y} = y(z = z_1, t = t_1) = A\cos(\omega t_1 - kz_1)$$
$$\bar{y} = y(z = z_2, t = t_1 + \Delta t) = A\cos[\omega(t_1 + \Delta t) - kz_2]$$

and since $\bar{y}$ is the same for any instant for the flagged point, we have (See Figure 2.9)

$$\cos(\omega t_1 - kz_1) = \cos[\omega(t_1 + \Delta t) - kz_2]$$

Now, we have assumed that the wave is traveling at speed $v$, so $z_2 = z_1 + v\Delta t_1$. Then

$$\cos(\omega t_1 - kz_1) = \cos[\omega t_1 + \omega\Delta t - k(z_1 + v\Delta t)]$$
$$= \cos(\omega t_1 - kz_1 + \omega\Delta t - kv\Delta t)$$

For this last equation to hold, we realize that

$$\omega\Delta t - kv\Delta t = 0$$

or

$$v = \frac{\omega}{k}$$

which is called the phase velocity of the wave. The flagged point (fixed on the wave) is called a point of constant phase. ▲

## 2.2 BASIC NOTATION AND METHODS

Figure 2.10 summarizes the basic model we will analyze for the remainder of the book. The TL voltage and current are $\mathscr{V}$ and $\mathscr{I}$, which are the superposition of

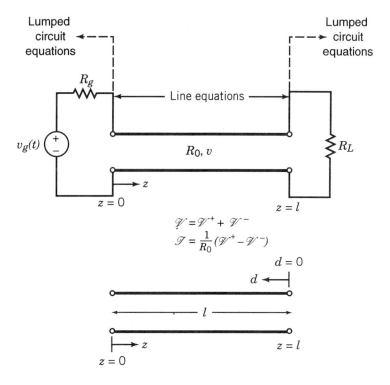

**FIGURE 2.10** The schematic for a transmission line with source and load connections. Notice that the two spatial coordinates $z$ and $d$ have different points of origin as well as direction.

waves traveling in both directions. The length of the TL is important, and it is usually designated on the sketch. On the ends of the line, lumped circuits are attached via filamentary conductors of ordinary circuit theory. The lengths and orientation of these wires have no significance. The dotted lines at the ends represent source and load planes wherein we merge the traveling waves of $\mathscr{V}$ and $\mathscr{I}$ on the TL to the ordinary currents and voltages of the connected lumped elements. In the lumped regions, Ohm's law is used; for example, at the source plane, with $\mathscr{V}_s$ the voltage at this plane, we have

$$\mathscr{V}_s = v_g(t) - \mathscr{I}(0,t)R_g \tag{2.11a}$$

Notice that $\mathscr{I}(0,t)$ is the current in the lumped elements. A second expression for $\mathscr{V}_s$ is

$$\mathscr{V}_s = \mathscr{V}^+(0,t) + \mathscr{V}^-(0,t) \tag{2.11b}$$

which is the superposition of the traveling waves at $z = 0$. At the load ($z = l$),

$$\mathscr{V} = \mathscr{V}_L = \mathscr{V}^+(l,t) + \mathscr{V}^-(l,t) \quad (2.12a)$$

and

$$\mathscr{V}_L = \mathscr{I}(l,t)R_L \quad (2.12b)$$

The functional forms of $\mathscr{V}^+$ and $\mathscr{V}^-$ (both shape and amplitude) depend on boundary and initial conditions at the ends of the line, as well as those on the line prior to throwing a switch. A word about coordinate systems is in order. In the literature of TLs, one finds several frames. The origin in Figure 2.10 is at the source plane, but in Chapter 5, we will move it to the load plane. Sometimes, another system is used wherein the origin is at the load plane with the positive direction toward the source (coordinate $d$ as shown). Thus, for the geometry shown, $d = l - z$ for any point on the line. For now, we will use the source plane as the origin because it is convenient for time domain (pulse) applications. The equations for $\mathscr{V}$, $\mathscr{I}$, and other quantities have slightly different expressions depending on the choice of origin, so one must be careful when mixing formulas from various sources (check the coordinate system).

The characteristic impedance will most often be chosen as $50\,\Omega$. This value is a compromise between those that yield either minimum loss or provide high power levels prior to arcing. Remember, the $50\,\Omega$ is a characteristic impedance and applicable only in time-varying situations. At dc, the line is just two wires, so an ohmmeter measures zero resistance for a shorted line and infinity for an opened section of $50$-$\Omega$ line.

## 2.3 WAVE LAUNCHING BY GENERATOR

With reference to Figure 2.11, we consider the launching of a wave by the generator. The line is terminated in $R_0$, so no reflections occur at the load plane. Therefore, only forward waves $\mathscr{V}^+$ and $\mathscr{I}^+$ exist, and $\mathscr{V}^- = \mathscr{I}^- = 0$. Apply

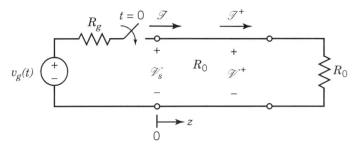

**FIGURE 2.11** The launching of a single forward wave on a matched ($R_L = R_0$) line.

Kirchoff's voltage law (KVL) at the source plane:

$$\mathscr{V}_s = v_g(t) - \mathscr{I}(0,t)R_g \qquad (2.13)$$

and on the line,

$$\mathscr{V}_s = \mathscr{V}^+(0,t), \qquad \mathscr{I}(0,t) = \mathscr{I}^+(0,t)$$

so

$$\mathscr{V}^+(0,t) = v_g(t) - \mathscr{I}^+(0,t)R_g \qquad (2.13a)$$

Now, the line forces

$$\mathscr{I}^+(0,t) = \frac{1}{R_0}\mathscr{V}^+(0,t)$$

Then

$$\mathscr{V}^+(0,t) = v_g(t) - \frac{1}{R_0}\mathscr{V}^+(0,t)R_g$$

or

$$\mathscr{V}^+(0,t) = v_g(t)\frac{R_0}{R_0 + R_g} \qquad (2.14)$$

So the line acts like a simple voltage divider for the source $v_g(t)$. Thus, $\mathscr{V}^+(0,t)$ follows $v_g(t)$ in time and moves away down the line. At the point $z$ down the line, we have

$$\mathscr{V}(z,t) = \mathscr{V}^+\left(t - \frac{z}{v}\right) = v_g\left(t - \frac{z}{v}\right)\frac{R_0}{R_0 + R_g} \qquad (2.15a)$$

and

$$\mathscr{I}(z,t) = \frac{1}{R_0}\mathscr{V}^+(z,t) = \frac{v_g(t - z/v)}{R_0 + R_g} \qquad (2.15b)$$

***Example 2.6*** Consider a generator with $R_g = R_0$ launching a pulse onto a line with characteristic impedance $R_0$ and terminated in $R_0$. Let $v_g(t) = 2p(t)$; then $\mathscr{V}^+(0,t) = p(t)$ is launched onto the line [apply Eq. (2.14)]. Let $p(t)$ be of the form:

$$p(t) = \frac{27}{4}t(1-t)^2 W(t)$$

where

$$W(t) = U(t) - U(t-1)$$

Then $p(t)$ appears as in Figure 2.12a. The signal at $z = z_0$ is

$$p(z_0, t) = p\left(t - \frac{z_0}{v}\right) = \frac{27}{4}\left(t - \frac{z_0}{v}\right)\left[1 - \left(t - \frac{z_0}{v}\right)\right]^2 W\left(t - \frac{z_0}{v}\right)$$

**66** THE MATHEMATICS OF TRAVELING WAVES

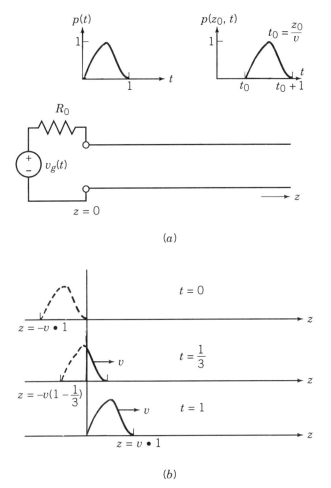

**FIGURE 2.12** The early stages of a pulse being launched onto a TL. (a) Time waveforms at $z = 0$ and $z = z_0$.

which is shown in Figure 2.12a. Figure 2.12b illustrates the signal at various instants.

Suppose that now the line extends to the left as shown in Figure 2.13. Assume that $v = 2\,\text{m/s}$; then the expression for the signal is

$$p(z,t) = \frac{27}{4}\left(t + \frac{z}{2}\right)\left[1 - \left(t + \frac{z}{2}\right)\right]^2 W\left(t + \frac{z}{2}\right)$$

since it moves to the left. Notice, in both figures, the method of sketching the pulse as a dotted waveform in the regions off the line. This is useful for quick sketching during the early instants when the pulse is only partially launched. ▲

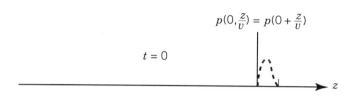
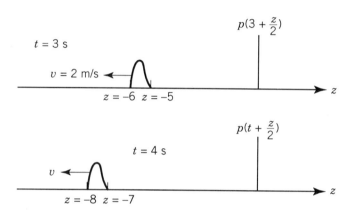

**FIGURE 2.13** A pulse launched at the load end of the line.

## 2.4 ANALYSIS AT THE LOAD PLANE

Assume that the wave launched by the generator $\mathscr{V}^+$ is about to hit the load plane and $R_L \neq R_0$, so a reflection occurs. The portion reflected is determined as follows. At the load plane $z = l$, we have

$$\mathscr{V}(l, t) = \mathscr{V}^+(l, t) + \mathscr{V}^-(l, t) \qquad (2.16\text{a})$$

$$\mathscr{I}(l, t) = \frac{1}{R_0}[\mathscr{V}^+(l, t) - \mathscr{V}^-(l, t)] \qquad (2.16\text{b})$$

and Ohm's law at the load is

$$\mathscr{V}(l, t) = R_L \mathscr{I}(l, t) \qquad (2.16\text{c})$$

## 68   THE MATHEMATICS OF TRAVELING WAVES

Define the load voltage reflection coefficient $K_L^V$ by

$$K_L^V = \frac{\mathscr{V}^-}{\mathscr{V}^+}$$

and employ this in Eq. (2.16c) using (2.16a) and (2.16b):

$$\left(\mathscr{V}^+ + K_L^V \mathscr{V}^+\right) = R_L \frac{1}{R_0}\left[\mathscr{V}^+ - K_L^V \mathscr{V}^+\right]$$

Rearrange

$$K_L^V = \frac{\mathscr{V}^-}{\mathscr{V}^+} = \frac{R_L - R_0}{R_L + R_0} \qquad (2.17)$$

which then gives $\mathscr{V}^-$ in terms of $\mathscr{V}^+$, $R_0$, and $R_L$. It is convenient to define a load current reflection coefficient by

$$\mathscr{I}^- = K_L^I \mathscr{I}^+$$

and since

$$\mathscr{I}^- = \frac{-1}{R_0}\mathscr{V}^- = \frac{-1}{R_0}K_L^V \mathscr{V}^+ = \frac{-K_L^V}{R_0}R_0 \mathscr{I}^+$$

comparing with the above shows

$$K_L^I = -K_L^V \qquad (2.18)$$

***Example 2.7***  Consider a long-duration 50-V pulse on a 50-Ω line terminated in 30 Ω (see Figure 2.14). What are $K_L^V$, $K_L^I$, and power absorbed by the load?

The incident voltage $\mathscr{V}^+ = 50$ V (given) has associated with it an incident current

$$\mathscr{I}^+ = \frac{1}{R_0}\mathscr{V}^+ = \frac{1}{50}(50) = 1 \text{ A}$$

The power available over an interval (short with respect to the pulse duration) is 50 W. The reflection coefficients at the load are

$$K_L^V = \frac{R_L - R_0}{R_L + R_0} = \frac{30 - 50}{30 + 50} = -\frac{1}{4}$$

$$K_L^I = -K_L^V = +\frac{1}{4}$$

so

$$\mathscr{V}^- = (-\tfrac{1}{4})(50) = -12.5 \text{ V}$$

$$\mathscr{I}^- = (+\tfrac{1}{4})(1) = \tfrac{1}{4} \text{ A}$$

Then the load voltage and current are

$$\mathscr{V}(l, t) = \mathscr{V}^+ + \mathscr{V}^- = 50 + (-12.5) = 37.5 \text{ V}$$

$$\mathscr{I}(l, t) = \mathscr{I}^+ + \mathscr{I}^- = 1 + \tfrac{1}{4} = 1.25 \text{ A}$$

The third sketch in Figure 2.14 gives incident, reflected, and load values. The power absorbed in the load is (37.5 V)(1.25 A) = 46.875 W. We can deduce this in another manner. The incident power is $P_{\text{inc}} = (50 \text{ V})(1 \text{ A}) = 50 \text{ W}$, and the reflected power is $P_{\text{refl}} = (-12.5)(-\tfrac{1}{4}) = +3.125 \text{ W}$. The $(-\tfrac{1}{4})$ is used since we want

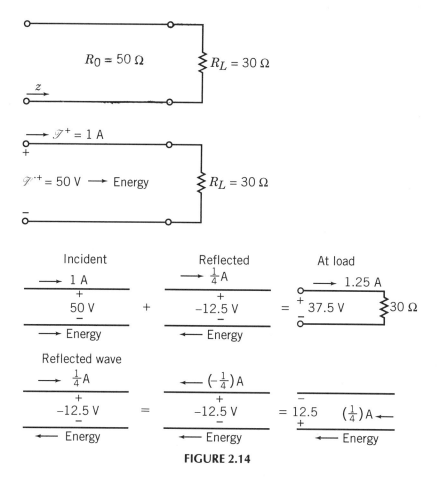

FIGURE 2.14

the current entering the positive node of the "effective circuit-element" that "absorbs" the reflected energy. Thus, the absorbed power is $P_{abs} = P_{inc} - P_{refl} = 50 - 3.125 = 46.875$ W. The bottom sketch shows three equivalent forms for the reflected wave. Recall that the convention for $\mathscr{I}^+$ and $\mathscr{I}^-$ is directed to the right in the upper wire (positive $z$ direction). Their signs are dictated by each specific problem. ▲

## 2.5 REFLECTION AT THE SOURCE PLANE

Consider the instant when the leading edge of a pulse that has been reflected at the load returns to the source plane. Assume that the time for the leading edge to propagate the length of the line is $T = l/v$. Then if the leading edge was launched at $t = 0$, the instant in question is $2T$. At the source plane, we have

$$v_g(2T) - \mathscr{I}(0, 2T)R_g = \mathscr{V}^+(0, 2T) + \mathscr{V}^-(0, 2T) \tag{2.19}$$

$$\mathscr{I}(0, 2T) = \frac{1}{R_0}[\mathscr{V}^+(0, 2T) - \mathscr{V}^-(0, 2T)]$$

Solve for $\mathscr{V}^+(0, 2T)$.

$$\mathscr{V}^+(0, 2T) = \frac{R_0}{R_0 + R_g} v_g(2T) + \frac{R_g - R_0}{R_g + R_0} \mathscr{V}^-(0, 2T) \tag{2.20}$$

which we interpret as follows. First, write this as

$$\mathscr{V}^+(0, 2T) = A + \mathscr{V}_r^+$$

and note that $A = v_g(2T)R_0/(R_0 + R_g)$ is just the voltage divider process discussed in Section 2.3. The second term $\mathscr{V}_r^+$ can be looked as a reflection of $\mathscr{V}^-(0, 2T)$ with reflection coefficient:

$$K_g^V = \frac{\mathscr{V}_r^+}{\mathscr{V}^-} = \frac{R_g - R_0}{R_g + R_0} \tag{2.21}$$

Notice that this reflection is independent of $v_g$, since $v_g$ could be zero at this instant [see Eq. (2.20)]. Thus, $K_g^V$ is called the source reflection coefficient. Thus, $\mathscr{V}^+(0, 2T)$ is the superposition of voltage $A$ leaking away due to the voltage divider on $v_g$, as well as $\mathscr{V}_r^+$, which is a reflection of the wave impinging on the source plane from the line. Now notice that

$$\mathscr{V}^-(0, 2T) = K_L^V \mathscr{V}^+(l, T) = \mathscr{V}^-(l, t)$$

since the line is lossless. Also, $\mathscr{V}^+(l, T) = \mathscr{V}^+(0,0)$ for the same reason. Thus,

$$\mathscr{V}^-(0, 2T) = K_L^V \mathscr{V}^+(0,0)$$

and we know from Eq. (2.14) that

$$\mathscr{V}^+(0,0) = v_g(0)\frac{R_0}{R_0 + R_g}$$

which leads to [use these in Eq. (2.20)]

$$\mathscr{V}^+(0, 2T) = \frac{R_0}{R_0 + R_g}v_g(2T) + K_g^V K_L^V \left[\frac{R_0}{R_0 + R_g}v_g(0)\right] \quad (2.22)$$

which relates $\mathscr{V}^+(0, 2T)$ to $v_g$ at $t = 0$ and $t = 2T$. Observe that $\mathscr{V}^+(0, 2T)$ is the leading edge of the net wave of voltage traveling to the right. It depends on the generator at times 0 and $2T$. It also depends on both the generator impedance $R_g$ as well as the load impedance $R_L$ (via $K_g^V$ and $K_L^V$). Therefore, in general, $\mathscr{V}^+(z, t)$ represents the *net* voltage wave moving to the right, and it depends on boundary conditions at *both ends* of the TL. It is *not* the value launched by the generator. It *is* just the value launched by the generator in the *special case* when either or both $K_L^V$ and $K_g^V$ are zero, as can be seen from Eq. (2.22). Similar statements can be made for $\mathscr{V}^-(z, t)$. Said in another way, the amplitudes and forms of $\mathscr{V}^+$ and $\mathscr{V}^-$ are the arbitrary coeffcients of the solutions of the differential equations. They depend on boundary and initial conditions on *both* ends of the line in the general case.

## 2.6 BOUNCE DIAGRAM

Consider the process of achieving a dc steady state with a battery, TL with a switch, and a load resistor $R_L$. The process is best described with reference to Figure 2.15. In Figure 2.15a, the voltage on the line is

$$\mathscr{V}_1^+ = V_0 \frac{R_0}{R_0 + R_g}$$

where the subscript 1 designates the first forward wave. After the reflection at the load, the net voltage on the line is indicated in Figure 2.15b. Here,

$$\mathscr{V}_1^- = K_L^V \mathscr{V}_1^+ = \frac{R_L - R_0}{R_L + R_0}\mathscr{V}_1^+$$

and notice that the net voltage on the line in the intervals $0 \leq z \leq (l - d)$ and

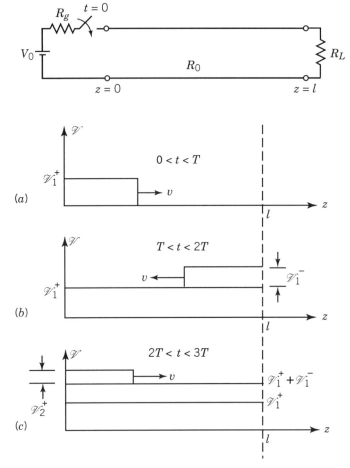

**FIGURE 2.15** At $t = 0$, a voltage source is applied to a TL with a load $R_L$. The line is $l$ meters long, and the propagation time is $T = l/v$.

$(l - d) \leq z \leq l$ is $\mathscr{V}_1^+$ and $\mathscr{V}_1^+ + \mathscr{V}_1^-$, respectively. In Figure 2.15c the first reflection at the source plane has occurred and $\mathscr{V}_2^+ = K_g^V \mathscr{V}_1^-$. Notice that for all time the source is continuously leaking the value $\mathscr{V}_1^+$ onto the line. Equilibrium occurs when all the overlapping waves sum to the steady final value of $\mathscr{V}_{\text{final}} = V_0 \dfrac{R_L}{R_L + R_g}$, for then the TL is just two wires and the capacity of the line has been charged to $\mathscr{V}_{\text{final}}$ volts. The steady current in the loop is just $V_0/(R_g + R_L)$.

As a second example, consider the circuit of Figure 2.16. The switch is initially at position 1. It then moves to 2 at $t = 0$ and remains there for $\tau$ seconds before returning to 1. We have therefore a pulse of duration $\tau$ launched. Assume that

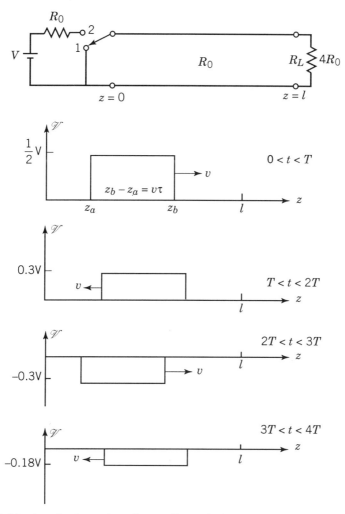

**FIGURE 2.16** A pulse bouncing along a line. The load reduces the amplitude 0.6, whereas the source inverts it.

$\tau < T$, where $T$ is the propagation time $l/v$. When the switch is at position 2, the incident wave is

$$\mathscr{V}_1^+ = V\frac{R_0}{R_0 + R_0} = \frac{1}{2}V$$

The reflection coefficient at the load is

$$K_L^V = \frac{4R_0 - R_0}{4R_0 + R_0} = 0.6$$

74   THE MATHEMATICS OF TRAVELING WAVES

and that for the leading edge of the return wave is (switch back at position 1)

$$K_g^V = \frac{0 - R_0}{0 + R_0} = -1$$

That is, a short produces a reflection coefficient of $-1$. The pulse winds its way back and forth on the line as shown. The load reduces the incoming amplitude by 0.6, whereas the source reflection changes the polarity and sends the pulse back toward the load.

A final example of a pulse on a line is depicted in Figures 2.17, 2.18, and 2.19. The generator votage is a pulse of 200 V that lasts $\tau = 1\mu s$. The load is a short (which could be realized as a large metal plate for an open wire line, or a metal

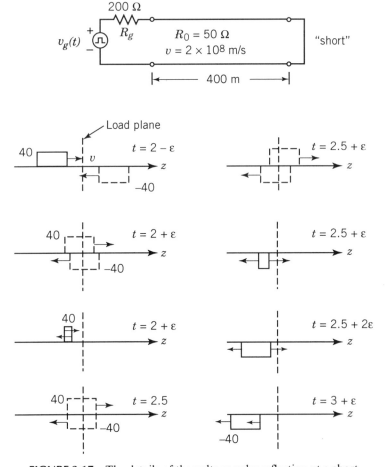

**FIGURE 2.17**   The details of the voltage pulse reflection at a short.

plug in a coax), which means all the incident energy is reflected. The amplitude of the initial pulse is

$$\mathscr{V}_1^+ = 200\frac{50}{250} = 40\text{ V}$$

The reflection coefficients are

$$K_L^V = \frac{0-50}{0+50} = -1$$

$$K_g^V = \frac{200-50}{200+50} = 3/5$$

The time to travel the line $T = l/v = 400\text{ m}/(2 \times 10^8 \text{ m/s}) = 2\,\mu\text{s}$. The 40-V pulse travels to the load and reaches it in $2\,\mu\text{s}$. The lower sketches of Figure 2.17 indicate the reflection process at the load. It is a convenient technique to imagine a pulse of amplitude $K_L^V \mathscr{V}_1^+$ approaching the load plane as shown. This artificial pulse hits the load plane when the real pulse does. At the instant (in microseconds) $t = 2 - \varepsilon$, the 40-V pulse is close to the load plane. At $t = 2 + \varepsilon$, part of it has crashed into the load and the reflected portion is modeled by the part of the artifical pulse that has moved out onto the line. Both pulses are shown dotted for clarity. In the sketch immediately below that just discussed (also at $t = 2 + \varepsilon$), the net voltage is shown. Notice in the overlap region that the net voltage is zero and the net voltage elsewhere appears as a pulse with both walls moving so as to narrow the pulse to zero width. The voltage at the load plane is zero, as there can be none across a short. The next sketch is the instant $t = 2.5$ wherein both pulses completely overlap. Then the net voltage is instantaneously zero over all portions of the line. After $3\,\mu\text{s}$, the reflected pulse is completely on the line as shown in the last sketch.

The $-40$-V pulse arrives at the source at $4\,\mu\text{s}$ and generates a reflected wave of $(-40)(3/5) = -24\text{ V}$. The details are shown in Figure 2.18. Also shown is the sketch of the total voltage at the source plane over the interval 0 to $6\,\mu\text{s}$. This represents the oscilloscope trace across the input terminals. The current pulses can be treated the same way, with reflection coefficients that are the negatives of those for voltage. Figure 2.19 shows the voltage at the center of the line for two different cases of pulse width. The lower one is for a pulse duration of $6\,\mu\text{s}$. Here, the duration is longer than the time to go up and back on the line. Thus, the leading edge can return to the generator before the trailing edge emerges from the source plane. At the center of the line, this overlap is handled by simply stretching each of the original pulses to the right such that each is $6\,\mu\text{s}$ in duration (dotted). Then these stretched pulses are combined to indicate the net voltage with time (solid line).

A systematic method to study pulse transients is called the "bounce diagram." Figure 2.20 shows the principal ideas. The left side illustrates a pulse of amplitude

**76** THE MATHEMATICS OF TRAVELING WAVES

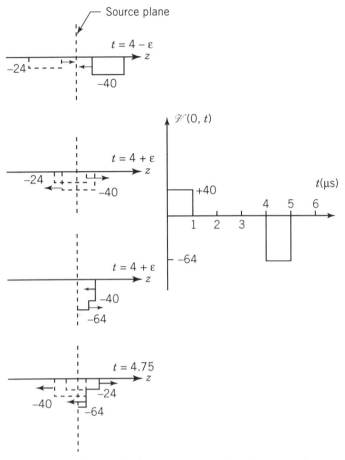

**FIGURE 2.18** Details of the reflection process at specific instants at the source plane.

unity leaving the source and arrives at the load in $T$ seconds. The time axis is vertical and increases downward. The pulse is reflected with coefficient $(K_L^V = \frac{1}{3})\frac{1}{3}$ and returns to the source at $t = 2T$. The source reflects this and sends the value $(\frac{1}{3})(\frac{1}{2})$ back to the load, which arrives there at $t = 3T$. The process continues indefinitely as shown. The sketch on the right is for a unit current pulse with the corresponding reflection coefficients shown at the bottom. Suppose that the generator has amplitude 100 V; then the initial amplitude on the line is $100(R_0/4R_0) = 25$ V. Consequently, the amplitude over the interval $(3T, 4T)$ is $(\frac{1}{3})^2(\frac{1}{2})(25) = 1.39$ V, which is moving from load to source. The net voltage at the load at $t = 3T$ is the superposition of incident and reflected waves: $25[(\frac{1}{3})(\frac{1}{2}) + (\frac{1}{3})^2(\frac{1}{2})] = 5.56$ V. This value lasts $\tau$ seconds (pulse width). The current diagram is used similarly. The incident value on the line is $\mathscr{I}^+ = \mathscr{V}^+/R_0 = 25/R_0$, and for calculation purposes, let $R_0 = 50\,\Omega$. Then the net current at $t = T$ is

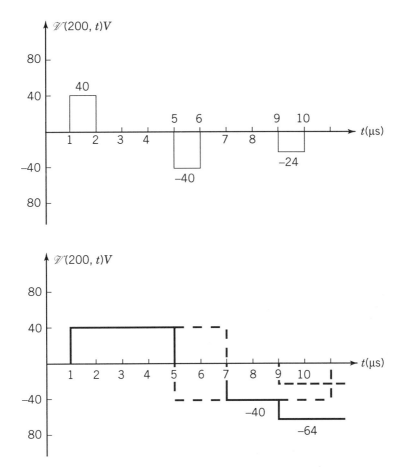

**FIGURE 2.19** The voltages at the center of the line for the cases where $\tau = 1\mu s$ (upper) and $\tau = 6\mu s$ (lower).

$\frac{1}{2}[1 + (-\frac{1}{3})] = (\frac{1}{3})$ A at the load. The current moving from load to source over the interval $(3T, 4T)$ is $\frac{1}{2}[(-\frac{1}{3})^2(-\frac{1}{2})] = -27.8$ mA. Some of the chapter problems will show the additional benefits of using bounce diagrams.

*Example 2.8* Consider the process of reaching steady state for a battery charging a line terminated in a resistor.

The initial wave is designated $V_1^+$ and the first reflected one $V_1^- = K_L V_1^+$. The next two are $V_2^+ = K_g V_1^-$ and $V_2^- = K_L V_2^+$. Now listing all the left to right $(+)$ waves and all the right to left $(-)$ waves, we have

Left to right

$$V_1^+ + V_2^+ + V_3^+ + \cdots = V_1^+ + K_g K_L V_1^+ + K_g^2 K_L^2 V_1^+ + \cdots$$

# 78 THE MATHEMATICS OF TRAVELING WAVES

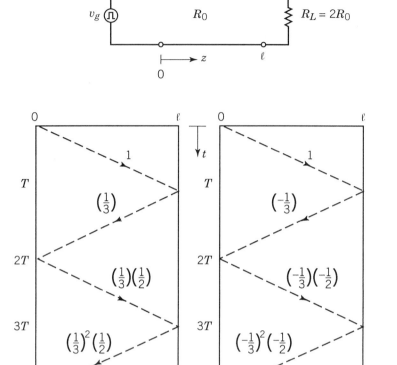

**FIGURE 2.20** A bounce diagram that systematically follows voltage and current pulses with time along a TL.

Right to left

$$V_1^- + V_2^- + V_3^- + \cdots = K_L V_1^+ + K_L(K_g K_L)V_1^+ + K_L(K_g K_L)^2 V_1^+ + \cdots$$

All left to right sum to (using $1 + x + x^2 + \cdots = 1/(1-x)$, $|x| < 1$)

$$V_1^+[1 + (K_g K_L) + (K_g K_L)^2 + (K_g K_L)^3 + \cdots] = V_1^+ \frac{1}{1 - K_g K_L} \equiv V_f^+$$

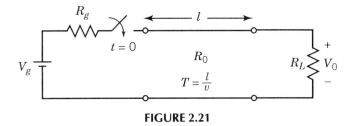

**FIGURE 2.21**

and those from right to left give

$$V_1^+ K_L[1 + (K_g K_L) + (K_g K_L)^2 + \cdots] = V_1^+ K_L \frac{1}{1 - K_g K_L} \equiv V_f^-$$

The total steady voltage $V_0$ on the line is

$$V_0 = V_1^+ \left( \frac{1}{1 - K_g K_L} + \frac{K_L}{1 - K_g K_L} \right) = V_1^+ \left( \frac{1 + K_L}{1 - K_g K_L} \right)$$

and using $V_1^+ = V_g R_0/(R_0 + R_g)$, we have

$$V_0 = V_g \frac{R_L}{R_L + R_g}$$

which we recognize is the solution once the "line" is just two wires (after the transient). Figure 2.21 indicates the decomposition of a steady voltage across the line into a "forward" $V_f^+$ and "reflected" $V_f^-$, where the subscript $f$ means "final."

$V_0 =$ final load voltage $= [V(t \to \infty)]$

$$V_f^+ = V_1^+ \frac{1}{1 - K_g K_L} = \text{final net "incident" wave } (+ \text{ wave})$$

$$V_f^- = V_1^+ \frac{K_L}{1 - K_g K_L} = \text{final net "reflected" wave } (- \text{ wave})$$

$$V_1^+ = V_g \frac{R_0}{R_0 + R_g} \qquad \blacktriangle$$

## 2.7 CAPACITIVE AND INDUCTIVE LOADS

Figure 2.22 consists of a generator and TL terminated by an arbitrary load. The equations at the load plane $z = l$ are

$$\mathscr{V}_L(t) = \mathscr{V}^+(l, t) + \mathscr{V}^-(l, t) \qquad (2.23)$$

$$\mathscr{I}_L(t) = \frac{1}{R_0} [\mathscr{V}^+(l, t) - \mathscr{V}^-(l, t)] \qquad (2.24)$$

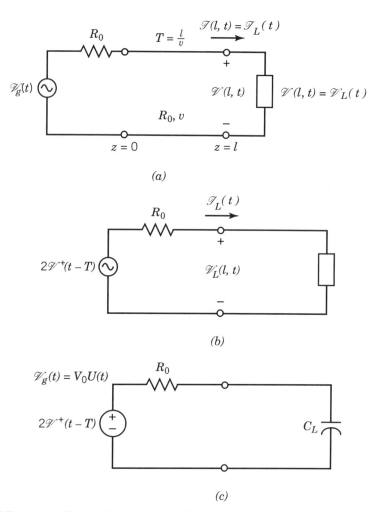

**FIGURE 2.22** (a) Standard TL notation. (b) Lumped equivalent at the load plane. (c) Specific case of a capacitor for the load.

and after elimination of $\mathscr{V}^-(l, t)$, we have

$$2\mathscr{V}^+(l, t) = \mathscr{V}_L(t) + R_0 \mathscr{I}_L(t)$$

This expression may be written as

$$2\mathscr{V}^+(0, t - T) = \mathscr{V}_L(t) + R_0 \mathscr{I}_L(t) \qquad (2.25)$$

where we have used $\mathscr{V}^+(l, t) = \mathscr{V}^+(0, t - T)$, that is, the incident wave at $(l, t)$ is that which occurred at $z = 0$ and $(t - T)$ seconds earlier. Note that the generator

CAPACITIVE AND INDUCTIVE LOADS   81

must be matched to the line. Here, $T$ is the propagation time from source to load. This last equation is that of a simple lumped network at the plane $z = l$, given in Figure 2.22b. Therefore, if $\mathscr{V}^+(t - T)$ is known, one can quickly determine $\mathscr{V}_L(t)$ and $\mathscr{I}_L(t)$. For example, consider a capacitor terminating the line. Assume zero initial current through the load. Then

$$\mathscr{I}_L(t) = C_L \frac{d\mathscr{V}_L(t)}{dt}$$

and Equation (2.25) is

$$2\mathscr{V}^+(t - T) = \mathscr{V}_L(t) + R_0 C_L \frac{d\mathscr{V}_L}{dt}$$

Assume a step input at $t = 0$. Then $2\mathscr{V}^+(t - T) = V_0 U(t - T)$ [i.e., $\mathscr{V}^+(t) = \mathscr{V}_g(t) R_0/(R_0 + R_0) = \frac{1}{2}\mathscr{V}_g(t) = \frac{1}{2}V_0 U(t)$], and we have

$$R_0 C_L \frac{d\mathscr{V}_L}{dt} + \mathscr{V}_L(t) = V_0 U(t - T) \quad (2.26)$$

The solution is

$$\mathscr{V}_L(t) = V_0[1 - e^{-(t-T)/R_0 C_L}]U(t - T) \quad (2.27)$$

which will permit the following relationship:

$$\mathscr{V}^-(l, t) = \mathscr{V}_L(t) - \mathscr{V}^+(l, t)$$
$$= \mathscr{V}_L(t) - \mathscr{V}^+(t - T)$$
$$= \frac{V_0}{2} U(t - T) - V_0 e^{-(t-T)/R_0 C_L} U(t - T) \quad (2.28)$$

And $\mathscr{V}^-(t)$ is sketched in Figure 2.23. Notice that the ratio $\mathscr{V}^-(t)/\mathscr{V}^+(t)$ changes with time, so the reflection coefficient is time-varying in the general case. The expression for $\mathscr{V}^-(z, t)$ is found from Eq. (2.28) as follows. Physically, we know at a point $z$, the value there is that at $z = l$ delayed by $d/v$ seconds, where $d$ is $(l - z)$. Thus, replace $t$ by $t - (l - z)/v = t - T + z/v$ in Eq. (2.28):

$$\mathscr{V}^-(z, t) = \frac{V_0}{2} U\left(t + \frac{z}{v} - 2T\right) - V_0 e^{-(t+z/v - 2T)/R_0 C_L} U\left(t + \frac{z}{v} - 2T\right) \quad (2.28a)$$

At any instant, we can write the total line voltage as

$$\mathscr{V}(z, t) = \mathscr{V}^+(z, t) + \mathscr{V}^-(z, t)$$

$$= \frac{V_0}{2} U\left(t - \frac{z}{v}\right) + \frac{V_0}{2} U\left(t + \frac{z}{v} - 2T\right)$$
$$- V_0 e^{-(t+z/v - 2T)/R_0 C_L} U\left(t + \frac{z}{v} - 2T\right) \quad (2.29)$$
$$= \frac{V_0}{2} U\left(t - \frac{z}{v}\right) + V_0 U\left(t + \frac{z}{v} - 2T\right)\left[\frac{1}{2} - e^{-(t+z/v - 2T)/R_0 C_L}\right]$$

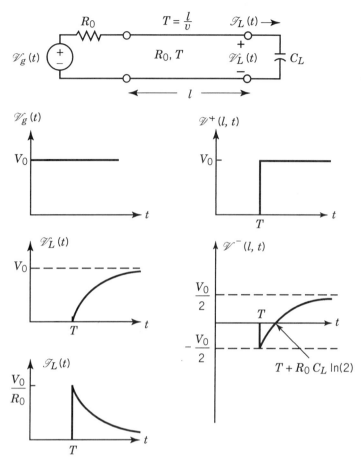

**FIGURE 2.23** The step response for a line with propagation time $T$ terminated in a capacitor.

The point where $\mathscr{V}^-(z,t) = 0$ occurs where the exponential equals unity (by inspection). Thus,

$$\mathscr{V}^-(z,t) = 0 \quad \text{when} \quad \left(t + \frac{z}{v} - 2T\right) = 0 \tag{2.30}$$

Sketches of the waveform on the line at various instants are shown in Figure 2.24. The sketch during the interval $(T, 2T)$ is easily found by sliding $\mathscr{V}^-(l,t)$ out onto the line and adding to $\mathscr{V}^+(z,t)$. Recall that $\mathscr{V}^-$ waves are not spun around when sketched vs. $z$.

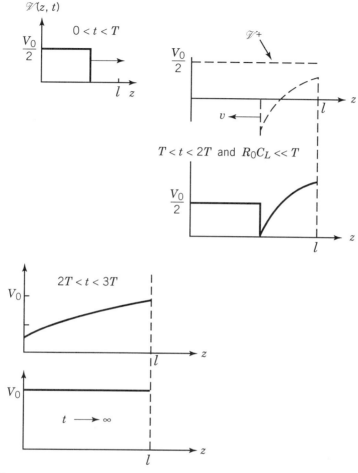

**FIGURE 2.24** The evolution of the total voltage vs. distance for the capacitively terminated line.

An inductive load is treated similarly:

$$2\mathscr{V}^+(t-T) = L_L \frac{d\mathscr{I}_L}{dt} + R_0 \mathscr{I}_L(t) + \text{ic}$$

where the initial condition is designated ic. The solutions for $\mathscr{I}_L$ and $\mathscr{V}_L$ are duals as we know, so pure reactive loads are easily sketched.

**Example 2.9** Find a lumped equivalent at the source end of a TL of length $l$ and delay $T$.

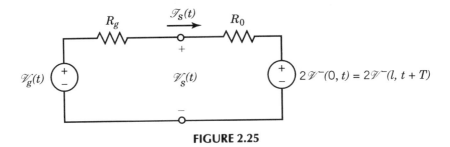

FIGURE 2.25

Using

$$\mathscr{V}_s(t) = \mathscr{V}^+(0,t) + \mathscr{V}^-(0,t)$$

$$\mathscr{I}_s(t) = \frac{1}{R_0}[\mathscr{V}^+(0,t) - \mathscr{V}^-(0,t)]$$

eliminate $\mathscr{V}^+(0,t)$:

$$2\mathscr{V}^-(0,t) = \mathscr{V}_s(t) - R_0 \mathscr{I}_s(t)$$

The lumped equivalent is shown in Figure 2.25. ▲

## 2.8 LAPLACE TRANSFORMS

A systematic and powerful way to obtain solutions for complex problems is to use the Laplace transform. When such factors as frequency dependence of the line parameters (R, L, C, G) and/or repetitive signals are included, transform analysis is essential. Here, we briefly introduce the technique, and the reader who may have use for such an approach can start with two articles (Refs. 1 and 2) in the literature. In general, for real-world problems, a computer solution is needed.

Start with Eqs. (1.60) and (1.61) and eliminate the current to obtain

$$\frac{\partial^2 \mathscr{V}}{\partial z^2} = LC\frac{\partial^2 \mathscr{V}}{\partial t^2} + (LG + RC)\frac{\partial \mathscr{V}}{\partial t} + RG\mathscr{V} \tag{2.31}$$

Take the transform

$$\frac{d^2 V(z,s)}{dz^2} = LC\left[s^2 V(z,s) - sV(z,0) - \frac{d\mathscr{V}(z,0)}{dt}\right]$$

$$+ (LG + RC)[sV(z,s) - \mathscr{V}(z,0)] + RGV(z,s)$$

$$= [s^2 LC + s(LG + RC) + RG]V(z,s) - LCs\mathscr{V}(z,0)$$

$$- LC\frac{d\mathscr{V}(z,0)}{dt} - (LG + RC)\mathscr{V}(z,0) \tag{2.31a}$$

Define
$$\Gamma^2(s) = [s^2LC + s(LG + RC) + RG] \tag{2.32}$$
Then
$$\frac{d^2V}{dz^2} = \Gamma^2(s)V - LC\frac{d\mathscr{V}(z,0)}{dt} - (sLC + LG + RC)\mathscr{V}(z,0)$$
or
$$\frac{d^2V}{dz^2} - \Gamma^2(s)V = F(0) \tag{2.33}$$

where we have used
$$V(z,s) = \int_0^\infty e^{-st}\mathscr{V}(z,t)\,dt$$
$$F(0) = -LC\frac{d\mathscr{V}(z,0)}{dt} - (sLC + LG + RC)\mathscr{V}(z,0)$$

***Example 2.10*** Find the solution for the voltage for a lossless, uncharged line using the transform results.

The first expression is
$$\Gamma^2 = s^2 LC$$
since the line is lossless. Also $F(0) = 0$ since it was uncharged. Then
$$\frac{d^2V}{dz^2} - \Gamma^2 V = 0$$

The solution is just
$$V = A_1(s)e^{-sz\sqrt{LC}} + A_2(s)e^{sz\sqrt{LC}}, \qquad v = \frac{1}{\sqrt{LC}}$$

and these transforms are easily recognized as
$$\mathscr{V}(z,t) = a_1\left(t - \frac{z}{v}\right) + a_2\left(t + \frac{z}{v}\right)$$
$$= \mathscr{V}^+\left(t - \frac{z}{v}\right) + \mathscr{V}^-\left(t + \frac{z}{v}\right)$$

which are just our familiar forward and reflected waves. ▲

## 86 THE MATHEMATICS OF TRAVELING WAVES

***Example 2.11*** Find the general solution for the distributed network in Figure 2.26 using the Laplace transform method:

First, transform the network and replace the capacitor by an arbitrary impedance as in Figure 2.27:

The equations for the transformed variables $V(z,\omega)$ and $I(z,\omega)$ are

$$\frac{d^2 V(z,\omega)}{dz^2} - \Gamma^2 V(z,\omega) = 0 \tag{1}$$

$$\frac{d^2 I(z,\omega)}{dz^2} - \Gamma^2 I(z,\omega) = 0 \tag{2}$$

The solutions are

$$V = V_+(\omega)e^{-\Gamma(\omega)z} + V_-(\omega)e^{\Gamma(\omega)z} \tag{3}$$

$$I = \frac{1}{R_0}[V_+(\omega)e^{-\Gamma(\omega)z} - V_-(\omega)e^{\Gamma(\omega)z}] \tag{4}$$

$$\Gamma = j\omega\sqrt{LC} = j\frac{\omega}{v}, \qquad R_0 = \sqrt{\frac{L}{C}} \tag{5}$$

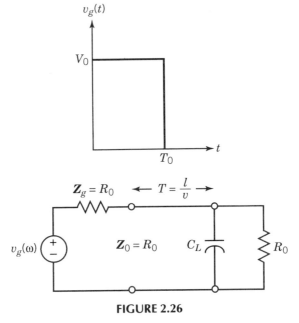

**FIGURE 2.26**

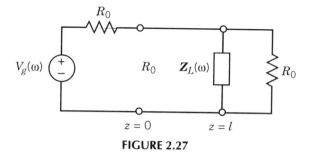

**FIGURE 2.27**

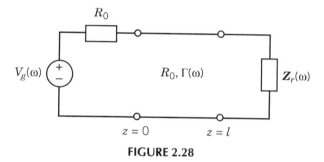

**FIGURE 2.28**

Now redraw the transformed network into a form suitable for any load (In Figure 2.28 the subscript $r$ refers to the receiving end):

The conditions at the source and load (receiving) planes are

$$V(0, \omega) = V_g(\omega) - I(0, \omega)R_0 \tag{6a}$$

$$V(l, \omega) = I(l, \omega)Z_r(\omega) \tag{6b}$$

which result in

$$V_+ + V_- = V_g - \frac{R_0}{R_0}(V_+ - V_-)$$

$$V_+ e^{-\Gamma l} + V_- e^{\Gamma l} = \frac{R_0}{R_0}(V_+ e^{-\Gamma l} - V_- e^{\Gamma l})$$

Define

$$\Gamma_g = \frac{Z_g - R_0}{Z_g + R_0}, \quad \Gamma_r = \frac{Z_r - R_0}{Z_r + R_0}$$

## THE MATHEMATICS OF TRAVELING WAVES

Then

$$V_+ = \frac{V_g}{2}\frac{1}{1-\Gamma_g\Gamma_r e^{-2\Gamma l}}$$

$$V(z,\omega) = \frac{V_g}{2}\frac{e^{-\Gamma z}+\Gamma_r e^{-2\Gamma l}e^{\Gamma z}}{1-\Gamma_g\Gamma_r e^{-2\Gamma l}} \tag{7a}$$

$$I(z,\omega) = \frac{V_g}{2}\frac{e^{-\Gamma z}-\Gamma_r e^{-2\Gamma l}e^{\Gamma z}}{1-\Gamma_g\Gamma_r e^{-2\Gamma l}} \tag{7b}$$

which are the general solutions for the voltage and current. For our special case,

$$Z_r(\omega) = Z_L(\omega) \| R_0$$

Note

$$\Gamma_g = 0, \qquad \Gamma = \frac{j\omega}{v}, \qquad T = \frac{l}{v}$$

Then

$$V(z,\omega) = \frac{V_g(\omega)}{2}[e^{-j\omega z/v}+\Gamma_r(\omega)e^{-j\omega(2T-z/v)}] \tag{8a}$$

$$I(z,\omega) = \frac{V_g(\omega)}{2}[e^{-j\omega z/v}-\Gamma_r(\omega)e^{-j\omega(2T-z/v)}] \tag{8b}$$

Here,

$$Z_L(\omega) = \frac{1}{j\omega C_L}, \qquad \therefore Z_r = \frac{R_0}{1+j\omega R_0 C_L}, \qquad \Gamma_r(\omega) = \frac{-j\omega R_0 C_L}{2+j\omega R_0 C_L}$$

The generator is expressed as

$$v_g(t) = V_0 p(t)$$

where

$$p(t) = U(t) - U(t-T_0)$$

$$V_g(\omega) = 2V_0 e^{-j\omega T_0/2}\frac{\sin \omega T_0/2}{\omega}$$

Use

$$f(t) \Leftrightarrow F(\omega)$$
$$f(t-T) \Leftrightarrow F(\omega)e^{-j\omega T}$$

$$V(z,\omega) = V_0 e^{-j\omega T_0/2}\frac{\sin \omega T_0/2}{\omega}\left[e^{-j\omega z/v} - \frac{j\omega R_0 C_L}{2+j\omega R_0 C_L}e^{-j\omega(2T-z/v)}\right]$$

Use

$$\sin\frac{\omega T_0}{2} = \frac{(e^{j\omega T_0/2} - e^{-j\omega T_0/2})}{2j}$$

$$V(z,\omega) = \left\{V_0 e^{-j\omega T_0/2}\frac{\sin\omega T_0/2}{\omega}\right\}e^{-j\omega z/v}$$

$$-\frac{V_0}{2}\left(\frac{1}{j\omega + \dfrac{2}{R_0 C_L}}\right)\left\{e^{-j\omega(2T-z/v)} - e^{-j\omega(T_0+2T-z/v)}\right\}$$

$$\mathscr{V}(z,t) = \tfrac{1}{2}v_g\!\left(t - \frac{z}{v}\right) - \frac{V_0}{2}\Big\{e^{-(2/R_0 C_L)(t-2T+z/v)}U\!\left(t - 2T + \frac{z}{v}\right)$$

$$- e^{-(2/R_0 C_L)(t-T_0-2T+z/v)}U\!\left(t - T_0 - 2T + \frac{z}{v}\right)\Big\}$$

which is the solution at all points on the line. The values at the source and load ends are

$$\mathscr{V}(0,t) = \tfrac{1}{2}v_g(t) - \frac{V_0}{2}\{e^{-(2/R_0 C_L)(t-2T)}U(t-2T) - e^{-(2/R_0 C_L)(t-T_0-2T)}U(t-T_0-2T)\}$$

The voltage at the source is shown in Figure 2.29, where we have assumed

$$\tau = R_0 C_L < T_0 \ll T$$

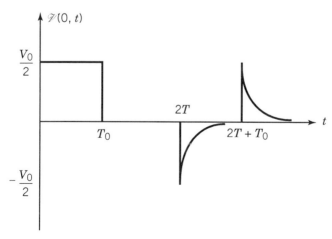

**FIGURE 2.29**

# THE MATHEMATICS OF TRAVELING WAVES

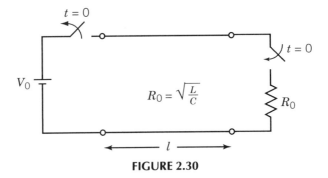

**FIGURE 2.30**

and

$$\mathscr{V}(l,t) = \tfrac{1}{2}v_g(t-T) - \frac{V_0}{2}\{e^{-(2/\tau)(t-T)}U(t-T) - e^{-(2/\tau)(t-T_0-T)}U(t-T_0-T)\}$$

▲

***Example 2.12*** Find and sketch $\mathscr{V}(l,t)$ for the circuit in Figure 2.30. This is a standard problem demonstrating the pulse-forming properties of an initially charged line:

The equation for the voltage $V(z,s)$ is found from Eq. (2.33) with the initial condition:

$$\frac{d^2V(z,s)}{dz^2} - \Gamma^2 V(z,s) = -sLCV_0 \tag{1}$$

$$\Gamma = s\sqrt{LC} \tag{2}$$

The homogeneous solution is

$$V = K_1(s)e^{-\Gamma z} + K_2(s)e^{\Gamma z} \tag{3}$$

For the particular integral, try a solution $V = A(s)$ that yields

$$0 - \Gamma^2 A(s) = -sLCV_0$$
$$-s^2 LC A(s) = -sLCV_0$$

or

$$A(s) = \frac{V_0}{s}$$

Then the solutions for $V(z, s)$ and $I(z, s)$ are

$$V = K_1 e^{-\Gamma z} + K_2 e^{\Gamma z} + \frac{V_0}{s}$$

$$I = \sqrt{\frac{C}{L}}(K_1 e^{-\Gamma z} - K_2 e^{\Gamma z})$$

Notice that the current has no particular integral since the initial condition on the current (by inspection) $\mathscr{I}(z, 0) = 0$ gives no term on the right-hand side for the current equation corresponding to Eq. (2.33). To find the constants, we use

$$I(0, s) = 0 \quad \text{for} \quad t > 0$$

Then

$$0 = K_1 - K_2 \qquad \therefore K_1 = K_2 = K$$

Now at $z = l$,

$$V(l, s) = R_0 I(l, s) = \sqrt{\frac{L}{C}} I(l, s)$$

$$\therefore K = -\frac{V_0}{2s} e^{-\Gamma l}$$

$$\therefore I(z, s) = \frac{1}{2}\sqrt{\frac{C}{L}} V_0 \left[ \frac{e^{-s\sqrt{LC}(l-z)}}{s} - \frac{e^{-s\sqrt{LC}(l+z)}}{s} \right]$$

Then

$$I(l, s) = \frac{V_0}{2R_0 s}[1 - e^{-2sl\sqrt{LC}}]$$

$$\mathscr{I}(l, t) = \frac{V_0}{2R_0}[U(t) - U(t - 2l\sqrt{LC})]$$

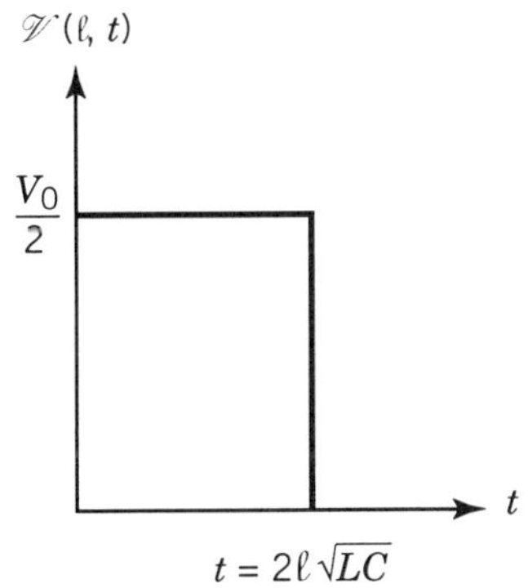

$\mathscr{V}(l, t)$

$\frac{V_0}{2}$

$t = 2l\sqrt{LC}$

**FIGURE 2.31**

Finally (See Figure 2.31),

$$\mathscr{V}(l,t) = \frac{V_0}{2}[U(t) - U(t - 2l\sqrt{LC})]$$

▲

## SUPPLEMENTARY EXAMPLES

**2.1** Find the general expression for the voltage and current at load and source ends when a line is connected to a battery and terminated in a resistor:

$$\Gamma_s = \frac{R_s - R_0}{R_s + R_0}$$

$$\Gamma_L = \frac{R_L - R_0}{R_L + R_0}$$

In Figure 2.32, $T$ is the propagation time for a signal on the line $T = l/v$. We observe that the load voltage changes at times $(2n - 1)T$, $n = 1, 2, 3, \ldots$, whereas at the source plane, changes occur at times $2(n - 1)T$, $n = 1, 2, 3, \ldots$ The $n$th wave leaving the source is

$$\mathscr{V}_n^+ = \frac{R_0}{R_0 + R_s}V_0 + \Gamma_s\mathscr{V}_{n-1}^- \equiv \Gamma_0 V_0 + \Gamma_s\mathscr{V}_{n-1}^- \qquad (1)$$

which is the superposition of the battery contribution and the partial reflection of the wave coming from the line to the source plane. We know

$$\mathscr{V}_{n-1}^- = \Gamma_L \mathscr{V}_{n-1}^+ \qquad (2)$$

and when substituted into Eq. (1), this gives

$$\mathscr{V}_n^+ - \Gamma_s\Gamma_L\mathscr{V}_{n-1}^+ = \Gamma_0 V_0 \qquad (3)$$

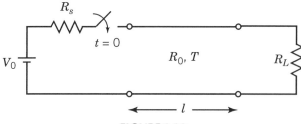

FIGURE 2.32

which is a constant coefficient difference equation relating successive forward waves. The particular solution is $\mathscr{V}_n^+ = C$, a constant:

$$C - \Gamma_s \Gamma_L C = \Gamma_0 V_0 \Rightarrow C = \frac{\Gamma_0 V_0}{1 - \Gamma_s \Gamma_L} \tag{4}$$

The homogeneous equation is

$$\mathscr{V}_n^+ - \Gamma_s \Gamma_L \mathscr{V}_{n-1}^+ = 0$$

Try the solution

$$\mathscr{V}_n^+ = A \lambda^n$$

$$A\lambda^n - \Gamma_s \Gamma_L A \lambda^{n-1} = 0 \tag{5}$$

or

$$A\lambda^{n-1}(\lambda - \Gamma_s \Gamma_L) = 0$$

and for a nontrivial solution,

$$\lambda = \Gamma_s \Gamma_L$$

So

$$\mathscr{V}_n^+ = \frac{\Gamma_0 V_0}{1 - \Gamma_s \Gamma_L} + A(\Gamma_s \Gamma_L)^n \tag{6}$$

is the complete solution. We determine $A$ by applying the known condition:

$$\mathscr{V}_1^+ = \Gamma_0 V_0 = \frac{\Gamma_0 V_0}{1 - \Gamma_s \Gamma_L} + A(\Gamma_s \Gamma_L)$$

so

$$A = \frac{-\Gamma_0 V_0}{(1 - \Gamma_s \Gamma_L)} \tag{7}$$

Using this in Eq. (6) yields

$$\mathscr{V}_n^+ = \left( \frac{\Gamma_0 V_0}{1 - \Gamma_s \Gamma_L} \right)[1 - (\Gamma_s \Gamma_L)^n] \tag{8}$$

Then the total voltage after $n$ reflections at times $(2n-1)T$, $n = 1, 2, 3, \ldots$ (at the load) is

$$\mathscr{V}_{n_L} = \mathscr{V}_n^+ + \mathscr{V}_n^- = \frac{V_0 \Gamma_0}{1 - \Gamma_s \Gamma_L}[1 - (\Gamma_s \Gamma_L)^n] + \Gamma_L \mathscr{V}_n^+$$

which simplifies to

$$\mathscr{V}_{n_L} = \frac{R_L}{R_L + R_s} V_0 [1 - (\Gamma_s \Gamma_L)^n] \tag{9a}$$

Then

$$\mathscr{I}_{n_L} = \frac{\mathscr{V}_n}{R_L} \tag{9b}$$

At the source plane, the times of change are $2(n-1)T$, $n = 1, 2, 3, \ldots$, and after $n$ reflections

$$\mathscr{V}_{n_s} = \mathscr{V}_n^+ + \mathscr{V}_{n-1}^- = \left(\frac{\Gamma_0 V_0}{1 - \Gamma_s \Gamma_L}\right)[1 - (\Gamma_s \Gamma_L)^n] + \Gamma_L \left(\frac{\Gamma_0 V_0}{1 - \Gamma_s \Gamma_L}\right)[1 - (\Gamma_s \Gamma_L)^{n-1}] \tag{10a}$$

and the source current is

$$\mathscr{I}_{n_s} = \frac{V_0 - \mathscr{V}_{n_s}}{R_s} \tag{10b}$$

Since $R_s$ and $R_L$ are positive (passive terminations are always assumed throughout text),

$$|\Gamma_s \Gamma_L| < 1$$
$$\therefore \lim_{n \to \infty} (\Gamma_s \Gamma_L)^n = 0 \tag{11}$$

Then the steady-state values at the load are

$$V_L = \frac{R_L}{R_L + R_s} V_0$$

$$I_L = \frac{V_0}{R_L + R_s}$$

and at the source

$$V_s = \frac{\Gamma_0 V_0}{1 - \Gamma_s \Gamma_2} + \Gamma_L \frac{\Gamma_0 V_0}{1 - \Gamma_s \Gamma_L} = \frac{R_L}{R_L + R_s} V_0 = V_L$$

$$I_s = \frac{V_0 - V_s}{R_s} = \frac{V_0}{R_L + R_s} = I_L$$

which are to be expected when the "line" is just two wires.

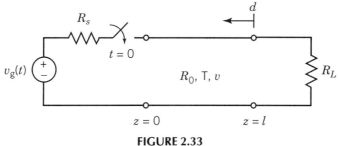

**FIGURE 2.33**

**2.2** Write the general expression for the response of voltage and current on a lossless line as shown in Figure 2.33:

For $0 < t \leq l/v$,

$$\mathscr{V}(z,t) = \frac{R_0}{R_0 + R_s} v_g\left(t - \frac{z}{v}\right) U\left(t - \frac{z}{v}\right)$$

$$\mathscr{I}(z,t) = \frac{1}{R_0 + R_s} v_g\left(t - \frac{z}{v}\right) U\left(t - \frac{z}{v}\right)$$

which are the first left to right waves. For $l/v \leq t \leq 2l/v$,

$$\mathscr{V}(z,t) = \frac{R_0}{R_0 + R_s}\left[v_g\left(t - \frac{z}{v}\right) U\left(t - \frac{z}{v}\right) + \Gamma_L v_g\left(t - \frac{2l-z}{v}\right) U\left(t - \frac{2l-z}{v}\right)\right]$$

$$\mathscr{I}(z,t) = \frac{1}{R_0 + R_s}\left[v_g\left(t - \frac{z}{v}\right) U\left(t - \frac{z}{v}\right) - \Gamma_L v_g\left(t - \frac{2l-z}{v}\right) U\left(t - \frac{2l-z}{v}\right)\right]$$

where now the first leftward wave exists. The first terms are those produced by the generator, whereas the second ones are caused by the load reflection. For these reflected waves, the leading edge is the replica of $v_g(t=0)$ delayed by the total time the edge has "accumulated." The edge has gone the distance $l + d$, where $d$ is measured from the load. But since $d = (l - z)$, the net distance accumulated is $l + l - z = 2l - z$, so the time delay is $(2l - z)/v$. For $2l/v \leq t \leq 3l/v$,

$$\mathscr{V}(z,t) = \frac{R_0}{R_0 + R_s}\left[v_g\left(t - \frac{z}{v}\right) U\left(t - \frac{z}{v}\right) + \Gamma_L v_g\left(t - \frac{2l-z}{v}\right) U\left(t - \frac{2l-z}{v}\right)\right.$$
$$\left. + \Gamma_s \Gamma_L v_g\left(t - \frac{2l+z}{v}\right) U\left(t - \frac{2l+z}{v}\right)\right]$$

$$\mathscr{I}(z,t) = \frac{1}{R_0 + R_s}\left[v_g\left(t - \frac{z}{v}\right) U\left(t - \frac{z}{v}\right) - \Gamma_L v_g\left(t - \frac{2l-z}{v}\right) U\left(t - \frac{2l-z}{v}\right)\right.$$
$$\left. + \Gamma_s \Gamma_L v_g\left(t - \frac{2l+z}{v}\right) U\left(t - \frac{2l+z}{v}\right)\right]$$

**96** THE MATHEMATICS OF TRAVELING WAVES

where the second set of forward waves has been generated. Notice that the delay is $[l + l$ (up and back), and $z]/v = (2l + z)/v$.

The process continues as

$$\mathscr{V}(z,t) = \frac{R_0}{R_0 + R_s}\left[v_g\left(t - \frac{z}{v}\right)U\left(t - \frac{z}{v}\right) + \Gamma_L v_g\left(t - \frac{2l-z}{v}\right)U\left(t - \frac{2l-z}{v}\right)\right.$$
$$+ \Gamma_s\Gamma_L v_g\left(t - \frac{2l+z}{v}\right)U\left(t - \frac{2l+z}{v}\right) + \Gamma_L^2\Gamma_s v_g\left(t - \frac{4l-z}{v}\right)$$
$$\cdot U\left(t - \frac{4l-z}{v}\right) + \Gamma_L^2\Gamma_s^2 v_g\left(t - \frac{4l+z}{v}\right)U\left(t - \frac{4l+z}{v}\right) + \cdots\Bigg]$$

$$\mathscr{I}(z,t) = \frac{1}{R_0 + R_s}\left[v_g\left(t - \frac{z}{v}\right)U\left(t - \frac{z}{v}\right) - \Gamma_L v_g\left(t - \frac{2l-z}{v}\right)U\left(t - \frac{2l-z}{v}\right)\right.$$
$$+ \Gamma_s\Gamma_L v_g\left(t - \frac{2l+z}{v}\right)U\left(t - \frac{2l+z}{v}\right) - \Gamma_L^2\Gamma_s v_g\left(t - \frac{4l-z}{v}\right)$$
$$\cdot U\left(t - \frac{4l-z}{v}\right) + \Gamma_L^2\Gamma_s^2 v_g\left(t - \frac{4l+z}{v}\right)U\left(t - \frac{4l+z}{v}\right) + \cdots\Bigg]$$

**2.3** For the lossless line shown in Figure 2.34, sketch the value of $\mathscr{V}(z)$ and $\mathscr{I}(z)$ for the instants shown on the sets of axes. Label the ordinates in terms of $V_b$ and $R_0$.

By quick inspection, we see that

$$K_L^V = \Gamma_L = \frac{0 - R_0}{0 + R_0} = -1, \qquad K_L^I = +1$$

$$K_g^V = \Gamma_s = \frac{0 - R_0}{0 + R_0} = -1, \qquad K_g^I = +1$$

Notice that this problem (Fig. 2.34) will run into trouble at $t \to \infty$, for we have a short across a battery!

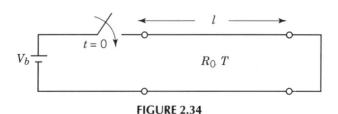

**FIGURE 2.34**

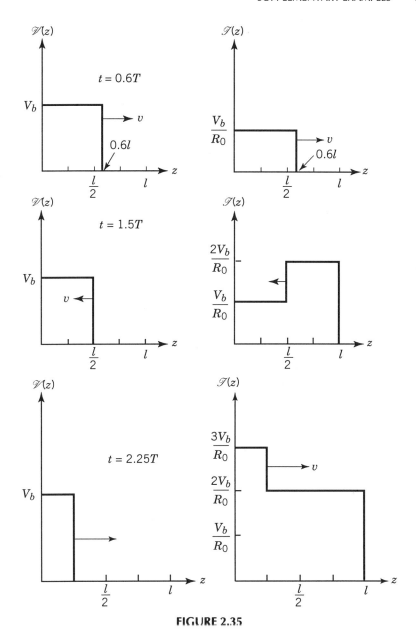

**FIGURE 2.35**

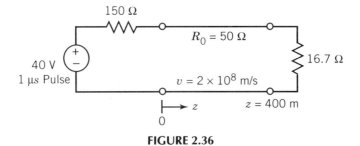

**FIGURE 2.36**

**2.4** Find the sending-end voltage and current over the interval $(0 \leqslant t \leqslant 7\,\mu s.)$ Using Figure 2.36.

The delay from one end to the other is $T = l/v = 2\,\mu s$. The reflection coefficients are

$$K_L^V = \frac{16.7 - 50}{16.7 + 50} = -\frac{1}{2}, \quad K_L^I = +\frac{1}{2}$$

$$K_g^V = \frac{150 - 50}{150 + 50} = +\frac{1}{2}, \quad K_g^I = -\frac{1}{2}$$

Construct bounce diagrams as in Figure 2.37:

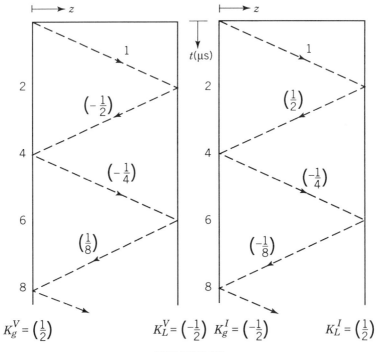

**FIGURE 2.37**

The initial voltage on the line is

$$\mathscr{V}_1^+ = 40\frac{50}{150+50} = 10\text{ V}$$

The initial current is

$$\mathscr{I}_1^+ = \frac{\mathscr{V}_1^+}{50} = 200\text{ mA}$$

At 4 μs, the first reflected wave reaches the source. Then the net voltage is

$$(-5)[(-\tfrac{1}{2})+(-\tfrac{1}{4})] = +3.75\text{ V}$$

The current at that time is

$$100[(\tfrac{1}{2})+(-\tfrac{1}{4})] = 25\text{ mA}$$

Then we can sketch $\mathscr{V}(0,t)$ and $\mathscr{I}(0,t)$ over the specified interval as in Figure 2.38.

**2.5** Consider the problem of two lines connected in tandem and terminated as shown. The initial condition is no current or voltage at any point. At $t=0$, a voltage wave of 120-V amplitude arrives at plane 1 − 1′. What is the initial current in the load?

The assumption we make in Figure 2.39 is that the variables $\mathscr{V}(z,t)$ and $\mathscr{I}(z,t)$ are continuous across junctions of lines with different characteristic impedances. At plane 1 − 1′, the total voltage is $\mathscr{V}_1^+ + \mathscr{V}_1^-$ on the 30-Ω line. It is also $\mathscr{V}_2^+$ (that launched onto the 60-Ω line). Thus, we may state

$$\mathscr{V}_1^+ + \mathscr{V}_1^- = \mathscr{V}_2^+ \tag{1}$$

or

$$\mathscr{V}_1^+(1 + K_{Lj}^V) = \mathscr{V}_2^+ \tag{2}$$

where

$$K_{Lj}^V - \frac{60-30}{60+30} = \frac{1}{3}$$

Equation (2) is also written

$$\mathscr{V}_2^+ = \mathscr{V}_1^+(1 + K_{Lj}^V) = \tau \mathscr{V}_1^+ \tag{3}$$

where $\tau$ is the transmission coefficient. Here, $\tau = 1 + \tfrac{1}{3} = \tfrac{4}{3}$. Notice that

**100** THE MATHEMATICS OF TRAVELING WAVES

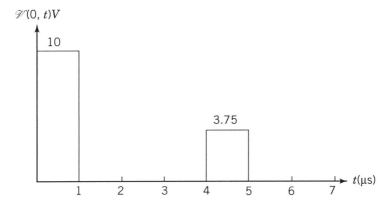

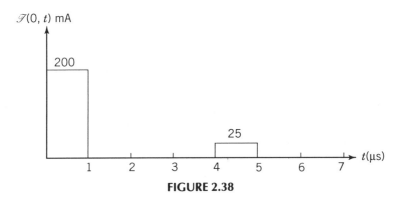

**FIGURE 2.38**

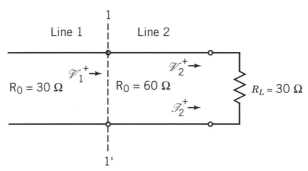

**FIGURE 2.39**

a larger voltage wave can be produced on the second line. The current launched is, however, less than that of the first line. That is,

$$\mathcal{I}_2^+ = \frac{\mathcal{V}_2^+}{60} = \left(\frac{4}{3}\right)\frac{120}{60} = \frac{8}{3}A$$

whereas

$$\mathcal{I}_1^+ = \frac{\mathcal{V}_1^+}{30} = \frac{120}{30} = 4 \text{ A}$$

Also, the net energy launched onto the second line must be less than that available on the first one. For this specific problem, the ratio is

$$\frac{\mathcal{W}_2}{\mathcal{W}_1} = \frac{\mathcal{V}_2 \mathcal{I}_2}{\mathcal{V}_1 \mathcal{I}_1} = \frac{[(4/3)120](8/3)}{(120)(4)} = \frac{8}{9}$$

or only $\frac{1}{9}$ of the available energy is reflected. Thus,

$$\mathcal{V}_2^+ = \tfrac{4}{3}(120) = 160 \text{ V}, \qquad \mathcal{I}_2^+ = \tfrac{8}{3} \text{ A}$$

reach the load. The net current is determined as follows:

$$K_L^V = \frac{30 - 60}{30 + 60} = -\frac{1}{3}$$

Then

$$K_L^I = +\frac{1}{3}$$

so

$$\mathcal{I}_{\text{Tot}} = \mathcal{I}_2^+ (1 + K_L^I) = (\tfrac{8}{3})(\tfrac{4}{3}) = \tfrac{32}{9} \text{ A}$$

**2.6** Consider two lines in tandem as shown in Figure 2.40. For simplicity, we assume that the wave velocity is the same on both lines. In general, at the junction of two lines, the change in geometry causes a reflection. Quite often, the effect of the discontinuity can be modeled by lumped elements at the plane of the junction. For example, in Figure 2.41 the junction could appear as shown where the shunt element diverts some

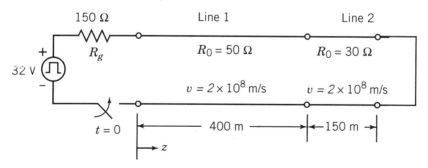

**FIGURE 2.40**

## 102 THE MATHEMATICS OF TRAVELING WAVES

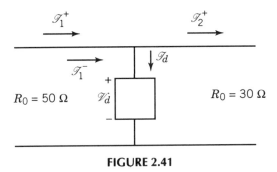

**FIGURE 2.41**

of the incident current. Using Kirchoff's current law at the junction gives

$$(\mathcal{I}_1^+ + \mathcal{I}_1^-) = \mathcal{I}_d + \mathcal{I}_2^+$$

that is, the net current into the junction equals the net leaving. The other constraint would be

$$(\mathcal{V}_1^+ + \mathcal{V}_1^-) = \mathcal{V}_d = \mathcal{V}_2^+$$

where $\mathcal{V}_d$ is the voltage across the element. Once the functional relationship between $\mathcal{V}_d$ and $\mathcal{I}_d$ is known, the junction is described. The determination of this functional relationship is a very tough problem in most cases. The actual modeling is a combination of analysis and experimentation. For this problem, we neglect any discontinuity effects, so $\mathcal{I}_d \equiv 0$.

The sequence of events is as follows. At $t = 0$, an 8-V pulse is launched onto line 1:

$$\mathcal{V}_1^+ = 32 \frac{50}{50 + 150} = 8 \text{ V}$$

When the 8-V wave reaches the junction, a 6-V pulse is launched into line 2 and a $-2$-V pulse returns to the source:

$$K_L^V = \frac{30 - 50}{30 + 50} = -\frac{1}{4}$$

$$\mathcal{V}_2^+ = \tau \mathcal{V}_1^+ = [1 + (-\tfrac{1}{4})]8 = 6 \text{ V}$$

$$\mathcal{V}_1^- = (-\tfrac{1}{4})8 = -2 \text{ V}$$

The $\mathcal{V}_2^+ = 6$ V pulse is reflected by the short and returns to the junction. There, it encounters a reflection coefficient:

$$K_j^V = \frac{50 - 30}{50 + 30} = +\frac{1}{4}$$

and a wave is launched onto line 1 of value

$$\mathscr{V}_1^- = (1 + K_j^V)\mathscr{V}_2^- = (1 + \tfrac{1}{4})(-6) = -7.5 \text{ V}$$

Thus, each line has several pulses traveling in both directions and launching new waves on the other each time a pulse contacts the junction. A bounce diagram is useful as shown in Figure 2.42. The transit times on lines 1 and 2 are 2 and 0.75 µs, respectively. The diagram gets very cluttered, but can be instructive in pin-pointing a particular event. For example, notice that pulses from both lines meet at the junction for the first time at 7.5 µs. Thus, the waves moving away from the junction at that

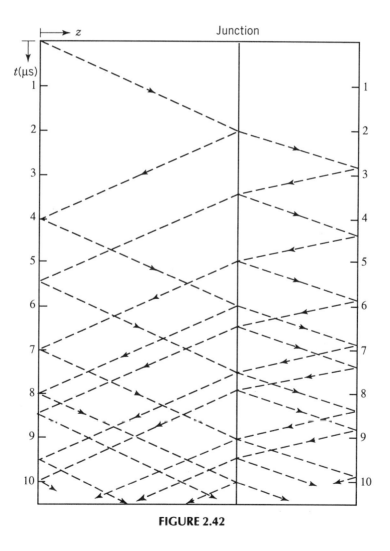

**FIGURE 2.42**

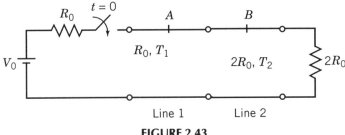

**FIGURE 2.43**

instant are composed of both the reflected as well as transmitted portions of the incident waves.

**2.7** Sketch the voltage and current at planes $A$ and $B$ (midpoints), as functions of time using Figure 2.43:

The sequence of events is as follows. The incident wave on line 1 is $V_0/2$ and it arrives at $A$ at $T_1/2$. The leading edge gets reflected at the junction with coefficient:

$$K_j^V = \frac{2R_0 - R_0}{2R_0 + R_0} = \frac{1}{3}$$

so

$$\mathscr{V}_1^- = \left(\frac{V_0}{2}\right)\left(\frac{1}{3}\right) = V_0/6$$

passes $A$ at $3T_1/2$. The source is matched so no reflections occur there. The wave launched onto line 2 is

$$\mathscr{V}_2^+ = \mathscr{V}_1^+(1 + K_j^V) = \left(\frac{V_0}{2}\right)\left(\frac{4}{3}\right) = \frac{2V_0}{3}$$

and it passes $B$ at $T_1 + (T_2/2)$. The load completely absorbs this incident signal since it is matched to line 2. In Figure 2.44, the currents are

$$\mathscr{I}^+(A,t) = \frac{\mathscr{V}^+(A,t)}{R_0}, \qquad \mathscr{I}^+(B,t) = \frac{\mathscr{V}^+(B,t)}{2R_0}$$

On the first line, $\mathscr{V}_1^- = V_0/6$ so $\mathscr{I}_1^- = -V_0/6R_0$. Thus, the net current for $t > 3T_1/2$ is $V_0/2R_0 - V_0/6R_0 = V_0/3R_0$.

**2.8** Sketch $\mathscr{V}(A,t)$, $\mathscr{I}(A,t)$ and the same at point $B$ (midpoints) as a function of time using Figure 2.45:

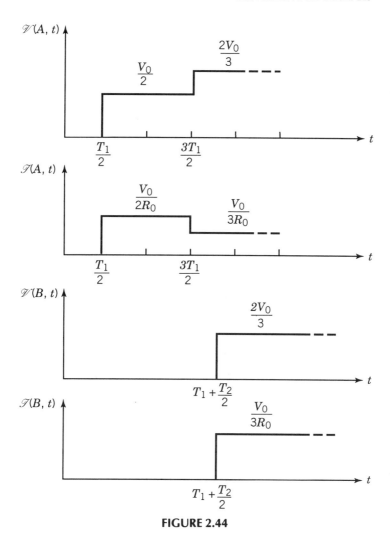

**FIGURE 2.44**

By inspection, $\mathscr{V}_1^+ = V_0$, and at the junction it sees the characteristic value $2R_0$ in parallel with the $R_0$ element. Thus, the effective load is $2R_0 \| R_0 = 2R_0/3$ with reflection coefficient:

$$K_j^V = \frac{2R_0/3 - R_0}{2R_0/3 + R_0} = -\frac{1}{5}$$

The net voltage across the junction is

$$V_0\left[1 + \left(-\frac{1}{5}\right)\right] = \frac{4V_0}{5}$$

**106** THE MATHEMATICS OF TRAVELING WAVES

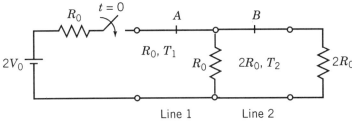

FIGURE 2.45

Therefore, $\mathscr{V}_2^+ = 4V_0/5$ is launched onto line 2 and absorbed completely by the load $2R_0$. The reflected wave on line 1 is absorbed by the generator that matches the line. The values at $A$ are at $T_1/2$

$$\mathscr{V}_1^+ = V_0, \qquad \mathscr{I}_1^+ = \frac{V_0}{R_0}$$

$$\mathscr{V}_1^- = -\frac{V_0}{5}, \qquad \mathscr{I}_1^- = \frac{V_0}{5R_0}$$

and at $t = 3T_1/2$

$$\mathscr{V} = V_0 - \frac{V_0}{5} = \frac{4V_0}{5}$$

$$\mathscr{I} = \frac{V_0}{R_0} + \frac{V_0}{5R_0} = \frac{6V_0}{5R_0}$$

and at $B$

$$\mathscr{V}_2^+ = \frac{4V_0}{5}, \qquad \mathscr{I}_2^+ = \frac{2V_0}{5R_0}, \qquad t > T_1 + \frac{T_2}{2}$$

Notice that at the junction the net current from line 1 is

$$\mathscr{I}_1 = \frac{V_0}{R_0} + \frac{V_0}{5R_0} = \frac{6V_0}{5R_0}$$

The net current in line 2 is $\mathscr{I}_2 = \mathscr{I}_2^+ = 2V_0/5R_0$. The current through the shunt resistor $R_0$ is

$$\mathscr{I}_s = \frac{4V_0}{5R_0}$$

and notice that $\mathscr{I}_1 = \mathscr{I}_s + \mathscr{I}_2$ so Kirchoff's law is satisfied (Figure 2.46).

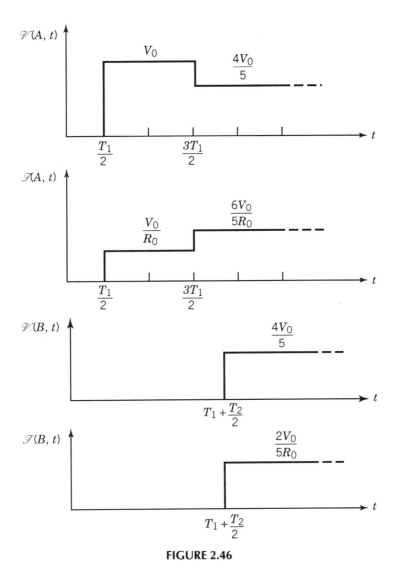

**FIGURE 2.46**

**2.9** Using Figure 2.47, consider the initiation of a pulse by closing a switch at the load:

Prior to the switch closing, the voltage at the load plane is $V_0$, that is, $\mathscr{V}_L = V_0$ for $t < 0$. This means that the line is charged to $V_0$ volts. At $t = 0$, a voltage wave is induced and it should satisfy

$$V_0 + \mathscr{V}_0^- = \mathscr{V}_L = \mathscr{I}_L R_L \tag{1}$$

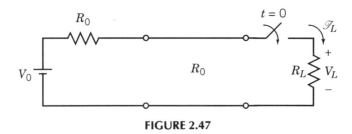

**FIGURE 2.47**

The $\mathscr{V}_1^-$ wave is accompanied by the current wave $\mathscr{I}_1^- = -\mathscr{V}_1^-/R_0$. At the load plane, the total current from the line into the plane is $\mathscr{I}^-$ (recall that $\mathscr{I}^-$ is referenced positive to the right in the upper conductor). Then $\mathscr{I}^- = \mathscr{I}_L$ so Eq. (1) becomes

$$V_0 + \mathscr{V}_1^- = \mathscr{I}^- R_L = -\frac{\mathscr{V}_1^-}{R_0} R_L$$

or

$$\mathscr{V}_1^- = -\frac{V_0 R_0}{R_0 + R_L} \qquad (2)$$

and

$$\mathscr{I}_1^- = \frac{V_0}{R_0 + R_L}$$

For this problem, these waves are completely absorbed in the matched source resistor.

**2.10** Consider a precharged line connected to a resistor as shown in Figure 2.48. Describe the discharge process.

This is a case of generating a pulse in $R$ by discharging a line through it. At $t = 0$, our equations are

$$V_R = \mathscr{I}_R R = -\mathscr{I}_1^+ R \qquad (1)$$

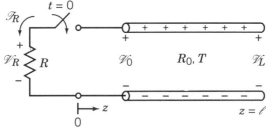

**FIGURE 2.48**

and
$$V_R = V_0 + \mathscr{V}_1^+ \tag{2}$$

where we assume a wave is initiated to permit the satisfaction of boundary conditions. We also have $\mathscr{V}_1^+ = R_0 \mathscr{I}_1^+$, and when combined with the above, it yields

$$\mathscr{V}_1^+ = -\frac{V_0 R_0}{R_0 + R} \tag{3}$$

$$\mathscr{I}_1^+ = -\frac{V_0}{R_0 + R} \tag{4}$$

which are the initial waves induced on the line. When these waves reach the open, the conditions are

$$\mathscr{I}_1^+ + \mathscr{I}_1^- = 0$$

so
$$\mathscr{I}_1^- = -\mathscr{I}_1^+ = \frac{V_0}{R_0 + R} \tag{5}$$

then using
$$\mathscr{V}_1^- = -R_0 \mathscr{I}_1^-$$

$$\mathscr{V}_1^- = -\frac{R_0 V_0}{R_0 + R} \tag{6}$$

We determine that the net voltage across the open is thus

$$\mathscr{V}_L = \mathscr{V}_1^+ + \mathscr{V}_1^- + V_0 = V_0 \frac{R - R_0}{R + R_0} \tag{7}$$

Consider first the case $R = R_0$. Then

$$\mathscr{V}_1^+ = -\frac{V_0}{2}$$

$$\mathscr{I}_1^+ = -\frac{V_0}{2R_0}$$

and the reflected waves are

$$\mathscr{V}_1^- = -\frac{V_0}{2}$$

$$\mathscr{I}_1^- = \frac{V_0}{2R_0}$$

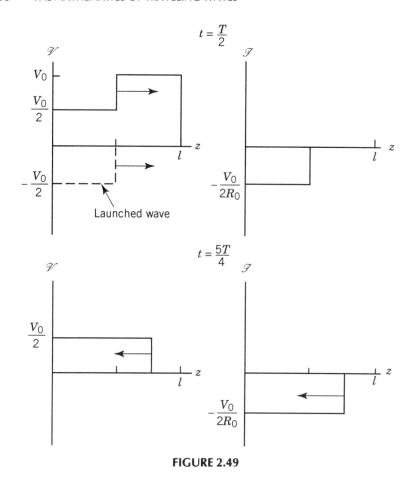

**FIGURE 2.49**

The total voltage and current on the line at $t = T/2$, $5T/4$ is sketched in Figure 2.49.

For $t < T$, the $\mathscr{V}_1^+$ wave has amplitude $-V_0/2$, which reduces the net voltage to $V_0/2$. When it reaches the open, the reflected wave $\mathscr{V}^-$ has amplitude $-V_0/2$ and the net value across the open, $\mathscr{V}_L$, is zero. The combination of $\mathscr{V}_1^+$ and $\mathscr{V}_1^-$ along with the initial value of $V_0$ produces zero volts in a region where both $\mathscr{V}_1^+$ and $\mathscr{V}_1^-$ exist. The reflected current cancels the incident value, so their superposition shows no net current. The current through the resistor is $-\mathscr{I}_1^+ = V_0/2R_0$ and it lasts $2T$ seconds. When $\mathscr{V}_1^-$ and $\mathscr{I}_1^-$ return to the source, they are absorbed ($K_g^V = K_g^I = 0$) so after $2T$ seconds the discharge is complete. The resistor current is shown in Figure 2.50.

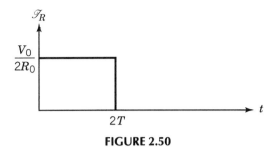

**FIGURE 2.50**

Consider the next case $R = R_0/2$:

$$\mathscr{V}_1^+ = -\frac{2}{3}V_0$$

$$\mathscr{I}_1^+ = -\frac{2V_0}{3R_0}$$

Now the $\mathscr{V}_1^+$ wave reduces the initial value by $\frac{2}{3}$, with the resultant equal to $V_0/3$. The reflected values are

$$\mathscr{V}_1^- = -\frac{2}{3}V_0$$

$$\mathscr{I}_1^- = \frac{2V_0}{3R_0}$$

and the net across the open is

$$\mathscr{V}_L = -\frac{V_0}{3}$$

The reflection coefficient at the source is

$$K_g^V = -\frac{1}{3}$$

so

$$\mathscr{V}_2^+ = -\frac{1}{3}\left(-\frac{2V_0}{3}\right) = \frac{2V_0}{9}$$

and the net voltage across the resistor at $t = 2T$ is

$$V_R = V_0 + \mathscr{V}_1^+ + \mathscr{V}_1^- + \mathscr{V}_2^+ = -\frac{V_0}{9}$$

**112** THE MATHEMATICS OF TRAVELING WAVES

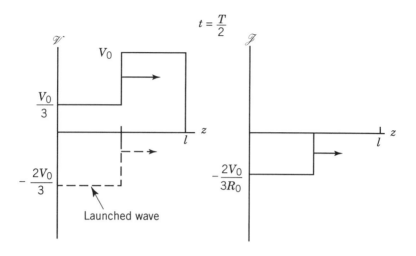

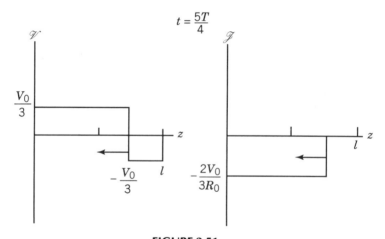

**FIGURE 2.51**

so

$$\mathscr{I}_R = \frac{V_R}{R} = -\frac{V_0}{9R} = -\frac{2V_0}{9R_0}$$

The process is continued in the above manner until the voltage approaches zero. Figure 2.51 indicates the process.

In Figure 2.52 the resistor current is shown,

and it jumps at $2T$ intervals.

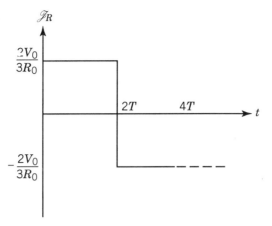

**FIGURE 2.52**

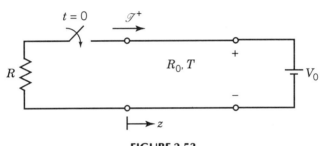

**FIGURE 2.53**

**2.11** Describe the discharge process for the case of a battery at the load:

In Figure 2.53, the line is again charged to $V_0$ volts and the equations at $t = 0$ are

$$-R\mathscr{I}^+ = V_0 + \mathscr{V}_1^+$$

$$\mathscr{I}^+ = \frac{\mathscr{V}_1^+}{R_0}$$

and thus,

$$\mathscr{V}_1^+ = -\frac{R_0}{R_0 + R} V_0 \tag{1}$$

$$\mathscr{I}_1^+ = -\frac{V_0}{R_0 + R} \tag{2}$$

**114**   THE MATHEMATICS OF TRAVELING WAVES

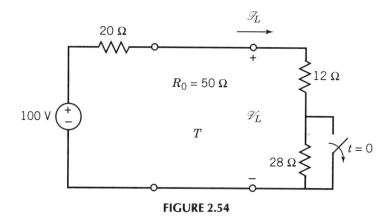

**FIGURE 2.54**

as in the previous problem. At the load, we have

$$V_0 + \mathscr{V}_1^+ + \mathscr{V}_1^- = V_0$$

or

$$\mathscr{V}_1^- = -\mathscr{V}_1 + \mathscr{V}_1^+ \tag{3}$$

and the process continues in a manner similar to the previous problem.

**2.12** Describe the transient response for Figure 2.54.

At $t = 0$, we expect that a wave will be generated at the load. Our equations read ($t = 0^+$):

$$\mathscr{V}_L = R_L \mathscr{I}_L, \qquad R_L = 40\,\Omega \tag{1}$$

or

$$\mathscr{V}^+ + \mathscr{V}^- = \frac{R_L}{R_0}(\mathscr{V}^+ - \mathscr{V}^-) \tag{2}$$

However, we need $\mathscr{V}^+$ to proceed. From physical reasoning, we notice $\mathscr{V}^+(0^-) = \mathscr{V}^+(0^+)$ as no changes in the steady-state $\mathscr{V}^+$ wave can occur immediately after the switch. We determine $\mathscr{V}^+(0^-)$ as follows. For $t < 0$, a dc steady state has been achieved, and

$$\mathscr{V}_L = \frac{12}{12 + 20} 100 = 37.5 \text{ V} \tag{3}$$

and

$$\mathscr{V}_L = \mathscr{V}^+ + \mathscr{V}^- = \mathscr{V}^+(1 + K_L^V)$$

and
$$K_L^V = \frac{12 - 50}{12 + 50} = -0.613$$

so
$$\mathscr{V}^+ = 96.875 \text{ V}$$

From Eq. (2), we find
$$\mathscr{V}^- = K_L^V \mathscr{V}^+ = (-0.6129)(96.875) = -59.375 \text{ V}$$

This wave will be reflected by the source in the normal manner, and the new forward wave sees the reflection coefficient:

$$K_L^V = \frac{42 - 50}{42 + 50} = -0.8696$$

and the process henceforth can be treated easily with a bounce diagram.

**2.13** Using Figure 2.55: **(a)** Find and sketch $\mathscr{V}_s$ and $\mathscr{V}_L$ over the interval $(0, 5T)$. **(b)** Do the same for the currents $\mathscr{I}_s$ and $\mathscr{I}_L$.

The initial voltage wave at $z = 0$ is

$$\mathscr{V}_1^+ = 200 \frac{50}{50 + 10} = 166.67 \text{ V}$$

When $t = T$,

$$\mathscr{V}_L = \mathscr{V}_1^+(1 + K_L^V) = 166.67(1 - \tfrac{3}{7}) = 95.24 \text{ V}$$

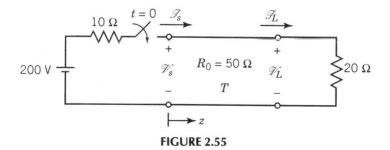

FIGURE 2.55

When $t = 2T$,

$$\mathscr{V}_s = \mathscr{V}^-(1 + K_g^V) = (-\tfrac{3}{7})(166.67)(1 - \tfrac{2}{3}) = -23.81 \text{ V}$$

The corresponding currents are

$$\mathscr{I}_1^+ = \frac{\mathscr{V}_1^+}{50} = 3.33 \text{ A}$$

$$\mathscr{I}_L(T) = \frac{95.24}{20} = 4.76 \text{ A}$$

$$\mathscr{I}_s(2T) = \frac{200 - \mathscr{V}_s(2T)}{10} = \frac{200 + 23.81}{10} = 22.38 \text{ A}$$

and so forth.

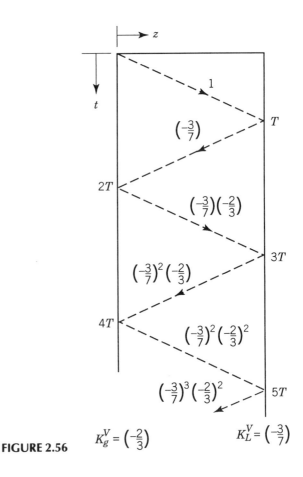

**FIGURE 2.56**

We will use the bounce diagram in Figure 2.56 to complete the problem. For voltage, we have

$$K_L^V = -\tfrac{3}{7}$$
$$K_g^V = -\tfrac{2}{3}$$

At the load (for a unit incident pulse),

$t = T,\qquad \mathscr{V}_L = 1 + (-\tfrac{3}{7}) = 0.5714$

$t = 3T,\qquad \mathscr{V}_L = 0.5714 + [(-\tfrac{3}{7})(-\tfrac{2}{3}) + (-\tfrac{3}{7})^2(-\tfrac{2}{3})] = 0.735$

$t = 5T,\qquad \mathscr{V}_L = 0.735 + [(-\tfrac{3}{7})^2(-\tfrac{2}{3})^2 + (-\tfrac{3}{7})^3(-\tfrac{2}{3})^2] = 0.7816$

At the source (for a unit incident pulse),

$t = 0^+,\qquad \mathscr{V}_s = 1$

$t = 2T,\qquad V_s = 1 + [(-\tfrac{3}{7}) + (-\tfrac{3}{7})(-\tfrac{2}{3})] = 0.8571$

$t = 4T,\qquad \mathscr{V}_s = 0.8571 + [(-\tfrac{3}{7})^2(-\tfrac{2}{3}) + (-\tfrac{3}{7})^2(-\tfrac{2}{3})^2] = 0.8163$

The actual (denormalized) values are therefore

| $t$ | $\mathscr{V}_s$ | $\mathscr{V}_L$ |
|---|---|---|
| $0^+$ | 166.67 | 0 |
| $T$ | 166.67 | (0.5714)(166.67) = 95.24 |
| $2T$ | 142.85 | 95.24 |
| $3T$ | 142.85 | (0.735)(166.67) = 122.50 |
| $4T$ | 136.05 | 122.50 |
| $5T$ | 136.05 | 130.27 |

The final values are

$$\mathscr{V}_L = \mathscr{V}_s = 200\frac{20}{20+10} = 133.33 \text{ V}$$

For the currents, we have

$$\mathscr{I}_1^+ = 3.33 \text{ A}$$

and the final values are (See Figure 2.57),

$$\mathscr{I}_s = \mathscr{I}_L = \frac{133.33}{20} = 6.67 \text{ A}$$

**118**   THE MATHEMATICS OF TRAVELING WAVES

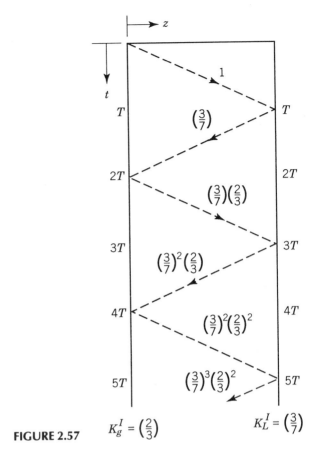

**FIGURE 2.57**

At the load (for a unit incident current),

$$t = T, \quad \mathscr{I}_L = 1 + \tfrac{3}{7} = 1.4286$$
$$t = 3T, \quad \mathscr{I}_L = 1.4266 + [(\tfrac{3}{7})(\tfrac{2}{3}) + (\tfrac{3}{7})^2(\tfrac{2}{3})] = 1.8368$$
$$t = 5T, \quad \mathscr{I} = 1.8368 + [(\tfrac{3}{7})^2(\tfrac{2}{3})^2 + (\tfrac{3}{7})^3(\tfrac{2}{3})^2] = 1.9534$$

At the source (for a unit incident current),

$$t = 0^+, \quad \mathscr{I}_s = 1$$
$$t = 2T, \quad \mathscr{I}_s = 1 + [(\tfrac{3}{7}) + (\tfrac{3}{7})(\tfrac{2}{3})] = 1.7143$$
$$t = 4T, \quad \mathscr{I}_s = 1.7143 + [(\tfrac{3}{7})^2(\tfrac{2}{3}) + (\tfrac{3}{7})^2(\tfrac{2}{3})^2] = 1.9184$$

The actual values are

| t | $\mathscr{I}_s$ | $\mathscr{I}_L$ |
|---|---|---|
| $0^+$ | 3.33 | 0 |
| $T$ | 3.33 | $(1.4286)(3.33) = 4.757$ |
| $2T$ | 5.709 | 4.757 |
| $3T$ | 5.709 | $(1.8368)(3.33) = 6.117$ |
| $4T$ | 6.388 | 6.117 |
| $5T$ | 6.388 | 6.505 |

which are sketched in Figure 2.58.

As you've noticed, the currents follow directly from the voltages as

$$\mathscr{I}_L = \frac{\mathscr{V}_L}{R_L} \quad (R_L = 20\,\Omega), \qquad \mathscr{I}_s = (200 - \mathscr{V}_s)/10\,\Omega$$

so the construction of the diagram is speeded up.

**2.14** For the network in Figure 2.59

(a) At $t = 0$, find $\mathscr{V}(0)$, $\mathscr{I}(0)$.
(b) At $t = T$, find the current in the 300 $\Omega$-resistor.
(c) At $t = 3T$, find the current and voltage associated with the 800-$\Omega$ load.
(d) At $t = 4T$, how much energy has been absorbed in the 300-$\Omega$ resistor?
(e) Sketch the sending-end voltage and current over the interval $(0, 7T)$.

(a) $\mathscr{V}(0) = 600 \dfrac{200}{200 + 400} = 200$ V

$\mathscr{I}(0) = \dfrac{200 \text{ V}}{200\,\Omega} = 1000$ mA

(b) The incident voltage wave $\mathscr{V}_1^+ (= 200$ V) sees an effective load of 300 $\Omega$ in parallel with the characteristic value (200 $\Omega$) of the second line. The reflection coefficient is

$$K_1^V = \frac{120 - 200}{120 + 200} = -\frac{1}{4}$$

and the net voltage at the junction is

$$\mathscr{V}_{\text{Tot}} = 200(1 - \tfrac{1}{4}) = 150 \text{ V}$$

**120**  THE MATHEMATICS OF TRAVELING WAVES

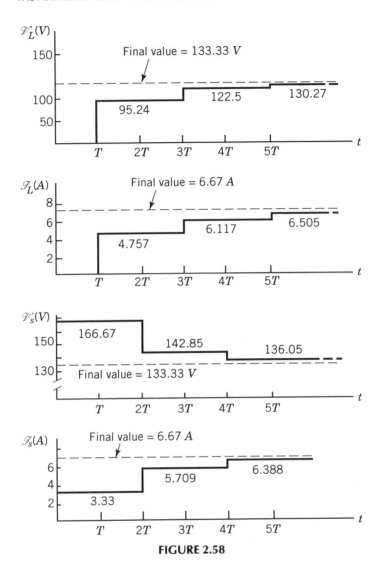

**FIGURE 2.58**

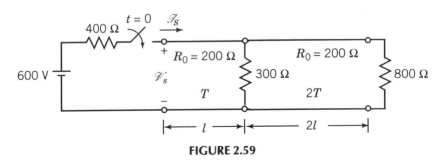

**FIGURE 2.59**

This is the value across the 300-Ω resistor as well as the value of $\mathscr{V}_2^+$ (that launched onto the next line). The resistor current is therefore $\mathscr{I} = 150\,\text{V}/300\,\Omega = 500\,\text{mA}$.

(c) Here,

$$K_2^V = \frac{800 - 200}{800 + 200} = 0.6$$

so

$$\mathscr{V}_L = 150(1 + 0.6) = 240\,\text{V}, \qquad \mathscr{I}_L = \frac{240}{800} = 300\,\text{mA}$$

(d) The reflection from the 800-Ω load has not arrived yet, so the voltage and current have stayed at 150 V and 500 mA for 2T seconds. The energy absorbed is

$$(150\,\text{V})(500\,\text{mA})(2T\,\text{s}) = 150\,T \quad \text{J}$$

(e) The first reflection from the 300-Ω resistor returns at $t = 2T$, whereas that from the 800-Ω load arrives at $t = 6T$. The subsequent reflections

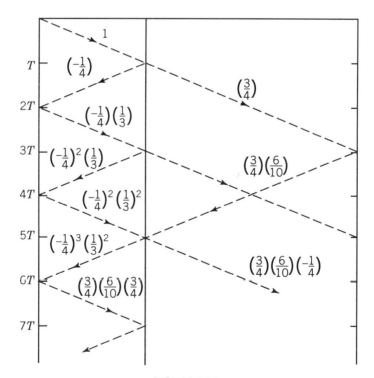

**FIGURE 2.60**

from the 800-$\Omega$ load arrive after $t = 7T$ and are not considered. A bounce diagram in Figure 2.60 is useful here.

At $t = 5T$, the wave from the 800-$\Omega$ load arrives back at the 300-$\Omega$ resistor. The load it sees is $300\,\Omega \parallel 200\,\Omega$ so the reflection coefficient is $(-\frac{1}{4})$ or a transmission coefficient of $(\frac{3}{4})$. Then over the interval $5T$ to $7T$, the net wave on the first line is composed of "ordinary" reflections as well as an input from the second line. Thus, the normalized sending-end voltage is

| $t$ | $\mathscr{V}_s$ |
|---|---|
| 0 | 1 |
| $2T$ | $(-\frac{1}{4}) + (-\frac{1}{4})(\frac{1}{3}) = -\frac{1}{3}$ |
| $3T$ | $-\frac{1}{3}$ |
| $4T$ | $-\frac{1}{3} + [(-\frac{1}{4})^2(\frac{1}{3}) + (-\frac{1}{4})^2(\frac{1}{3})^2] = -\frac{11}{36}$ |
| $5T$ | $-\frac{11}{36}$ |
| $6T$ | $-\frac{11}{36} + [(-\frac{1}{4})^3(\frac{1}{3})^2 + (\frac{3}{4})(\frac{6}{10})(\frac{3}{4})](1+\frac{1}{3}) = 0.1468$ |
| $7T$ | 0.1468 |

Find the source current from

$$\mathscr{I}_s = \frac{600 - \mathscr{V}_s}{400}$$

The sketch is shown in Figure 2.61.

**2.15** Sketch the load voltage and current over the interval $(0, 8T)$. The diode is assumed to be ideal as shown in Figure 2.62.

The diode permits any current $I_D$ if $V_D \geq 0.7$ V, and acts as an open circuit if $V_D < 0.7$ V. The first launched wave has amplitude

$$\mathscr{V}_1^+ = 8 \frac{50}{50 + 10} = 6\frac{2}{3}\text{ V}$$

and let us initially assume that the diode stays off ($I_D = 0$) when $\mathscr{V}_1^+$ is reflected. Under this assumption,

$$\mathscr{V}_L = \mathscr{V}_1^+ + \mathscr{V}_1^- = V_D + 5$$

$$\mathscr{I}_L = \frac{\mathscr{V}_1^+}{50} - \frac{\mathscr{V}_1^-}{50} = 0 \Rightarrow \mathscr{V}_1^+ = \mathscr{V}_1^-$$

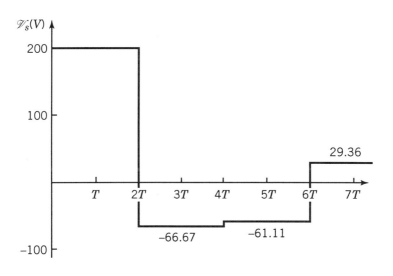

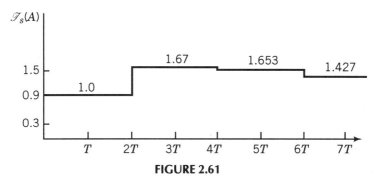

**FIGURE 2.61**

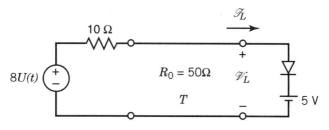

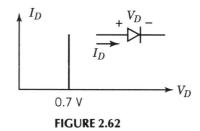

**FIGURE 2.62**

so

$$2\mathscr{V}_1^+ = V_D + 5 \Rightarrow V_D = 8\tfrac{1}{3}$$

but this implies that the diode would be on. Thus, next assume that the diode turns on, and we have

$$\mathscr{V}_1^+ + \mathscr{V}_1^- = 0.7 + 5 \Rightarrow \mathscr{V}_1^+ = -0.9666$$

and

$$\mathscr{I}_L = \frac{\mathscr{V}_L}{50} = 114\,\text{mA}$$

Therefore, the diode turns on at $t = T$ and draws 114 mA. The next incident wave has amplitude

$$\mathscr{V}_2^+ = (-0.9666)(-\tfrac{2}{3}) = 0.6444$$

Its reflected value is determined as follows:

$$\mathscr{V}_1^+ + \mathscr{V}_1^- + \mathscr{V}_2^+ + \mathscr{V}_2^- = 5.7 \Rightarrow \mathscr{V}_2^- = -\mathscr{V}_2^+$$

This means that the load has a reflection coefficient of $(-1)$ when the diode is on. The net current is then

$$\mathscr{I}_L = 114 + \frac{\mathscr{V}_2^+}{50} - \frac{\mathscr{V}_2^-}{50} = 139.8\,\text{mA}$$

The next incident wave has amplitude

$$\mathscr{V}_3^+ = (-0.6444)(-\tfrac{2}{3}) = +0.4296$$

and the load current becomes

$$\mathscr{I}_L = 139.8 + \frac{2(0.4296)}{50} = 156.98\,\text{mA}$$

The last incident wave of interest impinges on the load at $t = 7T$, and has amplitude

$$\mathscr{V}_4^+ = (-0.4296)(-\tfrac{2}{3}) = +0.2864$$

and the load current jumps to

$$\mathscr{I}_L = 156.98 + \frac{2(0.2864)}{50} = 168.44\,\text{mA.\ (See Figure 2.63)}.$$

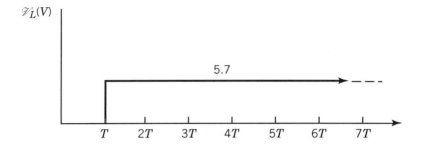

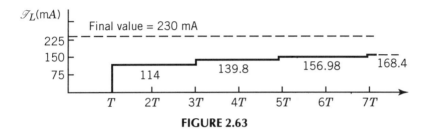

**FIGURE 2.63**

In this case, the diode turns on and stays on until steady state is reached. For various combinations of the parameters in this type of problem, the diode can turn off after several reflections. One must always check to ascertain if $\mathscr{I}_L > 0$. If $\mathscr{I}_L < 0$, then the diode is off, and the calculation must proceed accordingly.

**2.16** Find the reflection and transmission coefficients for the networks shown in Figures 2.64, 2.65 and 2.66.

**(a)**

Assume that an intial amplitude $\mathscr{V}_1^+$ arrives at the junction. The reflection coefficient is

$$K_L^V = \frac{R_{eq} - R_{01}}{R_{eq} + R_{01}}$$

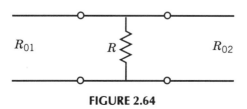

**FIGURE 2.64**

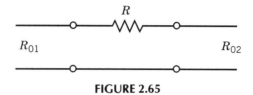

**FIGURE 2.65**

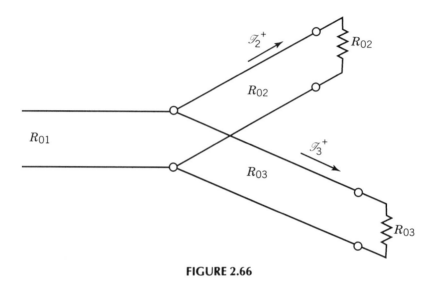

**FIGURE 2.66**

where $R_{eq} = R \| R_{02}$. The net voltage across $R$ is

$$\mathscr{V} = \mathscr{V}_1^+(1 + K_L^V) = \mathscr{V}_2^+$$

which is also the value launched onto the second line. Thus, $\mathscr{V}_2^+ = \tau \mathscr{V}_1^+$ so the transmission coefficient is $\tau = (1 + K_L^V)$.

**(b)**

Here, the effective load is $R_{eq} = R + R_{02}$ and $K_L^V$, $\tau$ follow immediately.

**(c)**

The effective load to the $R_{01}$ line is $R_{eq} = R_{02} \| R_{03}$. Then

$$K_L^V = \frac{R_{eq} - R_{01}}{R_{eq} + R_{01}}$$

and the voltage and currrent in line 1 at the junction are

$$\mathscr{V} = \mathscr{V}_1^+(1 + K_L^V)$$

$$\mathscr{I} = \frac{\mathscr{V}_1^+}{R_{01}}(1 - K_L^V)$$

The currents into the lines are (use current division)

$$\mathscr{I}_2^+ = \mathscr{I}\frac{R_{03}}{R_{03} + R_{02}}$$

$$\mathscr{I}_3^+ = \mathscr{I}\frac{R_{02}}{R_{03} + R_{02}}$$

with corresponding voltages:

$$\mathscr{V}_2^+ = \mathscr{I}_2^+ R_{02}$$
$$\mathscr{V}_3^+ = \mathscr{I}_3^+ R_{03}$$

## PROBLEMS

**2.1** Assume an infinitely long TL with $Z_0 = 50\,\Omega$ and phase velocity $v_p = 2 \times 10^8$ m/s. In Figure P2.1, a lightning stroke hits the line at $x = 0$ and induces the following voltage there:

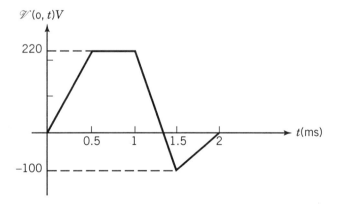

**FIGURE P2.1**

(a) Sketch $\mathscr{V}(z)$ on the line at $t = 0, 0.5, 1, 1.5, 2,$ and 3 ms.
(b) Sketch $\mathscr{V}(t)$ at $z = 100$ km and $z = -25$ km.

# 128   THE MATHEMATICS OF TRAVELING WAVES

**2.2** Which of the following functions are possible waveforms on a TL? Explain your answer.

(a) $f_1(z,t) = (A\cos\omega t)(B\sin\beta z)$
(b) $f_2(z,t) = (3t - 7z)^2$
(c) $f_3(z,t) = A\sin(\beta z - 16t)$

**2.3** A uniform plane wave in empty space is described by

$$\bar{E}(z) = \bar{a}_y 150 e^{-j\beta_0 z} \text{ mV/m}$$

where $f = 200$ MHz. What is the wave's amplitude, direction of travel, vector direction for $\bar{E}, \bar{H}$, and wavelength? Determine the phase velocity, $\beta_0$, and oscillation period.

**2.4** For the following voltage waveform on a 30-$\Omega$ TL,

$$\mathscr{V}(z,t) = 150\cos(3z - 100t) + 100\cos(3z + 100t) \text{ V}$$

(a) Does this represent a traveling or standing wave?
(b) What is the direction of energy travel (explain)?
(c) Find $\beta = 2\pi/\lambda$ and $\omega$.
(d) Find $\mathscr{I}^+(z,t), \mathscr{I}^-(z,t)$. Assume that the load is at $z = 0$.
(e) What is the voltage reflection coefficient and power absorbed in the load?
(f) Write $\mathscr{V}(z,t)$ in complex form.

**2.5** For the junction in Figure P2.5, the transmission coefficient is defined as $\tau \triangleq \mathscr{V}_L/\mathscr{V}^+$.

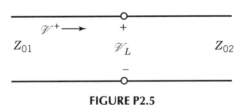

FIGURE P2.5

(a) Express $\tau$ in terms of $Z_{01}, Z_{02}$.
(b) Show that $1 = |\Gamma|^2 + |\tau|^2 (Z_{01}/Z_{02})$. Use $\langle P_L \rangle = \langle P^+ \rangle (1 - |\Gamma|^2)$, where $P^+ = Re\{V^+(I^+)^*\}$ is the incident power.

**2.6** Find the response for the network in Figure P2.6. Use a bounce diagram to describe the solution.

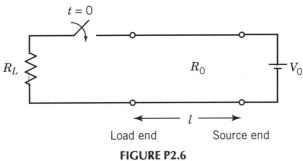

**FIGURE P2.6**

**2.7** Find the complete response for the network in Figure P2.7.

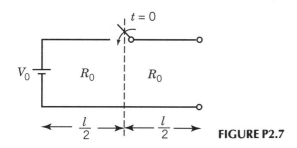

**FIGURE P2.7**

**2.8** For the network, in Figure P2.8, sketch $\mathscr{V}(z)$ along the line for $t = 0$, $l/2v, l/v, 3l/2v, 2l/v$, and $\infty$.

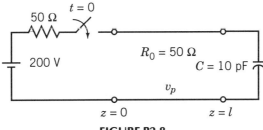

**FIGURE P2.8**

**130** THE MATHEMATICS OF TRAVELING WAVES

**2.9** Using Figure P2.9, find $\mathscr{V}^-$ and $\mathscr{V}^t$ in terms of $\mathscr{V}^+$:

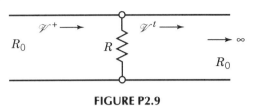

**FIGURE P2.9**

**2.10** Show that the wave produced at $t = 0$, in Figure P2.10 is given by

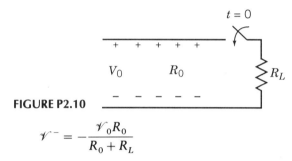

**FIGURE P2.10**

$$\mathscr{V}^- = -\frac{\mathscr{V}_0 R_0}{R_0 + R_L}$$

**2.11** Find the solution to the LR load shown in Figure P2.11, sketch $\mathscr{V}(0, t)$. Use Laplace transforms.

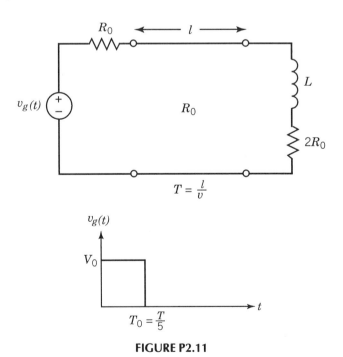

**FIGURE P2.11**

Assume

$$\frac{L}{2R_0} \ll T_0$$

## REFERENCES

1. Aitken, G. J. M., "A Fourier Transform Approach to Transmission-Line Analysis," *IEEE Trans. Education*, Vol. E-13, July 1970, pp. 50–52.
2. Shah, K. R., "On the Analysis of Transmission-Line Transients," *IEEE Trans. Education*, Vol. E-13, July 1970, pp. 52–54.

## BIBLIOGRAPHY

Johnson, W. C. *Transmission Lines and Networks* McGraw-Hill, New York, 1950

Brown, R. G., Sharpe, R. A., Hughes, W. L., and Post, R. E., *Lines, Waves, and Antennas*, Second Edition, John Wiley and Sons, New York, 1973.

Collin, R. E., *Foundations for Microwave Engineering*, McGraw-Hill, New York, 1966.

Lance, A. F. *Introduction to Microwave Theory and Measurements*, McGraw-Hill, New York, 1964.

CHAPTER THREE

# Coupled Lines

## 3.1 EQUATIONS FOR COUPLED LINES

When two transmission lines run parallel to one another and are separated by distances less than or on the order of the wavelength, a portion of the energy traveling on one will leak onto the other. This mutual coupling is the topic of the present chapter, and if it is unwanted, it is termed "crosstalk." Figure 3.1 shows two uniform conductors above a ground plane along with the schematics. All the fields exist in the same dielectric, so the TEM solution is the one assumed throughout the chapter. All the provisos of Chapter 1 still hold, and a few more will be added as we proceed. The mutual coupling is modeled by an inductor $L_m$ and a capacitor $C_m$ in a differential cell as shown. The differential equations relating $\mathscr{V}_1$, $\mathscr{I}_1$, $\mathscr{V}_2$, and $\mathscr{I}_2$ are developed in the same fashion as was done in Section 1.6. For line 1,

$$\mathscr{V}_1(z + \Delta z) = \mathscr{V}_1(z) - L_1 \Delta z \frac{\partial \mathscr{I}_1}{\partial t} - L_m \Delta z \frac{\partial \mathscr{I}_2}{\partial t}$$

or

$$\frac{\partial \mathscr{V}_1}{\partial z} = -L_1 \frac{\partial \mathscr{I}_1}{\partial t} - L_m \frac{\partial \mathscr{I}_2}{\partial t} \qquad (3.1)$$

and similarly for line 2,

$$\frac{\partial \mathscr{V}_2}{\partial z} = -L_2 \frac{\partial \mathscr{I}_2}{\partial t} - L_m \frac{\partial \mathscr{I}_1}{\partial t} \qquad (3.2)$$

The normal dot convention for coupled coils has been used as indicated on the figure. The equations for current are

$$\mathscr{I}_1(z + \Delta z) = \mathscr{I}_1(z) - C_1 \Delta z \frac{\partial \mathscr{V}_1}{\partial t} - C_m \Delta z \frac{\partial}{\partial t}(\mathscr{V}_1 - \mathscr{V}_2)$$

EQUATIONS FOR COUPLED LINES    133

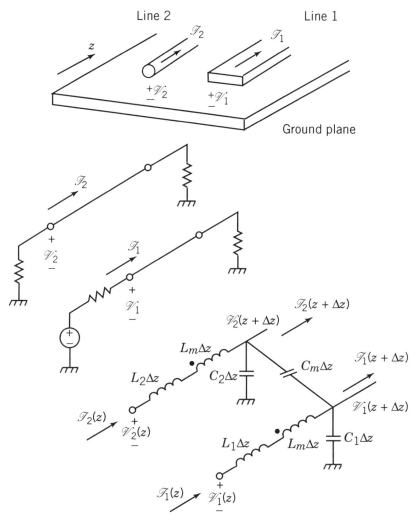

**FIGURE 3.1** Two conductors above a ground plane that propagate a three-conductor TEM wave (field energy between each conductor and ground, as well as between conductors). Below are schematics to indicate both excitation and the lumped equivalent.

or

$$\frac{\partial \mathscr{I}_1}{\partial z} = -(C_1 + C_m)\frac{\partial \mathscr{V}_1}{\partial t} + C_m \frac{\partial \mathscr{V}_2}{\partial t} \quad (3.3)$$

and

$$\frac{\partial \mathscr{I}_2}{\partial z} = -(C_2 + C_m)\frac{\partial \mathscr{V}_2}{\partial t} + C_m \frac{\partial \mathscr{V}_1}{\partial t} \quad (3.4)$$

When more than three conductors are involved, the representation is generally presented in a compact matrix formalization. Consider $N+1$ conductors ($N$

wires above the ground) and write the coupled equations as

$$\frac{\partial}{\partial z}[\mathscr{V}] = -[L]\frac{\partial}{\partial t}[\mathscr{I}] \tag{3.5}$$

$$\frac{\partial}{\partial z}[\mathscr{I}] = -[C]\frac{\partial}{\partial t}[\mathscr{V}] \tag{3.6}$$

where

$$[\mathscr{V}] = \begin{bmatrix} \mathscr{V}_1 \\ \mathscr{V}_2 \\ \vdots \\ \mathscr{V}_N \end{bmatrix}, \quad [\mathscr{I}] = \begin{bmatrix} \mathscr{I}_1 \\ \mathscr{I}_2 \\ \vdots \\ \mathscr{I}_N \end{bmatrix} \tag{3.7}$$

and the $N \times N$ matrices are

$$[L] = \begin{bmatrix} L_{11} & L_{12} & \cdots & L_{1N} \\ L_{21} & L_{22} & \cdots & L_{2N} \\ \vdots & & & \vdots \\ L_{N1} & \cdots & \cdots & L_{NN} \end{bmatrix} \tag{3.8}$$

$$[C] = \begin{bmatrix} C_{11} & C_{12} & \cdots & C_{1N} \\ C_{21} & C_{22} & \cdots & C_{2N} \\ \vdots & & & \vdots \\ C_{N1} & \cdots & \cdots & C_{NN} \end{bmatrix} \tag{3.9}$$

Here, $\mathscr{I}_k$ is the current in the $k$th conductor, and $\mathscr{V}_k$ is the voltage of that conductor with respect to ground. The elements of the matrices $[L]$ and $[C]$ are inductive and capacitive quantities that will be discussed at some length in the first chapter example. It turns out that the velocity of propagation is

$$[L][C] = \frac{1}{v^2}[I] \tag{3.10}$$

where

$$v = \frac{1}{\sqrt{\mu\varepsilon}} \tag{3.10a}$$

Here, $[I]$ is the identity matrix. For the special case that we will almost always treat in this book, that is, two conductors above ground, the above reduces to

$$\frac{\partial \mathscr{V}_1}{\partial z} = -L_{11}\frac{\partial \mathscr{I}_1}{\partial t} - L_{12}\frac{\partial \mathscr{I}_2}{\partial t} \tag{3.5a}$$

$$\frac{\partial \mathscr{V}_2}{\partial z} = -L_{21}\frac{\partial \mathscr{I}_1}{\partial t} - L_{22}\frac{\partial \mathscr{I}_2}{\partial t} \tag{3.5b}$$

$$\frac{\partial \mathscr{I}_1}{\partial z} = -C_{11}\frac{\partial \mathscr{V}_1}{\partial t} - C_{12}\frac{\partial \mathscr{V}_2}{\partial t} \tag{3.6a}$$

$$\frac{\partial \mathscr{I}_2}{\partial z} = -C_{21}\frac{\partial \mathscr{V}_1}{\partial t} - C_{22}\frac{\partial \mathscr{V}_2}{\partial t} \tag{3.6b}$$

Comparing these with those developed earlier, we find

$$L_1 = L_{11}, \qquad L_2 = L_{22}, \qquad L_m = L_{12} = L_{21} \tag{3.11a}$$

$$(C_1 + C_m) = C_{11}, \qquad (C_2 + C_m) = C_{22} \tag{3.11b}$$

$$C_m = -C_{12} = -C_{21} \tag{3.11c}$$

The velocity of propagation is therefore

$$v^2 = \frac{1}{L_1(C_1 + C_m) - L_m C_m} = \frac{1}{L_2(C_2 + C_m) - L_m C_m} \tag{3.12}$$

## 3.2 WEAK COUPLING ANALYSIS

The general solution for Eqs. (3.1) through (3.4) is not extremely useful for practical situations, and for our purposes the weak coupling approximation is adequate. Consider a signal on line 1 developed by a generator and line 2 weakly coupled. This means we assume that some energy in line 1 leaks onto line 2, but we neglect any reaction back on line 1 from line 2. Analytically, this means we neglect the terms $L_m$ and $C_m$ in Eqs. (3.1) and (3.3), while retaining them in Eqs. (3.2) and (3.4). The resulting system of equations is best solved using Laplace transforms. Assuming that all initial values are zero yields

$$\frac{dV_1(z)}{dz} + sLI_1(z) = 0, \qquad \text{that is, let } L_1 = L_2 = L \tag{3.13a}$$

$$\frac{dI_1(z)}{dz} + s(C_1 + C_m)V_1(z) = 0 \tag{3.13b}$$

$$\frac{dV_2(z)}{dz} + sLI_2(z) + sL_m I_1(z) = 0 \tag{3.13c}$$

$$\frac{dI_2(z)}{dz} + s(C_1 + C_m)V_2(z) - sC_m V_1(z) = 0, \qquad \text{that is, let } C_2 = C_1 \tag{3.13d}$$

**136** COUPLED LINES

In this approximation, we assume that the velocity of propagation is

$$v^2 \doteq \frac{1}{LC_1} \tag{3.14a}$$

and define

$$\gamma = \frac{C_m}{C_1}, \quad K = \frac{L_m}{L}\left(\frac{C_1}{C_m}\right) \tag{3.14b}$$

We also drop the mutual capacity term $C_m$ with respect to $C_1$, which after differentiating and rearranging gives

$$\frac{d^2 V_1}{dz^2} - \frac{s^2}{v^2} V_1 = 0 \tag{3.15a}$$

$$\frac{d^2 V_2}{dz^2} - \frac{s^2}{v^2} V_2 = \frac{s^2}{v^2} \gamma(K-1) V_1 \tag{3.15b}$$

which are ordinary differential equations whose solutions are obtained straightforwardly. Notice that the Laplace transform reduces partial differential equations to ordinary ones, where solutions are often more easily recognized. The first equation has just a homogeneous solution:

$$V_1 = A e^{s(z/v)} + B e^{-s(z/v)} \tag{3.16}$$

The second has a forcing function (proportional to $V_1$) and we write the solution as

$$V_2 = A \frac{sz\gamma(K-1)}{2v} e^{s(z/v)} - B \frac{sz\gamma(K-1)}{2v} e^{-s(z/v)} + D e^{s(z/v)} + E e^{-s(z/v)} \tag{3.17}$$

where the last two terms are the homogeneous solution and the first two form the particular integral. Since the forcing term has the same form as the homogeneous solution, we have to multiply the exponentials by $z$. The solution for $I_2$ is (Ref. 1)

$$Z_0 I_2 = -e^{s(z/v)}\left\{D - K\gamma A + \frac{sz\gamma(K-1)}{2v} A + \frac{\gamma(K-1)}{2} A\right\}$$

$$+ e^{-s(z/v)}\left\{E - K\gamma B - \frac{sz\gamma(K-1)}{2v} B + \frac{\gamma(K-1)}{2} B\right\} \tag{3.18}$$

where

$$Z_0 \triangleq \sqrt{\frac{L}{C_1}}$$

With reference to Figure 3.1, let us terminate all lines with $\mathbf{Z}_0$ and therefore assume that only a forward wave exists on line 1 (then $A = 0$). This is a reasonable approximation since line 1 is unaware of line 2, and its characteristic impedance (if line 2 were removed) is just $\mathbf{Z}_0$. To obtain the coefficients $B$, $D$, and $E$, we apply the following boundary conditions:

Line 1

$$z = 0, \quad V_1(0, s) = B \qquad (3.19a)$$
$$z = l, \quad V_1(l, s) = I_1(l, s)\mathbf{Z}_0 \qquad (3.19b)$$

where $V_1(0, s)$ is our known input signal:

Line 2

$$z = 0, \quad V_2(0, s) = -I_2(0, s)\mathbf{Z}_0 \qquad (3.19c)$$
$$z = l, \quad V_2(l, s) = I_2(l, s)\mathbf{Z}_0 \qquad (3.19d)$$

From Eq. (3.19c), we find

$$E = \frac{\gamma}{4}(K + 1)B$$

From Eq. (3.19d), we obtain

$$D = -\frac{\gamma(K + 1)}{4} e^{-2s(l/v)} B$$

For convenience, let $B(s) \triangleq V_s$. Then the solutions are

$$V_1(z, s) = V_s e^{-s(z/v)} \qquad (3.20a)$$

$$V_2(z, s) = V_s \left[ \frac{\gamma(K + 1)}{4} e^{-s(z/v)} - \frac{sz\gamma(K - 1)}{2v} e^{-s(z/v)} - \frac{\gamma(K + 1)}{4} e^{-s(2l/v) + s(z/v)} \right]$$

$$\triangleq V_2^a + V_2^b + V_2^c \qquad (3.20b)$$

where

$$V_2^a = \frac{\gamma(K + 1)}{4} V_s e^{-s(z/v)} \qquad (3.21a)$$

$$V_2^b = -\frac{z\gamma(K - 1)}{2v} sV_s e^{-s(z/v)} \qquad (3.21b)$$

$$V_2^c = -\frac{\gamma(K + 1)}{4} V_s e^{-s(\frac{2l-z}{v})} \qquad (3.21c)$$

## 138 COUPLED LINES

Now perform the inverse transform:

$$\mathcal{L}^{-1}\{V_2^a\} = \frac{\gamma(K+1)}{4} \mathcal{V}_0\left(t - \frac{z}{v}\right) U\left(t - \frac{z}{v}\right)$$

$$\mathcal{L}^{-1}\{V_2^b\} = -\frac{z\gamma(K-1)}{2v} \frac{d}{dt}\left[\mathcal{V}_0\left(t - \frac{z}{v}\right)\right] U\left(t - \frac{z}{v}\right)$$

$$\mathcal{L}^{-1}\{V_2^c\} = -\frac{\gamma(K+1)}{4} \mathcal{V}_0\left(t + \frac{z}{v} - \frac{2l}{v}\right) U\left(t + \frac{z}{v} - \frac{2l}{v}\right)$$

where we have used

$$\mathcal{L}\{f(t)U(t)\} \to F(s)$$

$$\mathcal{L}\{f(t-T)U(t-T)\} \to e^{-sT}F(s)$$

$$\mathcal{L}\{f'(t-T)U(t-T)\} \to e^{-sT}sF(s)$$

The last result is verified as follows:

$$\mathcal{L}\{f'(t-T)U(t-T)\} = \int_0^\infty f'(t-T)U(t-T)e^{-st}\,dt = \int_{-T}^\infty f'(\lambda)U(\lambda)e^{-s(\lambda+T)}\,d\lambda$$

$$= \int_0^\infty f'(\lambda)e^{-s\lambda}e^{-sT}\,d\lambda = e^{-sT}\int_0^\infty f'(\lambda)e^{-s\lambda}\,d\lambda \stackrel{\Delta}{=} e^{-sT}sF(s)$$

The response on line 1 is

$$\mathcal{L}^{-1}\{V_1(z,s)\} = \mathcal{V}_1(z,t) = \mathcal{V}_0\left(t - \frac{z}{v}\right)U\left(t - \frac{z}{v}\right) \qquad (3.22)$$

which is our familiar response for a single line with input signal $\mathcal{V}_0(t)$ applied at $z=0$; see Figure 3.2.

The response on line 2 is composed of three signals that we consider in turn. The one labeled "a" is

$$\mathcal{V}_2^a(z,t) = \frac{\gamma(K+1)}{4} \mathcal{V}_0\left(t - \frac{z}{v}\right) U\left(t - \frac{z}{v}\right)$$

which is a reduced version of the wave on line 1. This wave moves along in step with the one on line 1. The "b" signal is

$$\mathcal{V}_2^b(z,t) = -\frac{z\gamma(K-1)}{2v}\frac{d}{dt}\left[\mathcal{V}_0\left(t-\frac{z}{v}\right)\right]U\left(t-\frac{z}{v}\right)$$

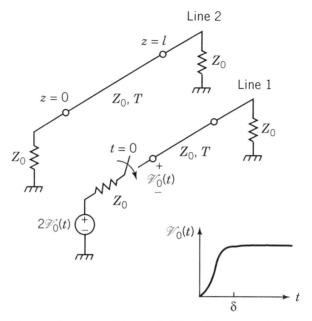

**FIGURE 3.2** The case for two weakly coupled lines. We assume that energy leaks from line 1 into line 2 and neglect any reverse transfer from line 2 back onto line 1.

which is a forward wave (travels toward the load) that increases in amplitude with $z$. It is not a replica of that on line 1, but rather has a shape that is proportional to the negative time derivative of $\mathscr{V}_0(t - z/v)$. The third wave is

$$\mathscr{V}_2^c(z,t) = -\frac{\gamma(K+1)}{4}\mathscr{V}_0\left(t + \frac{z}{v} - \frac{2l}{v}\right)U\left(t + \frac{z}{v} - \frac{2l}{v}\right)$$

which travels in the negative $z$ direction, with amplitude the negative of the "a" wave. By studying this wave at the midpoint of the line we observe that the signal does not appear until $t = 1.5T, (T = l/v)$. Recall that $U(\tau)$ is unity for $\tau > 0$, thus, at the midpoint

$$t + \frac{l}{2v} - \frac{2l}{v} > 0$$

or

$$t \geq 1.5\left(\frac{l}{v}\right)$$

Therefore, this wave is generated at the load end when the "a" wave arrives there.

Writing the complete response on line 2 gives

$$\mathscr{V}_2(z,t) = \frac{\gamma(K+1)}{4}\left[\mathscr{V}_0\left(t-\frac{z}{v}\right)U\left(t-\frac{z}{v}\right) - \mathscr{V}_0\left(t+\frac{z}{v}-\frac{2l}{v}\right)U\left(t+\frac{z}{v}-\frac{2l}{v}\right)\right]$$

$$-\frac{\gamma(K-1)}{2v}z\frac{d}{dt}\left[\mathscr{V}_0\left(t-\frac{z}{v}\right)\right]U\left(t-\frac{z}{v}\right) \qquad (3.23)$$

and evaluating this at the load $z = l$, we get

$$\mathscr{V}_2(l,t) = -\frac{\gamma(K-1)}{2}\left(\frac{l}{v}\right)\frac{d}{dt}[\mathscr{V}_0(t-T)]U(t-T) \qquad (3.23a)$$

so the "a" and "c" waves cancel at the load. With this result, we define a forward crosstalk constant:

$$K_f = -\frac{\gamma(K-1)}{2} = -\frac{1}{2}\left(\frac{L_m}{L} - \frac{C_m}{C_1}\right) = -\frac{v}{2}\left(\frac{L_m}{Z_0} - C_m Z_0\right) \qquad (3.24)$$

which is the portion of the amplitude into the load of line 2 that is independent of the line length $l$. Evaluating $\mathscr{V}_2(z,t)$ at the source $z = 0$ gives

$$\mathscr{V}_2(0,t) = \frac{\gamma(K+1)}{4}[\mathscr{V}_0(t)U(t) - \mathscr{V}_0(t-2T)U(t-2T)] \qquad (3.23b)$$

Thus, a signal exists immediately after $t = 0$ and it follows the shape of $\mathscr{V}_0(t)$ until $t = 2T$ (when the "c" wave arrives). After $2T$, we superpose $\mathscr{V}_0$ at $t$ with the negative of its value $2T$ seconds earlier. The scale factor wrt, the signal on line 1, is defined as $K_b$, the backward crosstalk constant:

$$K_b = \frac{\gamma(K+1)}{4} = \frac{v}{4}\left(\frac{L_m}{Z_0} + C_m Z_0\right) \qquad (3.25)$$

Rewriting $\mathscr{V}_2(z,t)$ now gives

$$\mathscr{V}_2(z,t) = K_f z \frac{d}{dt}\left[\mathscr{V}_0\left(t-\frac{z}{v}\right)U\left(t-\frac{z}{v}\right)\right]$$

$$+ K_b\left\{\mathscr{V}_0\left(t-\frac{z}{v}\right)U\left(t-\frac{z}{v}\right) - \mathscr{V}_0\left(t+\frac{z}{v}-2T\right)U\left(t+\frac{z}{v}-2T\right)\right\}$$

$$(3.26)$$

So in summary, we can state that when two lines are weakly coupled, the parasitic line has signals traveling in both directions when only a forward wave

WEAK COUPLING ANALYSIS **141**

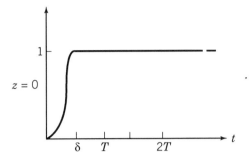

Line 1

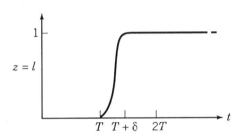

Line 2

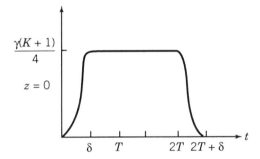

**FIGURE 3.3**

**142** COUPLED LINES

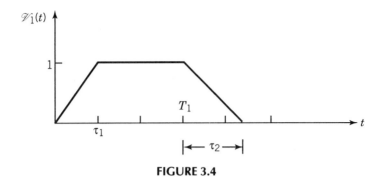

**FIGURE 3.4**

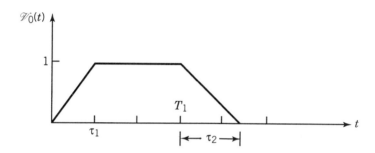

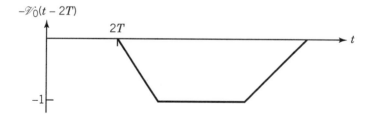

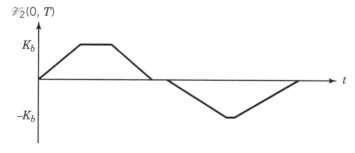

**FIGURE 3.5**

exists on the primary line. Two forward waves are induced: One is a reduced version of that on the primary, whereas the other has a shape corresponding to $-d/dt$ of the signal at $z = 0$, and its amplitude increases linearly with the distance traveled. When the second line is terminated, only the wave that increases linearly with $z$ produces a signal in the load. The above results are under the weak coupling assumptions, so the problem is perhaps even more involved than we might expect.

This case was considered in detail to illustrate the fact that the response in line 2 is not intuitively obvious. Now that we know what to expect with coupled lines, we can reexamine the problem from another (more intuitive) viewpoint. This is the thrust of the next section.

***Example 3.1*** For the input signal $\mathscr{V}_0(t)$ sketched in Figure 3.3 (top), sketch the signals at all the other ports. ▲

***Example 3.2*** For the signal applied to line 1 in Figure 3.4, find $\mathscr{V}_2(0, t)$.

Here, the total up and back delay $2T$ is less than $T_1$. Use Eq. (3.23b) which states that one just subtracts $\mathscr{V}_0(t - 2T)$ from $\mathscr{V}_0(t)$ and scales by the factor $K_b$. This is done graphically in Figure 3.5. ▲

## 3.3 LUMPED APPROACH TO COUPLED LINES

Two weakly coupled lines are shown in Figure 3.6 where a differential section of line 2 is shown for clarity. The section is located a distance $\xi$ from the source ($x = 0$) and is $d\xi$ long. We use $x$ rather than $z$ for the coordinate in this section. To find the signal developed in line 2, we use a lumped approach on the segment $d\xi$. Assume the signal on line 1 is such that $\mathscr{I}$ is positive with the direction shown. Also assume that only a forward wave exists on line 1 and the weak coupling assumptions hold. Of the net magnetic flux associated with the primary signal, that which links line 2 develops a voltage $d\mathscr{V}_L$ with polarity determined by Lenz's law. The subscript $L$ denotes inductive coupling. This value is

$$d\mathscr{V}_L = (L_m d\xi)\frac{\partial \mathscr{I}}{\partial t} \quad (3.27)$$

and the differential current developed (see the figure) is

$$d\mathscr{I}_L = \frac{d\mathscr{V}_L}{2\mathbf{Z}_0} \quad (3.28)$$

This current sets up waves in both directions $d\mathscr{V}_L^-, d\mathscr{V}_L^+$ as follows:

$$d\mathscr{V}_L^- = \mathbf{Z}_0 d\mathscr{I}_L = \tfrac{1}{2} d\mathscr{V}_L \quad (3.29a)$$

$$d\mathscr{V}_L^+ = -\mathbf{Z}_0 d\mathscr{I}_L = -\tfrac{1}{2} d\mathscr{V}_L \quad (3.29b)$$

**144** COUPLED LINES

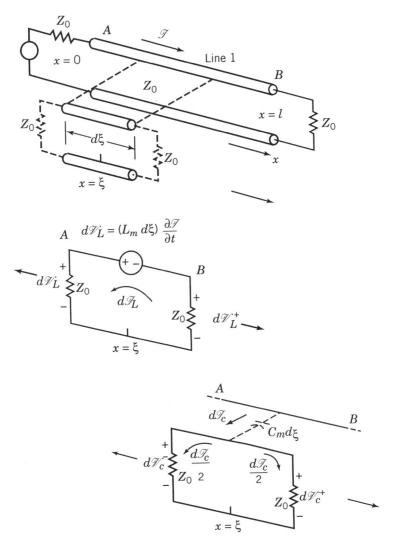

**FIGURE 3.6** Two weakly coupled lines studied from a lumped point of view. The primary signal on line 1 induces waves in both directions on line 2. This induction is considered the sum of inductive and capacitive coupling.

We can express these in terms of the voltage on line 1 since we assume that $\mathscr{I} = \mathscr{V}/\mathbf{Z}_0$, so

$$d\mathscr{V}_L^- = \frac{1}{2\mathbf{Z}_0} L_m d\xi \frac{d\mathscr{V}}{dt} \tag{3.30a}$$

$$d\mathscr{V}_L^+ = -\frac{1}{2\mathbf{Z}_0} L_m d\xi \frac{d\mathscr{V}}{dt} \tag{3.30b}$$

At the same time, since positive charge is accumulating in the upper conductor of line 1 ($\mathscr{I}$ increasing), it repels positive charge in the upper conductor on the $d\xi$ segment. Thus, the capacitive coupling produces currents moving downward in both $\mathbf{Z}_0$ elements in the figure. We write

$$d\mathscr{I}_C = C_m \, d\xi \, \frac{d\mathscr{V}}{dt} \tag{3.31}$$

and since this value splits equally, we write

$$d\mathscr{V}_C^- = d\mathscr{V}_C^+ = \tfrac{1}{2} \mathbf{Z}_0 C_m \, d\xi \, \frac{d\mathscr{V}}{dt} \tag{3.31a}$$

Now superposing the inductive and capacitive induced waves in each direction shows:

Toward $B$

$$d\mathscr{V}^+ = d\mathscr{V}_L^+ + d\mathscr{V}_C^+ = \frac{1}{2}\left(C_m \mathbf{Z}_0 - \frac{L_m}{\mathbf{Z}_0}\right) d\xi \, \frac{d\mathscr{V}}{dt} \tag{3.32a}$$

Toward $A$

$$d\mathscr{V}^- = d\mathscr{V}_L^- + d\mathscr{V}_C^- = \frac{1}{2}\left(C_m \mathbf{Z}_0 + \frac{L_m}{\mathbf{Z}_0}\right) d\xi \, \frac{d\mathscr{V}}{dt} \tag{3.32b}$$

Use the Laplace transform to find

$$dV^+ = \frac{1}{2}\left(C_m \mathbf{Z}_0 - \frac{L_m}{\mathbf{Z}_0}\right) d\xi \, sV, \qquad x > \xi \tag{3.33a}$$

$$dV^- = \frac{1}{2}\left(C_m \mathbf{Z}_0 + \frac{L_m}{\mathbf{Z}_0}\right) d\xi \, sV, \qquad x < \xi \tag{3.33b}$$

and the values on line 1 expressed as follows:

$$I = I_0 e^{-\Gamma \xi}, \qquad \Gamma = \frac{s}{v} \tag{3.34a}$$

$$V = \mathbf{Z}_0 I \tag{3.34b}$$

From the previous section's results, we may write the above as

$$dV^+ = \frac{1}{v} K_f \, d\xi \, sV, \qquad x > \xi \tag{3.35a}$$

$$dV^- = \frac{2}{v} K_b \, d\xi \, sV, \qquad x < \xi \tag{3.35b}$$

To obtain the signals at the source and load ends of line 2, we must sum contributions for all positions of $d\xi$. For a wave traveling toward $A$ (source), we know

$$\Delta V^- = \lambda_1 e^{\Gamma x}$$

and if we force $\Delta V^-$ to equal $dV^-$ at $x = \xi$, we have

$$\Delta V^- = \left(\frac{2}{v} K_b s V e^{-\Gamma \xi} d\xi\right) e^{\Gamma x} \qquad \text{for } x > \xi \qquad (3.36a)$$

Similarly, for a wave traveling toward $B$

$$\Delta V^+ = \lambda_2 e^{-\Gamma x}$$

and forcing $\Delta V^+ = dV^+$ at $x = \xi$ yields

$$\Delta V^+ = \left(\frac{K_f}{v} s V e^{\Gamma \xi} d\xi\right) e^{-\Gamma x}, \qquad x < \xi \qquad (3.36b)$$

Then the voltage at a point $x$ is

$$V(x,s) = \int_0^l \Delta V = \int_0^x \Delta V^+ + \int_x^l \Delta V^-$$

$$= \int_0^x \frac{K_f}{v} s e^{-\Gamma x} \mathbf{Z}_0 I_0 \, d\xi + \int_x^l \frac{2K_b}{v} s e^{\Gamma x}(\mathbf{Z}_0 I_0 e^{-2\Gamma \xi}) \, d\xi$$

$$= \left(\frac{K_f}{v} s \mathbf{Z}_0 I_0 e^{-\Gamma x}\right) x - \frac{1}{2\Gamma v}(2K_b s e^{\Gamma x} \mathbf{Z}_0 I_0)(e^{-2\Gamma l} - e^{-2\Gamma x}) \qquad (3.37)$$

Some algebra will show that this last expression is Eq. (3.20b). Now that we have some intuition concerning weakly coupled lines, we can gain more insight by studying an example and performing some short-cut calculations.

Consider the source voltage on line 1 as shown in Figure 3.7. The distribution of voltage with distance on line 1 is shown as part (a); notice that it is "spun around" as discussed in Chapter 2. Parts (b) and (c) show the two waves $\mathscr{V}_2^a$ and $\mathscr{V}_2^b$ at two instants $t_1$ and $t_2$. Here, we have assumed $\gamma(K-1) < 0$, so $\mathscr{V}_2^b$ has the polarity shown. We can calculate $\mathscr{V}_2(l,t)$ by using Eq. (3.32a) along with the following line of reasoning. First of all, we know $\mathscr{V}_2^c$ cancels $\mathscr{V}_2^a$ at the load, so $\mathscr{V}_2(l,t)$ is due to just $\mathscr{V}_2^b$. From the figure, we notice that the signal about $x_2$ for $\mathscr{V}_2^b(x,t_2)$ is composed of a wavelet developed by $d\mathscr{V}_1/dt$ about $x_2$, as well as all the previously developed wavelets at all points to the left of $x_2$. These previously developed wavelets all arrive at $x_2$ at the same instant (here, $t = t_2$) since all travel at speed $v$. Since the system is linear, it is not too surprising that the signal

LUMPED APPROACH TO COUPLED LINES    147

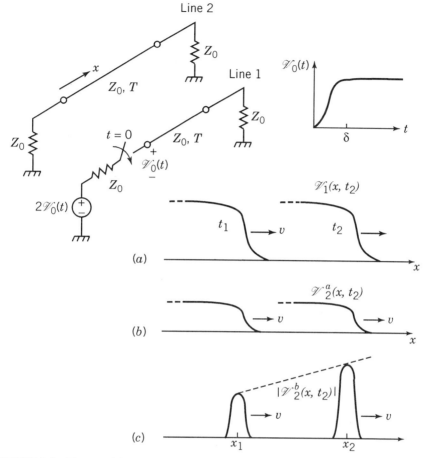

**FIGURE 3.7** Two weakly coupled lines showing the signals before the reflected wave on line 2 appears. In (a) the primary signal with distance at two instants $t_1$ and $t_2$ is shown. Parts (b) and (c) show the two waves induced on line 2. The polarity of $\mathscr{V}_2^b$ is not considered here, just its shape and growth with distance.

increases linearly with the distance traveled. Thus, the signal at $x = l$ is

$$\mathscr{V}^+(l,t) = \int_0^l d\mathscr{V}^+(\xi, t) = \int_0^l \frac{1}{2}\left(C_m Z_0 - \frac{L_m}{Z_0}\right) d\xi \frac{d\mathscr{V}_1}{dt} \qquad (3.38)$$

Now at each point $\xi$, $d\mathscr{V}_1/dt$ has the same value, and this allows $d\mathscr{V}_1/dt$ to come out of the integral. Thus,

$$\mathscr{V}^+(l,t) = \frac{1}{2}\left(C_m Z_0 - \frac{L_m}{Z_0}\right) l \frac{d\mathscr{V}_1}{dt} \qquad (3.39)$$

**148** COUPLED LINES

which is Eq. (3.23a), where Eq. (3.24) has been used. We are able to get the correct answer using this argument since all the wavelets developed by a particular point on the primary wave arrive at the load at the same instant. For the wavelets that arrive at the source plane, a simple argument is not possible. This is due to the fact that at any instant, wavelets arrive that have been developed at different points by different values of $d\mathscr{V}_1/dt$ at those points.

## 3.4 GENERAL SOLUTION FOR COUPLED LINES

We start the general case solution for two coupled lines by Laplace transforming equations (3.1) through (3.4):

$$\frac{dV_1}{dz} + sLI_1 + sL_m I_2 = 0 \tag{3.40a}$$

$$\frac{dV_2}{dz} + sL_m I_1 + sLI_2 = 0 \quad (L_2 = L_1 = L) \tag{3.40b}$$

$$\frac{dI_1}{dz} + sCV_1 - sC_m V_2 = 0 \quad (C = C_1 + C_m) \tag{3.40c}$$

$$\frac{dI_2}{dz} - sC_m V_1 + sCV_2 = 0 \quad (C_2 = C_1) \tag{3.4d}$$

After elimination, we have

$$\frac{d^2 V_1}{dz^2} = s^2(LC - L_m C_m)V_1 + s^2(CL_m - LC_m)V_2 \tag{3.41a}$$

$$\frac{d^2 V_2}{dz^2} = s^2(L_m C - LC_m)V_1 + s^2(LC - L_m C_m)V_2 \tag{3.41b}$$

and to simplify the solution of these coupled equations, we define two new variables:

$$V_E = \tfrac{1}{2}(V_1 + V_2) \tag{3.42a}$$

$$V_O = \tfrac{1}{2}(V_1 - V_2) \tag{3.42b}$$

where the subscripts $E$ and $O$ stand for even and odd. Their physical significance will be discussed shortly. Adding Eqs. (3.41a and 3.41b) yields

$$\frac{d^2 V_E}{dz^2} = \tfrac{1}{2}s^2\{LC(V_1 + V_2) + L_m C(V_1 + V_2) - L_m C_m(V_1 + V_2) - L_m C_m(V_1 + V_2)\}$$

$$= s^2(L + L_m)(C - C_m)V_E \tag{3.43a}$$

## GENERAL SOLUTION FOR COUPLED LINES

and after subtracting them, we find

$$\frac{d^2 V_O}{dz^2} = s^2(L - L_m)(C + C_m)V_O \qquad (3.43b)$$

which, conveniently, are two uncoupled equations. The propagation constants are

$$\Gamma_E^2 = s^2(L + L_m)(C - C_m)$$

or

$$\Gamma_E = \pm s\sqrt{(L + L_M)(C - C_M)} \qquad (3.44a)$$

and

$$\Gamma_O = \pm s\sqrt{(L - L_M)(C + C_m)} \qquad (3.44b)$$

The expressions for the corresponding even and odd currents are

$$I_E = \tfrac{1}{2}(I_1 + I_2), \qquad I_O = \tfrac{1}{2}(I_1 - I_2)$$

$$\frac{d^2 I_E}{dz^2} = s^2(L + L_m)(C - C_m)I_E$$

$$\frac{d^2 I_O}{dz^2} = s^2(L - L_m)(C + C_m)I_O$$

If we consider the even mode case, we know the solutions are

$$V_E = V_E^+ e^{-\Gamma_E z} + V_E^- e^{\Gamma_E z} \qquad (3.45a)$$

$$I_E = \frac{1}{Z_{0_E}}(V_E^+ e^{-\Gamma_E z} - V_E^- e^{\Gamma_E z}) \qquad (3.45b)$$

and we can determine $Z_{0E}$ as follows. Add Eqs. (3.40a) and (3.40b) to obtain

$$\frac{dV_E}{dz} = -s(L + L_m)I_E \qquad (3.46)$$

and if we set $V_E^- = 0$ in the expressions for $V_E$ and $I_E$ above, Eq. (3.46) yields

$$-\Gamma_E V_E^+ e^{-\Gamma_E z} = -s(L + L_m)\frac{V_E^+}{Z_{0_E}} e^{-\Gamma_E z}$$

or

$$Z_{0_E} = \frac{s(L + L_m)}{\Gamma_E} = \sqrt{\frac{(L + L_m)}{(C - C_m)}} \qquad (3.47)$$

**150** COUPLED LINES

Using the same procedure for the odd mode yields

$$Z_{o_o} = \frac{s(L - L_m)}{\Gamma_O} = \sqrt{\frac{(L - L_m)}{(C + C_m)}} \qquad (3.48)$$

After finding the solutions for $V_E$, $V_O$, $I_E$, etc. for a particular problem, the actual response is found from the following expressions:

$$V_1 = \tfrac{1}{2}(V_E + V_O) \qquad (3.49a)$$

$$V_2 = \tfrac{1}{2}(V_E - V_O) \qquad (3.49b)$$

$$I_1 = \tfrac{1}{2}(I_E + I_O) \qquad (3.49c)$$

$$I_2 = \tfrac{1}{2}(I_E - I_O) \qquad (3.49d)$$

## 3.5 PHYSICAL INTERPRETATION OF EVEN AND ODD MODES

With reference to Figure 3.8, we assume that the coupled lines are appropriately terminated such that no reflections exist. Thus, we can consider only forward waves. For the even mode case, we write

$$V_E = V_E^+ e^{-\Gamma_E z} \qquad (3.50a)$$

$$I_E = \frac{V_E^+}{Z_{0_E}} e^{-\Gamma_E z} \qquad (3.50b)$$

For the even mode, set the generators equal in magnitude and phase, so [see Eq. (3.42a)]

$$V_1 = V_2 = V_E$$

and similarly,

$$I_1 = I_2 = I_E$$

so

$$\frac{V_1}{I_1} = \frac{V_2}{I_2} = \frac{V_E}{I_E} = Z_{0_E} \qquad (3.51)$$

Consider the case wherein all fields exist in the same medium so a pure TEM mode exists. Then we know that the propagation velocity is $v_p = 1/\sqrt{\mu\varepsilon}$. Since $\Gamma = \omega/v_p$ for any wave, the condition of the same $v_p$ forces the following

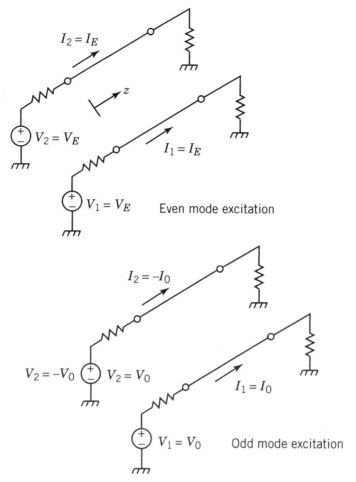

**FIGURE 3.8** Coupled lines excited in the even mode (upper) and odd (lower) modes, respectively.

conditions:

$$\Gamma_E = \frac{\omega}{v_{pE}} = \frac{\omega}{v_p} = \frac{\omega}{v_{po}} = \Gamma_O \tag{3.52}$$

From Eqs. (3.44a and 3.44b) the above implies

$$\frac{L}{C} = \frac{L_m}{C_m} \quad \text{or} \quad \frac{L_m}{L} = -\frac{C_m}{C} \quad \text{if } L_m < 0 \tag{3.53}$$

This allows us to eliminate $L_m$ in all the previous formulas. We have for future reference:

$$Z_{0_E} = \left[\frac{L}{C_1(C_1 + C_m)}(C_1 + 2C_m)\right]^{1/2} \quad (3.54)$$

$$Z_{0_O} = \left[\frac{L}{(C_1 + C_m)}\frac{C_1}{(C_1 + 2C_m)}\right]^{1/2} \quad (3.55)$$

Note that

$$Z_{0_E}Z_{0_O} = \frac{L}{(C_1 + C_m)} \triangleq Z_0^2 \quad (3.56)$$

It turns out that the terminations for zero reflections have the value $Z_0$. Since the velocities of either mode are the same,

$$\Gamma_E = \Gamma_O = s\sqrt{\frac{L}{C}(C + C_m)(C - C_m)}$$

$$v_p = \frac{\omega}{|\Gamma|} = \frac{1}{\sqrt{\frac{L}{C}(C + C_m)(C - C_m)}} = \frac{1}{\sqrt{Z_0^2(C_1 + 2C_m)C_1}} \quad (3.57)$$

The result in Eq. (3.53) is significant in that it shows $K_f = 0$ when a pure TEM mode is considered. In many practical cases, the TLs are metal strips on a dielectric board with air above. Thus, all field enegy is *not* in the same $\varepsilon_r$, and a quasi-TEM mode propagates with resulting forward crosstalk. For the case of a pure TEM mode, the expressions for $Z_{0_E}$ and $Z_{0_O}$ can be simplified as follows. We know

$$Z_{0_E} = \frac{1}{C_E v_p} \quad (3.58a)$$

$$Z_{0_O} = \frac{1}{C_O v_p} \quad (3.58b)$$

where this general relation between impedance, capacity, and phase velocity always holds. Here, $C_E$ and $C_O$ are effective capacities of the even and odd modes, respectively. Using Eqs. (3.55) and (3.57) yields [in (3.58a)]

$$\frac{L}{C_1(C_1 + C_m)}(C_1 + 2C_m) = \frac{1}{C_E^2}\frac{L}{C}(C + C_m)(C - C_m)$$

which reduces to

$$C_E = C_1 \quad (3.59a)$$

or

$$Z_{0_E} = \frac{1}{C_1 v_p} \quad (3.59b)$$

For the odd mode,

$$\frac{L}{(C_1 + C_m)} \frac{C_1}{(C_1 + 2C_m)} = \frac{1}{C_0^2} \frac{L}{C}(C + C_m)(C - C_m)$$

or

$$C_O = C_1 + 2C_m \quad (3.60a)$$

$$Z_{0_o} = \frac{1}{(C_1 + 2C_m) v_p} \quad (3.60b)$$

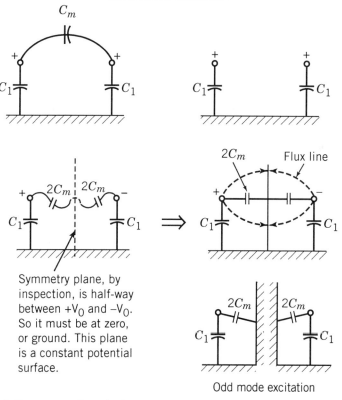

**FIGURE 3.9** For even mode excitation, both conductors are at the same potential, so $C_m$ is uncharged and not in the picture. For odd mode excitation, we can use symmetry arguments to show the effective capacity to be $(C_1 + 2C_m)$.

These latter expressions reduce the calculations of $\mathbf{Z}_{0_E}$ and $\mathbf{Z}_{0_O}$ to finding two capacities, since $v_p = 1/\sqrt{\mu\varepsilon}$, which can be determined once the dielectric is known. Referring to Figure 3.9, we can further expand on $\mathbf{Z}_{0_E}$ and $\mathbf{Z}_{0_O}$. For the even mode of excitation, we can bias each conductor at the same dc potential. (This is the case at any plane at microwave frequencies at any instant.) Since the mutual capacity is across equipotential points, it has no charge and can be eliminated. Then the effective capacity of either conductor to ground is $C_1$. Therefore, $C_E = C_1$ as shown earlier. For odd excitation, there exists a symmetry plane for which all electric flux lines are normal (shown as a vertical wall). We have also expressed $C_m$ as the sum of two $2C_m$ capacitors in series. The lower sketch shows that the total capacity to "ground" is now $C_O = C_1 + 2C_m$ as shown earlier.

Now the utility of the even and odd mode formalism is clear. Applying even or odd mode excitation allows one to effectively isolate the two lines. For the even case, one needs to solve a single transmission line wherein the capacity per unit length is just $C_E = C_1$. For odd excitation, again just a single line is analyzed, now with distributed capacity $C_O = C_1 + 2C_m$. After even and odd mode voltages and currents have been found, the total solution is the appropriate superposition of solutions, see Eqs. (3.49a–d).

## SUPPLEMENTARY EXAMPLES

3.1 Discuss the measurement and/or calculation schemes to determine the values for the parameters $L_1, L_2, L_m, C_1, C_2,$ and $C_m$ for coupled lines.

We will start with the analytical determination of the capacities $C_1, C_2,$ and $C_m$. It is necessary to review the topic of capacities in multiconductor systems, and for our purposes we will study at most three conductors above ground. From electrostatics, the charges, potentials, and capacities for multiconductor systems can be related in several ways, that is

$$\Phi = Pq \quad (1)$$

where $\Phi$ is the column vector of conductor potentials $\Phi_1, \Phi_2, \ldots, \Phi_N$ for $N$ conductors above a ground plane. The vector $q$ is that formed by the charges on each conductor $q_1, q_2, \ldots, q_N$. The array $P$ is called the coefficient of potential matrix. By inverting Eq. (1), we have

$$q = C^M \Phi \quad (2)$$

where $C^M = P^{-1}$ is the Maxwell capacitance matrix. For a three-wire system shown in Figure 3.10, this gives

$$\begin{bmatrix} q_1 \\ q_2 \\ q_3 \end{bmatrix} = \begin{bmatrix} C^M_{11} & C^M_{12} & C^M_{13} \\ C^M_{21} & C^M_{22} & C^M_{23} \\ C^M_{31} & C^M_{32} & C^M_{33} \end{bmatrix} \begin{bmatrix} \Phi_1 \\ \Phi_2 \\ \Phi_3 \end{bmatrix} \quad (2a)$$

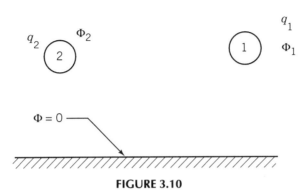

**FIGURE 3.10**

In this equation, $C_{11}^M$, $C_{22}^M$, $C_{33}^M$ are called coefficients of electrostatic capacitance (or coefficients of capacitance for short). The off-diagonal terms (all six of them) are called coefficients of electrostatic induction (coefficients of induction for short). To determine $C_{22}^M$, one proceeds as follows: refer to Figure 3.11,

Then
$$q_2 = C_{21}^M \Phi_1 + C_{22}^M \Phi_2 + C_{23}^M \Phi_3$$

$$C_{22}^M = \left. \frac{q_2}{\Phi_2} \right|_{\Phi_1 = \Phi_3 = 0} \tag{3}$$

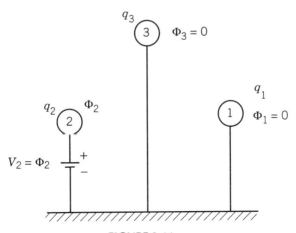

**FIGURE 3.11**

By definition, a positive charge on conductor 1 ($q_1$) will produce a positive potential there ($\Phi_1$). Therefore, $C_{11}^M$ is positive. Similarly, $C_{22}^M$ and $C_{33}^M$ are positive. The situation for the off-diagonal terms is more involved. For example,

$$q_1 = C_{11}^M \Phi_1 + C_{12}^M \Phi_2 + C_{13}^M \Phi_3$$

Then

$$C_{12}^M = \left. \frac{q_1}{\Phi_2} \right|_{\Phi_1 = \Phi_3 = 0} \tag{4}$$

Thus, we ground conductors 1 and 3 and apply a potential $\Phi_2$ to conductor 2. We then measure the resulting charge $q_1$ on the grounded conductor 1.

If $\Phi_2 > 0$ as shown above, then $q_2 > 0$. This will induce negative charges $q_1$ and $q_3$ on the other conductors. Therefore from Eq. (4), we see $C_{12}^M < 0$. Similar considerations show that all off-diagonal terms are negative.

The fact that all $C_{ij}^M$ ($i \neq j$) in Eq. (2) are negative has led to an alternate definition for $C^M$, namely [use Eq. (2a) specifically]

$$C' = \begin{bmatrix} C'_{11} & -C'_{12} & -C'_{13} \\ -C'_{21} & C'_{22} & -C'_{23} \\ -C'_{31} & -C'_{32} & C'_{33} \end{bmatrix} \tag{5}$$

where all the elements of $C'$ are positive (i.e., $-C'_{12} = C_{12}^M$; hence, $C'_{12} > 0$). This dual representation for the capacitance matrix $C$ is bothersome when attempting to correlate results from various sources in the literature. For the sake of clarity, we will call Eq. (5) the modified Maxwell capacitance matrix. Returning to Eq. (2), we can show (Ref. 2)

$$C_{ij}^M = C_{ji}^M$$
$$C_{ij}^M < 0, \quad i \neq j \tag{6}$$
$$C_{ii}^M \geq \sum_{j \neq i} C_{ji}^M > 0, \quad i = 1, 2, \ldots, N$$

The last expression has application later and can be demonstrated as follows. For our three-conductor example, ground 2 and 3 and apply $\Phi_1 > 0$. Then (See Figure 3.12)

$$q_1 = C_{11}^M \Phi_1$$
$$q_2 = C_{21}^M \Phi_1$$
$$q_3 = C_{31}^M \Phi_1$$

Then $q_1 > 0$, $q_2$ and $q_3 < 0$. Since the system is charge neutral,

$$q_1 + q_x + q_2 + q_3 = 0$$

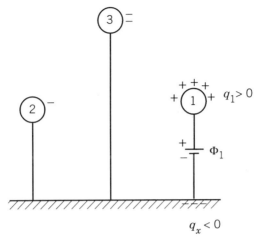

**FIGURE 3.12**

but $q_x < 0$ by inspection ($q_x$ is the charge in the ground below conductor 1). So dropping $q_x$ in the above expression gives

$$q_1 + q_2 + q_3 > 0$$

or

$$(C_{11}^M + C_{21}^M + C_{31}^M)\Phi_1 > 0$$

or

$$C_{11}^M + C_{21}^M + C_{31}^M > 0$$

which is the last expression in Eq. (6) for $i = 1, j = 2, 3$.

There is an alternate method of relating $q$ and $\Phi$, and it is

$$\begin{aligned} q_1 &= C_1\Phi_1 + C_{12}(\Phi_1 - \Phi_2) + C_{13}(\Phi_1 - \Phi_3) \\ q_2 &= C_{21}(\Phi_2 - \Phi_1) + C_2\Phi_2 + C_{23}(\Phi_2 - \Phi_3) \\ q_3 &= C_{31}(\Phi_3 - \Phi_1) + C_{32}(\Phi_3 - \Phi_2) + C_3\Phi_3 \end{aligned} \quad (7)$$

which is a rearrangement of Eq. (2). We notice that

$$\begin{aligned} C_1 + C_{12} + C_{13} &= C_{11}^M \\ C_{21} + C_2 + C_{23} &= C_{22}^M \\ C_{31} + C_{32} + C_3 &= C_{33}^M \end{aligned}$$

and in particular,

$$C_{12} = -C_{12}^M > 0$$
$$C_{13} = -C_{13}^M > 0$$

etc.

Thus, Eq. (7) is

$$\begin{bmatrix} q_1 \\ q_2 \\ q_3 \end{bmatrix} = \begin{bmatrix} (C_1 + C_{12} + C_{13}) & -C_{12} & -C_{13} \\ -C_{21} & (C_{21} + C_2 + C_{23}) & -C_{23} \\ -C_{31} & -C_{32} & (C_{31} + C_3 + C_{32}) \end{bmatrix} \begin{bmatrix} \Phi_1 \\ \Phi_2 \\ \Phi_3 \end{bmatrix}$$
(7a)

where all $C_{ij} > 0$. These elements are called direct capacities and can be placed on a diagram as shown in Figure 3.13.

We will call the matrix in Eq. (7a) the "conventional" capacitance matrix. In many applications, only the nearest-neighbor coupling between conductors is needed refer to Figure 3.14, and the above reduces to (neglect $C_{13}$)

$$\begin{bmatrix} C_1 + C_{12} & -C_{12} & 0 \\ -C_{12} & C_{12} + C_2 + C_{23} & -C_{23} \\ 0 & -C_{23} & C_3 + C_{23} \end{bmatrix} \quad (8)$$

For the special case of two lines above ground, the above reduces to

$$\begin{bmatrix} C_1 + C_m & -C_m \\ -C_m & C_2 + C_m \end{bmatrix} \quad (8a)$$

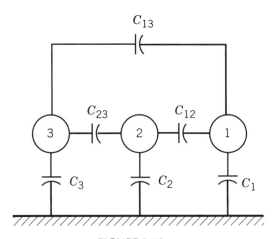

**FIGURE 3.13**

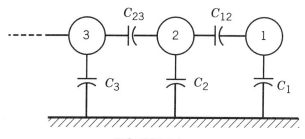

**FIGURE 3.14**

and comparison with Eqs. (3.9) and (3.11b–c) shows

$$C_{11} = C_{11}^M = C_1 + C_m$$
$$C_{22} = C_{22}^M = C_2 + C_m \qquad (8b)$$
$$C_m = C_{12}$$

To summarize, the diagonal elements in $[C]$ of Eq. (3.9) are the sums of the direct capacitors connected to a given conductor, that is $C_{22} = C_2 + C_{21} + C_{23}$. The remaining entries are the negatives of the direct capacitors between conductors. All direct capacitors are positive quantities.

For the inductors, we can write a flux-current relationship:

$$\psi_1 = L_{11}\mathscr{I}_1 + L_{12}\mathscr{I}_2$$
$$\psi_2 = L_{21}\mathscr{I}_1 + L_{22}\mathscr{I}_2 \qquad (9)$$

where $\psi_i$ is the flux through a surface that lies between a conductor and the ground plane (see Figure 3.15):

Using

$$\text{emf} = -\frac{d\psi}{dt}$$

$$\text{emf}_1 = -L_{11}\frac{\partial \mathscr{I}_1}{\partial t} - L_{12}\frac{\partial \mathscr{I}_2}{\partial t}$$

$$\text{emf}_2 = -L_{21}\frac{\partial \mathscr{I}_1}{\partial t} - L_{22}\frac{\partial \mathscr{I}_2}{\partial t}$$

and comparing with Eqs. (3.1) and (3.2), we observe

$$L_1 = L_{11}$$
$$L_2 = L_{22}$$
$$L_{12} = L_{21} = L_m$$

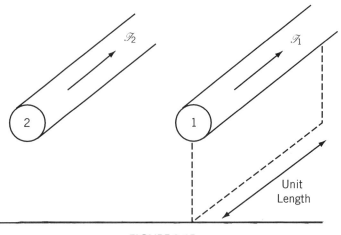

**FIGURE 3.15**

It is not necessary to solve the inductance problem, as it can be rather difficult. Instead, we use Eq. (3.10) to obtain the $L_{ij}$ from the $C_{ij}$.

Finally, this problem has attempted to introduce some of the terminology and notations used in the general literature. Sometimes, the elements in $[C]$ in Eq. (3.9) are called "capacities of one line *in the presence of the others.*" With this rich variety of terminology and notations, considerable confusion can arise if the reader is not careful to sort out exactly the details and definitions of a given article. For measurements of the various parameters, see Ref. 4.

**3.2** In Ref. 3, some capacitance coefficients are determined using numerical methods. Notice that their capacitance matrix is the Maxwell case [Eq. (2a) of Problem 3.1]. The thrust of the article is to show how multilayered dielectric regions can be handled. Then at microwave frequencies, these solutions for $L$, $C$, and thus $Z_0$ should be close to those of the actual quasi-TEM mode excited in a structure. Recall that, if more than one dielectric material is present, then only a quasi-TEM mode is possible. They calculate $C^M$ and determine $L$ by removing all dielectrics (thus, free space conditions exist). Then

$$L = \mu_0 \varepsilon_0 [C^M]^{-1}$$

Let us verify the entries for $L_{ij}$ given in Table III (Ref. 3). The system is the authors' Figure 6, reproduced below:

Their computations yield (subscript 0 for the free space condition)

$$C_0^M = (10^{-12} \, F/m) \begin{bmatrix} 23.91 & -8.427 \\ -8.427 & 20.42 \end{bmatrix}$$

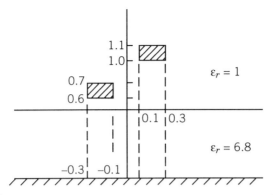

**FIGURE 3.16** Coupled lines in a nonhomogeneous situation. [Ref. 3] © 1984 IEEE.

Then using $\mu_0 \varepsilon_0 = (1/9) \times 10^{-16} \, s^2/m^2$ and for a general matrix, we know that if

$$A = \begin{bmatrix} a_{11} & a_{12} \\ a_{21} & a_{22} \end{bmatrix}$$

Then

$$A^{-1} = \frac{1}{\Delta} \begin{bmatrix} a_{22} & -a_{12} \\ -a_{21} & a_{11} \end{bmatrix}, \qquad \Delta = (a_{11}a_{22} - a_{12}a_{21})$$

Therefore,

$$L = \frac{(\frac{1}{9}) \times 10^{-6}(10^{12})}{417.228} \begin{bmatrix} 20.42 & 8.427 \\ 8.427 & 23.91 \end{bmatrix}$$

$$\therefore L = (10^{-6} \, H/m) \begin{bmatrix} 0.5437 & 0.2244 \\ 0.2244 & 0.6368 \end{bmatrix}$$

which agrees with their entries.

## PROBLEMS

**3.1** Assume two coupled transmission lines with the following parameters:

$$C_1 = C_2 = 1 \, pF/in., \qquad L_1 = L_2 = 20 \, nH/in.$$
$$C_m = 0.45 \, pF/in. \qquad L_m = 10.3 \, nH/in.$$

**(a)** Determine the phase velocity $v_p$.

(b) Determine $\Gamma_E$, $Z_{0_E}$ and $\Gamma_O$, $Z_{0_O}$, the even and odd mode propagation constants and characteristic impedances.

**3.2** Treat the coupled lines in the previous problem with the weak coupling approximation.

(a) Find the forward and backward crosstalk coupling constants $K_f$, $K_b$.

(b) Which mechanism dominates: electric or magnetic coupling of energy?

**3.3** Two coupled lines have the following parameters:

$$C_1 = C_2 = 0.85 \, \text{pF/in.}, \quad L_1 = L_2 = 15 \, \text{nH/in.}$$
$$C_m = 0.5 \, \text{pF/in.}, \quad L_m = 8 \, \text{nH/in.}$$

Assume that the input signal on line 1 of Figure 3.2 has the form of Figure P3.3

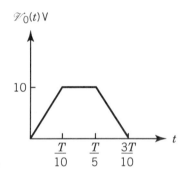

**FIGURE P3.3**

Sketch the signals on lines 1 and 2 at each end as a function of time.

## REFERENCES

1. Jarvis, D. B., "The Effects of Interconnections on High-Speed Logic Circuits," *IEEE Trans. Electronic Computers*, EC-12, Oct. 1963, pp. 476–487.
2. Carey, V. L., Scott, T. R., and Weeks, W. T., "Characterization of Multiple Parallel Transmission Lines Using Time Domain Reflectometry," *IEEE Trans. Instrumentation Measurement*, IM-18, Sept. 1969, pp. 166–171.
3. Wei, C., Harrington, R. F., Mautz, J. R., and Sarkar, T. K., "Multiconductor Transmission Lines in Multilayered Dielectric Media," *IEEE Trans. Microwave Theory Tech.*, MTT-32, April 1984, pp. 439–449.

## BIBLIOGRAPHY

Amemiya, H., "Time-Domain Analysis of Multiple Parallel Transmission Lines," *RCA Review*, Vol. 28, June 1967, pp. 241–276.

Amemiya, H., "Time-Domain Analysis of Multiple Parallel Transmission Lines by Means of Equivalent Circuits," Electronics Lett., 3(1), Jan. 1967, pp. 14–15.

DeFalco, J. A., "Reflection and Crosstalk in Logic Circuit Interconnections," *IEEE Spectrum*, Vol. 7, July 1970, pp. 44–50.

Isaacs, J. C., Jr. and Strakhov, N. A., "Crosstalk in Uniformly Coupled Lossy Transmission Lines," *Bell Syst. Tech. J.*, 52, Jan. 1973, pp. 101–115.

Plonsey, R. and Collin, R., *Principles and Applications of Electromagnetic Fields*, McGraw-Hill, New York, 1961, Section 3.5.

CHAPTER FOUR

# Time Domain Topics

### 4.1 TIME DOMAIN REFLECTOMETRY

Figure 4.1 displays a schematic of a time domain reflectometer (TDR) setup. It consists of a step generator, sampling oscilloscope, and a bridging tee. The output of the tee is connected to the terminated transmission line, which we consider comprises the device under test (DUT). The generator produces a step of voltage with amplitude $E_i$, and we show the instantaneous voltage distribution on the test line when the step edge is halfway down the line. The delay of the line is $T_0 = l/v$. Modern equipment can achieve a system rise time of about 30 ps that corresponds to a bandwidth of (BW $\simeq 0.35/t_r$) about 12 GHz. Thus, the system has a very broadband response. Its utility lies in the fact that it can separate reflections from different parts of the DUT, since reflections from these parts arrive back at the tee at different times.

The measurement concept is as follows. The first response of the scope occurs when the step passes through the bridging tee, and if the line and load are perfectly matched, no reflections exist and the scope trace appears as shown in Figure 4.1b. If, however, a reflection occurs at the termination, this reflected wave appears at the sampling point of the tee at $t = 2T_0$ seconds. In part (c), we show the ideal trace when the reflected signal $E_r$ subtracts from the incident one, $E_i$. Thus, the reflection coefficient can be determined, $K_L^V = E_r/E_i$, with the sign determined easily. The phase velocity on the line is simply $v_p = l/T_0$, and $T_0$ is directly determined by the trace (the length determined by proper calibration, i.e., actual reference plane of the line that excludes the connectors). In Figure 4.2, we show the ideal traces for various termination conditions. The line and the TDR system are both 50 Ω. At the present time, one can obtain time domain responses from Fourier analysis on swept frequency information. This capability is now available with Hewlett Packard's HP-8510B, so the diagram shown wherein step signals are actually applied to the line may become obsolete in the near future. We will use this model, however, for pedagogical purposes.

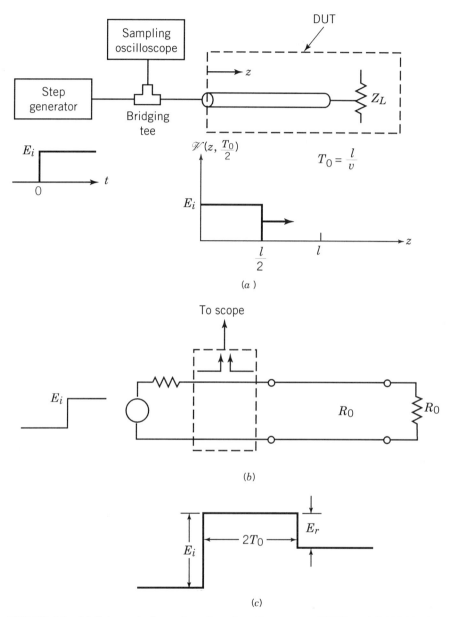

**FIGURE 4.1** (a) Schematic for a time domain reflectometer (TDR) and DUT (device under test). (b) The circuit diagram to indicate reference and reflected signals coupled into the oscilloscope. (c) The scope trace in a condition where the reflected signal subtracts from the incident level. The reflected signal returns after a delay of $2T_0$ (the up and return time on the line).

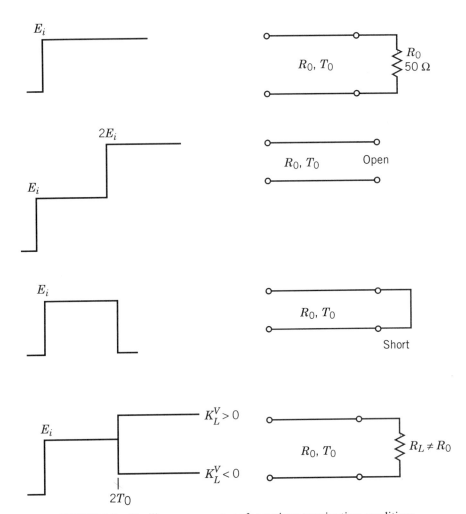

**FIGURE 4.2** Oscilloscope responses for various termination conditions.

The system can detect bad connectors between cascaded lines, as well as faults due to mechanical or electrical damage. It can also determine if line attenuation is predominantly series or shunt loss. The system has a large dynamic range and can resolve a reflection as small as 0.1%. The rise time is sufficient to resolve two reflections that are created only 1 to 2 mm apart on the line. It can be used to measure the variation of cable impedance and loss both with frequency and distance. For reasonably simple ideal loads, the scope response is easily calculated. In principle, the load impedance should be expressed as $\mathbf{Z}_L(s)$. Then the reflection coefficient is found:

$$K_L^V(s) = \frac{[\mathbf{Z}_L(s) - R_0]}{[\mathbf{Z}_L(s) + R_0]}, \quad [K_L^V(s) \equiv \rho(s)]$$

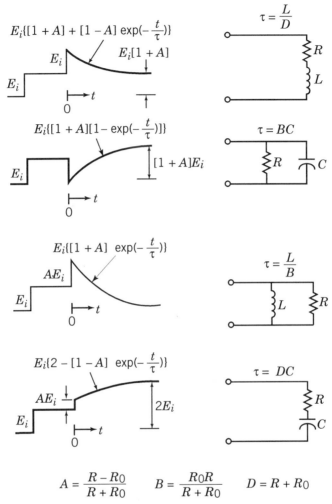

**FIGURE 4.3** Oscilloscope traces for single energy storage elements and resistors in various configurations. See Ref. 1. Reprinted with permission of Adam Microwave Consulting Inc.

Therefore, the input pulse is written $E_i/s$, so the final response at the load is $e_L(t) = \mathscr{L}^{-1}\{\rho(s)(E_i/s)\}$. Figure 4.3 displays the ideal traces for simple RC and RL loads. These responses are found using the following simple arguments. For the series RL load, the inductor appears as an open when the step arrives (inductor current cannot change). So the initial reflection is $+1$ for an open. The reflection coefficient changes with time as the coil permits more current, and after the transient, the inductor is a short. Then the reflection coefficient becomes $K_L^V = (R - R_0)/(R + R_0)$, so the final value of the trace must be $E_i(1 + K_L^V)$. The inductor charges through the effective resistance of $(R_0 + R)$, so the time constant

**168** TIME DOMAIN TOPICS

is $L/(R + R_0)$. Using the general form for an exponential change in a single energy storage element:

$$v(t) = v_f - (v_f - v_i)e^{-t/\tau} \qquad (4.1)$$

where

$v_f$ = final value of voltage
$v_i$ = initial value of voltage
$\tau$ = time constant

we have

$$v_f = E_i\left(1 + \frac{R - R_0}{R + R_0}\right)$$

$$v_i = E_i(1 + 1) = 2E_i$$

$$\tau = \frac{L}{(R + R_0)}$$

so

$$v(t) = E_i\left(1 + \frac{R - R_0}{R + R_0}\right) - \left[E_i\left(1 + \frac{R - R_0}{R + R_0}\right) - 2E_i\right]e^{-t/\tau}$$

$$= E_i\left\{\left(1 + \frac{R - R_0}{R + R_0}\right) + \left(1 - \frac{R - R_0}{R + R_0}\right)e^{-t/\tau}\right\} \qquad (4.2)$$

which is that shown in the figure. Similar reasoning is applicable to the other three cases. Other uses for TDR results will be discussed in the chapter examples.

***Example 4.1*** Consider a condition wherein two reflections occur as sketched below. From the trace information, determine the reflection coefficients $K_1^V$, $K_2^V$.

The first jump is that due to $K_1^V$. The reflection coefficient is negative as the resultant signal $(E_i - A) < E_i$, and its magnitude is $(A/E_i)$. Thus,

$$K_1^V = -\left(\frac{A}{E_i}\right)$$

The $E_2^+$ wave sent down the line is

$$E_2^+ = E_i(1 + K_1^V)$$

and, after reflection, it returns to the junction as

$$E_2^- = K_2^V E_2^+$$

The value transmitted to the 50-$\Omega$ line is therefore

$$E_1^- = E_2^-(1 + K')$$

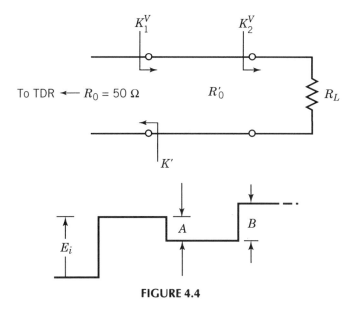

**FIGURE 4.4**

where $K'$ is the reflection coefficient for the $E_2^-$ wave. We know $K' = -K_1^V$, so

$$E_1^- = K_2^V E_2^+ (1 - K_1^V)$$

At the sampling tee, the net signal is $E_i + E_1^- + K_1^V E_i$ (the last term is that from the first bump). Thus,

$$E_i + B = E_i(1 + K_1^V) + K_2^V E_i(1 + K_1^V)(1 - K_1^V)$$

Since $B$, $E_i$, and $K_1^V$ are known, we can solve for $K_2^V$. ▲

## 4.2 DIGITAL APPLICATIONS

In high-speed digital circuits, the modeling of an interconnect wire as a small shunt capacitor is adequate as long as the length is less than the (rise time)·$(v_p)$ product. For high speeds, the interconnect is sometimes too long and should be treated as a transmission line. The effective impedance levels for digital lines can vary from about 30 to 200 Ω. The delay for a line $l$ units long is $T_D = l/v_p$, and the equivalent net capacity to ground is $C = T_D/Z_0$. The series net inductance is $L = Z_0 T_D$, where we use $Z_0$ for the characteristic impedance symbol. We notice then that low-impedance lines have high capacity, whereas high-impedance lines present large series inductance. A basic circuit for a digital switching application is shown in Figure 4.5, where the NOR gate at the load end is assumed not to load down (draw current) the resistor $R_L$. In other words, nearly all the line current is

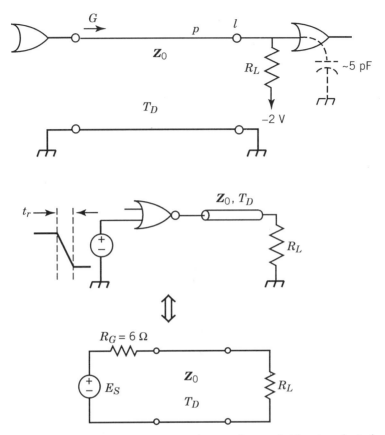

**FIGURE 4.5** Several schematic forms used to analyze switching transients in logic circuits.

delivered to $R_L$. The output of the generator side gate is often a low value ($\sim 5$ to $7\,\Omega$), so one does not want much reflection from $R_L$, or several bounces will occur for any pulse edge. Many of the pulse problems in Chapter 2 are easily extended to handle simple situations in these switching applications.

In many cases, the transmission lines are of the microstrip variety (a thin strip above a ground plane with a glass epoxy circuit board in between). For Motorola's older emitter-coupled logic families (10,000 for computers and communications, MECL III for instrumentation), the circuits are designed to drive 50-$\Omega$ lines and they function over the range of 50 to 120 $\Omega$. A rule of thumb is that if the rise time $t_r$ is less than 1.5 ns, then TL modeling is needed. The speed of propagation on epoxy glass boards is about 0.15 ns/in., which is given the symbol $t_{pd}$. In general, the delay per unit length is written as (for microstrip) $r_{pd} = 1.017\sqrt{\varepsilon_r/2}$ ns/ft, where $\varepsilon_r$ is that of the board.

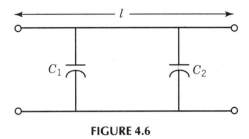

**FIGURE 4.6**

***Example 4.2*** Find the effective impedance and propagation delay for a TL with lumped capacitors attached as shown in Figure 4.6.

The approach taken is the assumption that the line is not too long, so it has a net capacity $C = C_0 l$, where $C_0$ is the distributed capacity in F/m. The lumped capacitors are made equivalent to $C_d = (C_1 + C_2)/l$ (distributed equivalent), so the characteristic impedance is

$$\mathbf{Z}_{eq} = \left(\frac{L}{C_0 + C_d}\right)^{1/2} = \frac{\sqrt{L/C_0}}{\sqrt{1 + C_d/C_0}} = \frac{\mathbf{Z}_0}{\sqrt{1 + C_d/C_0}} \quad (1)$$

The speed is $v_p = 1/(\mathbf{Z}_{eq} C_{eq})$, where $C_{eq} = C_0 + C_d$, so the delay/unit length is

$$t'_{pd} = \frac{1}{v_p} = \sqrt{LC_{eq}} = \sqrt{LC_0}\sqrt{1 + \frac{C_d}{C_0}} = t_{pd}\sqrt{1 + \frac{C_d}{C_0}} \quad (2)$$

▲

***Example 4.3*** For the network in Figure 4.7 assume that $\mathscr{V}_G$ increases 0.8 V at $t = 0$ (ideal step). Sketch the generator and load voltages and currents.

Here, the line has an initial dc steady state. The steady current is $(2 - 1.6)/100\,\Omega = 4\,\text{mA}$. Since superposition holds, we can add the step response to the

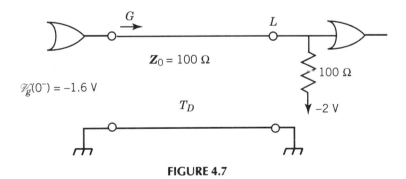

**FIGURE 4.7**

**172** TIME DOMAIN TOPICS

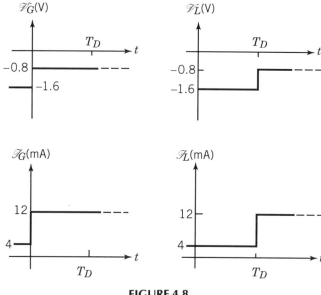

**FIGURE 4.8**

existing steady condition to find the complete response. The gate has a positive step of 0.8 V, so the current jumps to 4 mA + 0.8 V/100 Ω = 12 mA. The 0.8 V and 12-mA steps are completely absorbed in the matched load, so the results are quickly sketched as shown in Figure 4.8. ▲

***Example 4.4*** The output of an emitter-coupled logic curcuit is connected to the input of another by a TL with $Z_0 = 50\,\Omega$, $T = 2.5$ ns. A resistor is connected to the driver output to permit the effective output to be applied through 50 Ω, then into the line. The ECL circuit at the load plane is modeled by a 5-pF capacitor. Sketch the voltage and current waveforms at the load, generator, and a plane $3l/4$ from the generator (plane $p$) in Figure 4.5:

The circuit diagram in Figure 4.9 shows the added series resistor $R$ used to bring the effective generator resistance up to 50 Ω to prevent any reflections at the

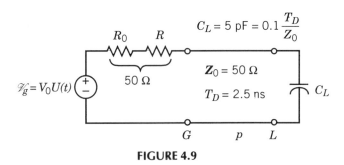

**FIGURE 4.9**

DIGITAL APPLICATIONS    173

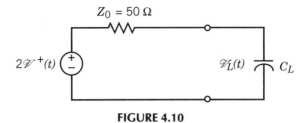

**FIGURE 4.10**

generator plane. The effective capacity of the line is $C = T_0/Z_0 = 50\,\text{pF}$, so the load may be represented as $C_L = C/10 = 0.1\,T_D/Z_0$. Since the line is matched to the generator, the initial step has amplitude $\mathscr{V}_g/2$. When this edge of the step reaches the initially uncharged load capacitor $C_L$, it is completely reflected since $C_L$ acts initially like a short $[\mathscr{V}_L(0^-) = \mathscr{V}_L(0^+) = 0]$. Since the generator is matched, the solution is straightforward. From Section 2.7, we know that the load plane has a lumped equivalent circuit as shown in Figure 4.10.

Here,

$$\mathscr{V}^+(t) = \frac{\mathscr{V}_g(t)}{2} = \frac{V_0}{2} U(t)$$

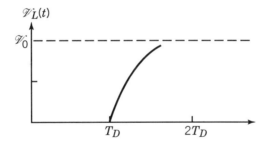

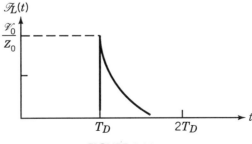

**FIGURE 4.11**

since no reflections occur at the generator. The solution is by inspection

$$\mathscr{V}_L(t) = V_0[1 - e^{-(t-T_D)/Z_0 C}]U(t - T_D)$$

and $Z_0 C = 0.1 T_D$. This along with the load current are sketched in Figure 4.11

The reflected waveform is determined as follows.

$$\mathscr{V}_L(t) = \mathscr{V}^+(t) + \mathscr{V}^-(t)$$

but $\mathscr{V}^+(t) = V_0/2\ U(t)$ for all time, so

$$\mathscr{V}^-(t) = -\frac{V_0}{2} U(t - T_D) + V_0[1 - e^{-(t-T_D)/(0.1 T_D)}]U(t - T_D)$$

$$= V_0[\tfrac{1}{2} - e^{-(t-T_D)/(0.1 T_D)}]U(t - T_D)$$

since $\mathscr{V}^-(t)$ doesn't start until $t = T_D$. Now this signal reaches plane $p$ at $t = 1.25 T_0$ and superposes with $V_0/2$ already present there ($\mathscr{V}^+ = V_0/2$). Thus, the signal at $p$ is

$$\mathscr{V}(p,t) = \frac{V_0}{2} U(t - 0.75 T_D) + V_0[\tfrac{1}{2} - e^{-(t-1.25 T_D)/(0.1 T_D)}]U(t - 1.25 T_D)$$

This and the current are sketched in Figure 4.12.

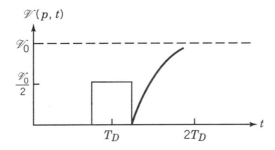

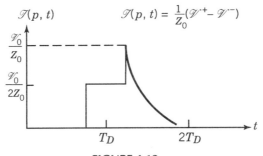

**FIGURE 4.12**

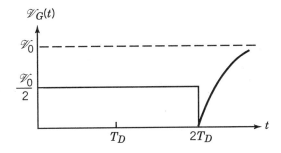

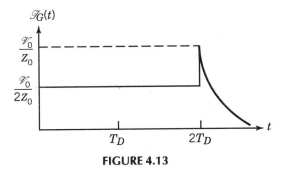

**FIGURE 4.13**

The signal at the generator is just $V_0/2\ U(t)$ superposed with $\mathscr{V}^{-}(t - 2T_D)$, which is shown in Figure 4.13.

The last problem was simplified by neglecting rise time, a resistor in shunt with $C_L$, and the real reflection that would occur at the generator. For a more realistic problem wherein the above are included, the use of Laplace transforms is a must. Then peaks of ringing, etc. are determined in a systematic way. The fact that the $I - V$ characteristics of logic gates are nonlinear further complicates the analysis, and computer modeling is the appropriate tool. The next section gives a brief introduction to this behavior.

▲

## 4.3 GRAPHICAL METHODS

The graphical procedure is a sequential application of simple load line constructions and particularly useful when the $I - V$ characteristics of the source and/or load are nonlinear. It is very useful for quickly studying the switching waveform details in high-speed digital circuits. See Refs. 2, 3, and 4. This insight is helpful before a more detailed software package is used to study a situation in depth. The assumptions inherent to the technique are:

1. The voltage waveforms for logic state transitions are essentially step functions.
2. The transient responses of the gates are adequately modeled by their static $I - V$ characteristics. In other words, we neglect any charge storage effects within the gates. Good agreement between measured results and those predicted using this method justifies its use (Ref. 2).

The technique consists of plotting the $I - V$ characteristcs of the source and load ports on the same set of axes. The source and load plane current and voltage values during the switch transient are found by constructing load lines of slopes $\pm 1/R_0$, where $R_0$ is the line's characteristic resistance. To introduce the procedure, we apply it to a simple linear system that could be easily solved using a bounce diagram. The circuit is given in Figure 4.14; take special care to notice that the reference directions for current and voltage must be strictly adhered to.

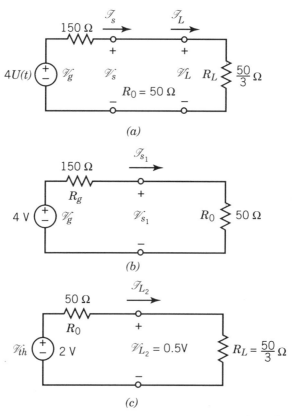

**FIGURE 4.14** A linear circuit to indicate reference directions for voltage and current when using the graphical solution technique.

GRAPHICAL METHODS 177

**TABLE 4.1** The source and load plane voltages and currents for the first five passes of the leading edge of the transition pulse. The steady values are $\mathscr{V}_S = \mathscr{V}_L = 0.4$ V, $\mathscr{I}_L = \mathscr{I}_S = 24$ mA.

| Time Interval | $\mathscr{V}_s$(V) | $\mathscr{I}_s$(mA) | $\mathscr{V}_L$(V) | $\mathscr{I}_L$(mA) |
|---|---|---|---|---|
| $0 < t < T$ | 1.0 | 20.0 | 0.0 | 0.0 |
| $T < t < 2T$ | 1.0 | 20.0 | 0.5 | 30.0 |
| $2T < t < 3T$ | 0.25 | 25.0 | 0.5 | 30.0 |
| $3T < t < 4T$ | 0.25 | 25.0 | 0.375 | 22.5 |
| $4T < t < 5T$ | 0.4375 | 23.75 | 0.375 | 22.5 |

The $I - V$ characteristics must be plotted using the convention. Be careful at the source end, where the current is defined as leaving the positive node. In other words, if the available $I - V$ curve for the source gate is defined with the current entering the positive node, then be sure to reflect the curve about the $V$ axis when plotting it for application of this technique. The source and load voltage and current for the example were determined using a bounce diagram and are tabulated in Table 4.1.

We start by investigating the source plane for the interval $(0, 2T)$: this is shown in Figure 4.14b. The network equation is

$$\mathscr{V}_{s_1} = \mathscr{V}_g - \mathscr{I}_{s_1} R_g = \mathscr{I}_{s_1} R_0 \tag{4.3}$$

and the subscript 1 denotes the first time interval of study. Now plot these two straight lines on the $I - V$ plane as shown in Figure 4.15. Their intersection is at point $A$, which gives the initial $\mathscr{V}$ and $\mathscr{I}$ values at the source. Excitation at the load starts in the interval $(T, 2T)$, and the lumped equivalent at that plane during $(T, 3T)$ is in part (c) of Figure 4.14. The Thévenin source value is determined as follows. The total voltage at the load is $\mathscr{V}^+(1 + K_L^V)$, where $\mathscr{V}^+ = \mathscr{V}_{s_1} = 1.0$ V. From the lumped equivalent, we see

$$\mathscr{V}_{L_2} = \mathscr{V}_{th} \frac{R_L}{R_L + R_0} = \mathscr{V}^+(1 + K_L^V) \tag{4.4}$$

or

$$\mathscr{V}_{th} = 2\mathscr{V}^+ = 2 \text{ V}$$

From the circuit, we write

$$\mathscr{V}_{L_2} = \mathscr{I}_{L_2} R_L = \mathscr{V}_{th} - R_0 \mathscr{I}_{L_2} \tag{4.5}$$

which yields point $B$ in Figure 4.15 (the initial $\mathscr{V}$ and $\mathscr{I}$ of the load). Next consider the wave returning to the source; see Figure 4.16. The Thévenin generator is

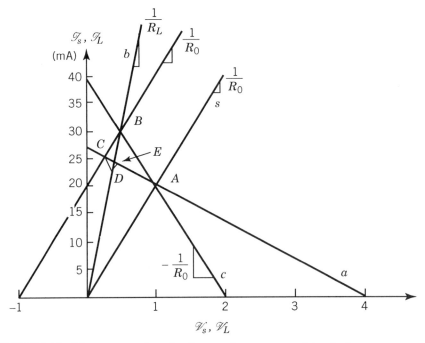

**FIGURE 4.15** The constructions leading to the source and load plane currents at successive intervals.

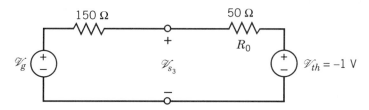

**FIGURE 4.16** The effective equivalent circuit over the interval $(2T, 4T)$.

found using

$$\mathscr{V}_s = \mathscr{V}^+ + \mathscr{V}^-(1 + K_s^V) \tag{4.6}$$

$$\mathscr{V}^+ = \mathscr{V}_g \frac{R}{R_0 + R_g} \tag{4.7}$$

$$\mathscr{V}_s = \mathscr{I} R_0 + \mathscr{V}_{th} \tag{4.8}$$

$$\mathscr{I} = \frac{\mathscr{V}_g - \mathscr{V}_{th}}{R_g + R_0} \tag{4.9}$$

or all together,

$$\mathscr{V}_s = \mathscr{V}_g \frac{R_0}{R_0 + R_g} + \mathscr{V}^-(1 + K_s^V) = \mathscr{V}_{th} + R_0\left(\frac{\mathscr{V}_g - \mathscr{V}_{th}}{R_0 + R_g}\right) \quad (4.10)$$

from which we extract $\mathscr{V}_{th} = 2\mathscr{V}^-$. The correct circuit for $(2T, 4T)$ is in Figure 4.16, and the equation from it is

$$\mathscr{V}_{s_3} = \mathscr{I}_{s_3} R_0 + \mathscr{V}_{th} = \mathscr{V}_g - \mathscr{I}_{s_3} R_g \quad (4.11)$$

which yields lines 'a' and 'c' in Figure 4.15. The intersection C gives $\mathscr{V}_s$ and $\mathscr{I}_s$ over $(2T, 4T)$. Continuing in this manner gives points D and E.

We see that the method is essentially the movement between curves 'a' and 'b' on lines of slope $\pm 1/R_0$. Figure 4.17 summarizes the procedure on a less cluttered graph. The basic lines constructed are the $I - V$'s for source and load. Establish the steady starting values of $\mathscr{V}$ and $\mathscr{I}$ in the system, then move to each curve alternately. Leave a point on lines of slope $\pm 1/R_0$. Here, the initial system conditions for $t < 0$ are $\mathscr{V} = \mathscr{I} = 0$. We thus start at the origin and move to the source curve and establish point A. We must use a $+1/R_0$ slope in this particular case or we cannot reach the source curve. From 'A', we move along a slope of $-1/R_0$ and intersect the load curve at 'B'. We leave 'B' on a positive slope and reach 'C'. Continuing this procedure yields points D and E. Ultimately, we will

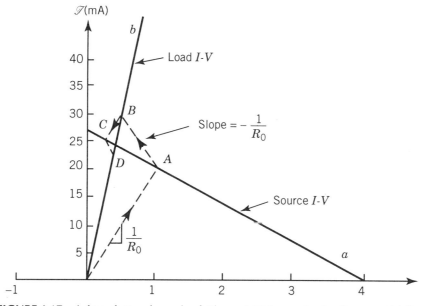

**FIGURE 4.17** A less cluttered graph of Figure 4.15 to emphasize the essential lines needed to obtain the desired information.

180   TIME DOMAIN TOPICS

stop this procedure when we reach the final steady value (the intersection of the source and load curves). Note that the intersections with the source $I - V$ give values at the source, whereas those with the load $I - V$ give load plane values. The axes in Figure 4.17 were chosen so that the slope corresponding to $1/R_0$ is inclined at 45°, and therefore, $-1/R_0$ is inclined at 135° to the voltage axis. For this special case, the lines with slope $\pm 1/R_0$ are perpendicular. For other arbitrary scales, this perpendicularity does not hold. This special axis situation is implied throughout Ref. 5.

***Example 4.5***  Consider a 50-Ω line with time delay $T_D$ connected to a source as shown below. The line is open-circuited. Analyze the circuit for the time interval $(0, 4T_D)$ in Figure 4.18.

The source $I - V$ is designated by '$a$' in Figure 4.19; the load's is the voltage axis. The three successive source values read from the graph are (in V and mA): (0.4, 8), $(-0.675, -2.5)$, (0.575, 1.0). The first two load sets are (0.8, 0) and (0.535, 0).  ▲

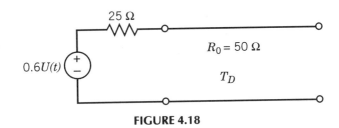

**FIGURE 4.18**

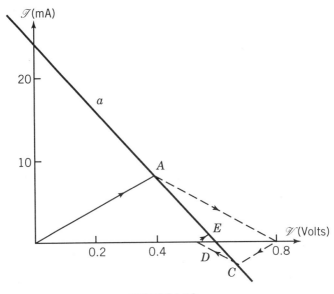

**FIGURE 4.19**

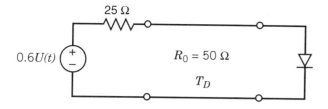

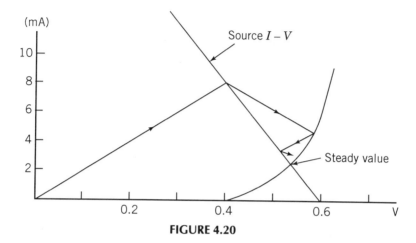

**FIGURE 4.20**

***Example 4.6*** A line with $R_0 = 50\,\Omega$ and a source given by $\mathscr{V}_g = 0.6\,\text{V}$, $R_g = 25\,\Omega$ is connected to a diode as shown in Figure 4.20. Study the initial transient using the $I - V$ curve.

The settling time is about $4T_D$. Note that we are neglecting the charging effects of the diffusion capacitance. ▲

***Example 4.7*** Figure 4.21 shows a connection of two inverters with a 50-$\Omega$ line. Assume that the source has two distinct piecewise linear curves for the 0 and 1 logic states that are shown below and the input $I - V$ at the load remains the same for both logic states. The transitions from 0 to 1 and back again are shown. The initial state for $\mathscr{V}_s$ corresponds to logical 0. The voltage across the line is about $\approx 0.8\,\text{V}$ and $\mathscr{I}_s \approx -1.2\,\text{mA}$. The dotted line indicates that the successive $I - V$ points passed through in reaching logic 1. The switch back to 0 is indicated by the arrowheads on the dot-dash trajectory. See the text by A. Barna (Ref. 6.) ▲

***Example 4.8*** For the ECL circuit driving a 50-$\Omega$ line with the load given in Figure 4.22, find the transient response both analytically and graphically.

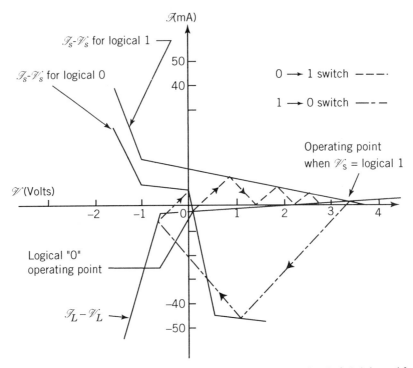

**FIGURE 4.21** The connection of two inverters with a 50-Ω line. See Ref. 6. Adapted from Barna, 1970. Reprinted by permission of John Wiley & Sons, Inc.

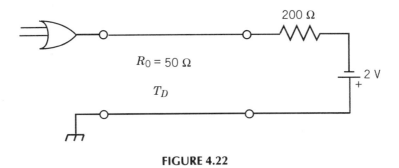

**FIGURE 4.22**

The equivalent circuit is shown in Figure 4.23.

Assume that the source $\mathscr{V}_s$ changes from the logic 0 level of $-1.6$ V to a logic 1 level of $-0.8$ V at $t = 0$. First establish the dc conditions on the line prior to the transition. A dc analysis shows $\mathscr{V}_s = -1.619$ V and $\mathscr{I}_s = 1.9048$ mA. We calculate the transient response by assuming no initial dc level, then use superposition to

GRAPHICAL METHODS   183

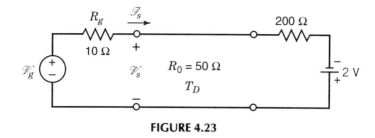

**FIGURE 4.23**

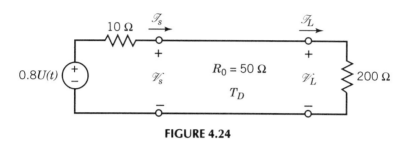

**FIGURE 4.24**

get the complete values. The transient problem is treated by using the circuit in Figure 4.24.

Table 4.2 gives source and load values for the above obtained with a bounce diagram and added to the dc result. The final steady dc values are $\mathscr{V} = -0.857$ V, $\mathscr{I} = 5.71$ mA. The voltage and current waveforms are shown in Figure 4.25.

The graphical method proceeds as shown in Figure 4.26. The source and load equivalent circuits and their $I - V$ plots are shown.

Notice that the source has two separate plots, one for each value of the source voltage. Starting from the logic '0,' we converge on the '1' point; see Figure 4.27. It takes about $6T_D$ for the complete transition.  ▲

**TABLE 4.2   Response for circuit in Figure 4.22.**

| Time Interval | $\mathscr{V}_s$(V) | $\mathscr{I}_s$(mA) | $\mathscr{V}_L$(V) | $\mathscr{I}_L$(mA) |
|---|---|---|---|---|
| $t < 0$ | −1.619 | 1.9048 | −1.619 | 1.9048 |
| $0 < t < T_D$ | −0.952 | 15.23 | −1.619 | 1.9048 |
| $T_D < t < 2T_D$ | −0.952 | 15.23 | −0.552 | 7.24 |
| $2T_D < t < 3T_D$ | −0.819 | 1.9048 | −0.552 | 7.24 |
| $3T_D < t < 4T_D$ | −0.819 | 1.9048 | −0.979 | 5.104 |
| $4T_D < t < 5T_D$ | −0.872 | 7.237 | −0.979 | 5.104 |
| $5T_D < t < 6T_D$ | −0.872 | 7.237 | −0.906 | — |

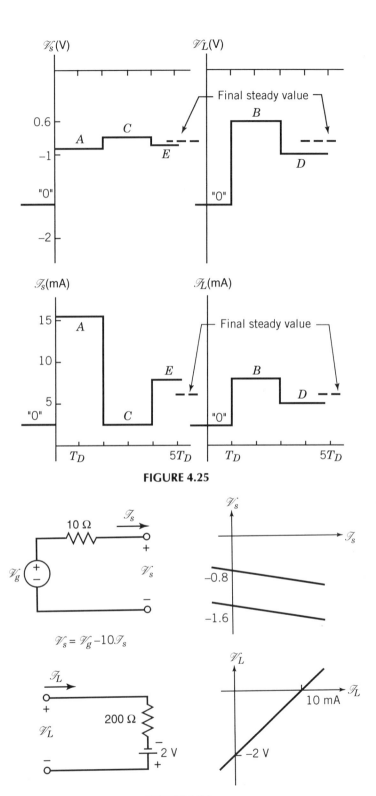

**FIGURE 4.25**

**FIGURE 4.26**

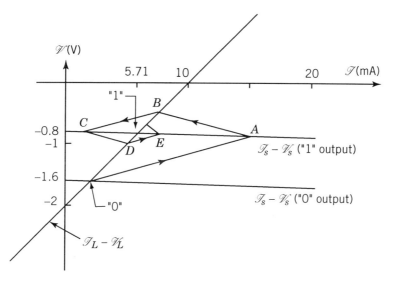

**FIGURE 4.27**

## PROBLEMS

**4.1.** Find the transient response for the network in Figure P4.1. Use graphical procedures.

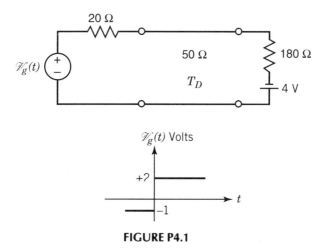

**FIGURE P4.1**

**186** TIME DOMAIN TOPICS

**4.2** For the TL, source, and load shown in Figure P4.2, sketch the transition from state *A* to state *B*. Plot both current and voltage excursions using the graphical approach.

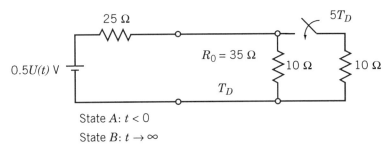

State *A*: $t < 0$
State *B*: $t \to \infty$

**FIGURE P4.2**

## REFERENCES

1. Adam, Stephen. F., *Microwave Theory and Applications*, Prentice-Hall, New Jersey, 1969.
2. Barna, A., *High Speed Pulse and Digital Techniques*, Wiley-Interscience, New York, 1980.
3. DeFalco, J., "Reflection and Crosstalk in Logic Circuit Interconnections," *IEEE Spectrum*, 7 July 1970, pp. 44–50.
4. Davidson, C. W., *Transmission Lines for Communications*, John Wiley & Sons, New York, 1978, Section 2.4.
5. Singleton, R. S., "No Need to Juggle Equations to Find Reflection—Just Draw Three Lines," *Electronics*, *41* Oct. 28, 1968, pp. 93–99.
6. Barna, A., *High-Speed Pulse Circuits*, Wiley-Interscience, New York, 1970.

## BIBLIOGRAPHY

Abdel-Latif, M. and Strutt, M. J. O., "Simple Graphical Method to Determine Line Reflections Between High-Speed-Logic Integrated Circuits," *Electronics Lett.*, *4*, Nov. 1968, pp. 496–498.

Blood, W. R., *MECL System Design Handbook*, Motorola, Inc., 1980.

Mohammadian, A. H. and Tai, C. T., "A General Method of Transient Analysis for Lossless Transmission Lines and Its Analytical Solution to Time-Varying Resistive Terminations," *IEEE Trans. Antennas Propagation*, *AP-32*, March 1984, pp. 309–312.

Mohammadian, A. H. and Tai, C. T., "Transients on Lossy Transmission Lines with Arbitrary Boundary Conditions," *IEEE Trans. Antenna Propagation*, *AP-32*, April 1984, pp. 418–422.

Tai, C. T., "Transients on Lossless Terminated Transmission Lines," *IEEE Trans. Antennas Propagation*, *AP-26*, July 1978, pp. 556–561.

CHAPTER FIVE

# Sinusoidal Steady State

## 5.1 SINUSOIDAL EXCITATION

Previously, we examined the achievement of a dc steady state after the application of a battery to a TL. The final steady value could be decomposed into two waves in opposite directions, such as done in Example 2.2. This same type of decomposition can be performed for the sinusoidal steady state. We write the counterstreaming waves as

$$\mathscr{V}^+(z,t) = A\cos(\omega t - \beta z + \varphi^+) \tag{5.1a}$$

$$\mathscr{V}^-(z,t) = B\cos(\omega t + \beta z + \varphi^-) \tag{5.1b}$$

where $\mathscr{V}^+(z,t)$ is the superposition of all the forward waves and $\mathscr{V}^-(z,t)$ is that for all waves moving from the load toward the generator. A single cosinusoidal wave is possible since sums of cosines are still a single cosine with appropriate amplitude and phase shift. The constants $A$, $B$, $\varphi^+$, $\varphi^-$ are determined by the boundary conditions at both ends of the line. The interference properties of these countersteaming waves are interesting and very important; consequently, we study them in depth. Figure 5.1 shows snapshots of $\mathscr{V}^+$, $\mathscr{V}^-$ and their sum at successive instants corresponding to 60° of the cycle. For this case, $A = 2B$ and the lower portions show a composite of the total voltage at the separate instants along with a dotted envelope of the maximum signal at any particular point. These envelopes for $\mathscr{V}$ and $\mathscr{I}$ are called "standing wave envelopes" for reasons to be made clearer later. The envelope is the quantity detected in many measurement schemes, and the properties of the load are calculated from a detailed investigation of this locus of points. As this envelope does not move [it is a locus of voltage (or current) extremes], it is called a standing wave pattern. The word "standing" comes from the special case when both waves have the same amplitude, then their superposition forms a series of alternate up and down loops that oscillate from positive to negative without any movement to the left or right. This

**188** SINUSOIDAL STEADY STATE

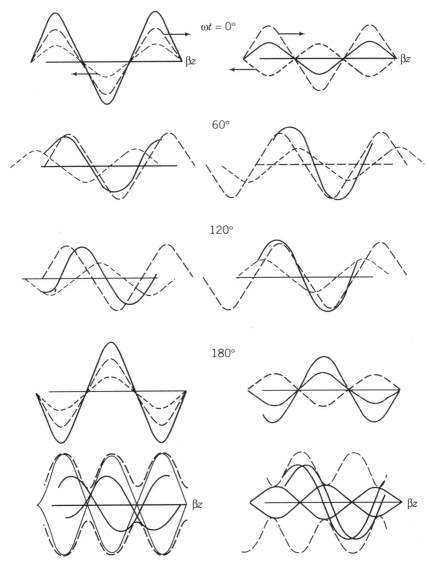

**FIGURE 5.1** On the left are two counterstreaming voltage waves, while those on the right are for current. The traveling waves are shown dotted, whereas their resultant is solid. Each sketch is delayed by 60° from the one directly above. We have chosen the reflected wave amplitude to be half the incident value.

occurs when the line is terminated in either a short or an open. In the general case wherein the load only reflects a portion of the incident wave ($|A| > |B|$), the pattern is more complicated; it appears to move toward the load in a nonuniform manner. A more detailed picture of the total voltage $\mathscr{V}_{tot} = \mathscr{V}^+ + \mathscr{V}^-$ is shown in Figure 5.2. The horizontal axis is related to time, whereas the vertical ones give

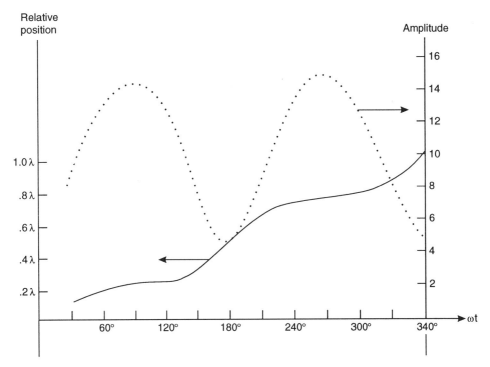

**FIGURE 5.2** The value of the crest of the net voltage on a TL where $\mathscr{V}_{tot}(z,t) = \mathscr{V}^+ + \mathscr{V}^-$, for the case $A = -2B$, $\varphi^+ = \varphi^- = 0$. Here, $A = 10$ and $B = -5$. The vertical axis on the left gives the crest position at various instants (converted to $\omega t$ in degrees). The right vertical axis shows the crest amplitude at the same instants.

the crest position (left) and amplitude (right) vs. time. Notice that the crest varies between $A + B$ and $A - B$ as we expect, but its instantaneous velocity (slope of solid line) is not uniform. The crest moves one wavelength during the time of one period ($T = 2\pi/\omega$), but its speed is not constant. It moves slower than average near its maximum and speeds up near the minimum. Its average speed is $v_p$ (recall Example 2.5). Observe in both Figures 5.1 and 5.3 that the position where $\mathscr{V}_{tot} = 0$ changes with time. If one focuses on a mechanical analog (driven string), the loops both bob up and down, as well as sway to and fro. When $|A| = |B|$ (no net flow of energy), the zero crossing point stays in the same place, and the loops strictly "stand" in a set interval and bob up and down (no to and fro motion). This is sketched in Figure 5.4. Observe that a loop does not appear to move sideways, as is the case when a net flow of energy occurs. As a matter of fact, when net energy flow occurs, one can visualize a loop moving in an "inch worm" manner toward the load. The "standing" concept has its roots in mechanical wave systems wherein one can see the difference between pure one-way energy flow vs. that with complete reflection at the load. For the complete reflection case, the transverse

# 190 SINUSOIDAL STEADY STATE

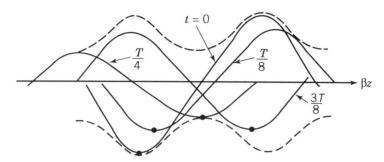

**FIGURE 5.3** The net voltage $\mathscr{V}_{\text{tot}}$ at four successive instants. Notice the nonuniform motion of the trough with respect to $z$. Notice also that the point where $\mathscr{V}_{\text{tot}}(z,t) = 0$ changes with time. The zero crossing point (if viewed for several periods) appears to rock back and forth.

displacement with time is easily distinguished visually from that for a one-way energy flow (Ref. 1). Many science/technology museums have wave machines as described in Ref. 1, wherein one can see the behavior. Some commercial software packages display the motion of standing waves in the general case and make for interesting displays.

***Example 5.1*** Find the expression for the total instantaneous voltage $\mathscr{V}_{\text{tot}}(z,t) = \mathscr{V}^+ + \mathscr{V}^-$ for two counterstreaming waves. Find the expression for the voltage standing wave ratio (VSWR) that is defined as

$$\text{VSWR} \triangleq S = \frac{\mathscr{V}_{\text{tot}}(\max)}{\mathscr{V}_{\text{tot}}(\min)}$$

The general expression is the addition of Eqs. (5.1a) and (5.1b):

$$\mathscr{V}_{\text{tot}}(z,t) = A\cos(\omega t - \beta z + \varphi^+) + B\cos(\omega t + \beta z + \varphi^-)$$
$$= A\{\cos \omega t \cos(\beta z - \varphi^+) + \sin \omega t \sin(\beta z - \varphi^+)\}$$
$$+ B\{\cos \omega t \cos(\beta z + \varphi^-) - \sin \omega t \sin(\beta z + \varphi^-)\}$$

Let

$$\alpha = \beta z - \varphi^+, \qquad \delta = \beta z + \varphi^-$$

Then

$$\mathscr{V}_{\text{tot}}(z,t) = [A\cos\alpha + B\cos\delta]\cos\omega t + [A\sin\alpha - B\sin\delta]\sin\omega t$$

Use

$$M\cos\omega t + N\sin\omega t = \sqrt{M^2 + N^2}\cos(\omega t - \psi)$$

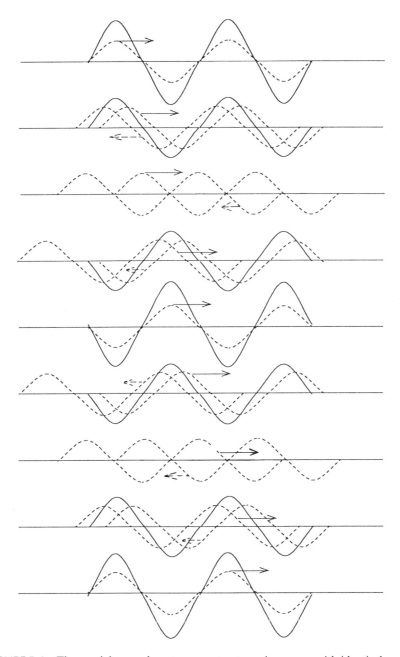

**FIGURE 5.4** The special case where two counterstreaming waves with identical amplitudes superpose to cause a standing wave pattern wherein the nodes remain at the same position. Each sketch differs by one-eighth of the period of either wave.

where $\tan\psi = N/M$. So

$$\mathscr{V}_{tot}(z,t) = \{A^2 + B^2 + 2AB[\cos\alpha\cos\delta - \sin\alpha\sin\delta]\}^{1/2}\cos(\omega t - \psi)$$
$$= \{A^2 + B^2 + 2AB\cos(\alpha + \delta)\}^{1/2}\cos(\omega t - \psi)$$
$$= \{A^2 + B^2 + 2AB\cos(2\beta z + \xi)\}^{1/2}\cos(\omega t - \psi)$$

where $\xi = \varphi^- - \varphi^+$. Thus, the amplitude varies with $z$ and has a maximum when $\cos(2\beta z + \xi) = 1$. The maximum and minimum amplitudes are

$$\mathscr{V}_{tot}(\max) = \sqrt{A^2 + B^2 + 2AB} = A + B$$
$$\mathscr{V}_{tot}(\min) = \sqrt{A^2 + B^2 - 2AB} = A - B \quad \text{when } \cos(2\beta z + \xi) = -1$$

Therefore,

$$\text{VSWR} = S = \frac{A+B}{A-B}$$

Notice that an rms voltmeter (if moved along the line) would record a voltage pattern with distance as

$$\mathscr{V}_{tot}(\text{rms}) = \frac{1}{\sqrt{2}}\{A^2 + B^2 + 2AB\cos(2\beta z + \xi)\}^{1/2}$$

The total waveform for the special case $\xi = 0$ is

$$\mathscr{V}_{tot}(z,t) = A[1 + \rho^2 + 2\rho\cos 2\beta z]^{1/2}\cos(\omega t - \psi)$$

The envelope of this waveform is sketched in Figure 5.5 for $A = 1$, $\rho = -\frac{1}{2}$. ▲

***Example 5.2*** Although using analogies is generally not a good idea (for often the systems have more differences than commonalities), we will discuss water waves in a canal and the situation of standing waves therein. For an infinitely long canal, the sinusoidal wave produced at the source corresponds to our wave. Assume we arrange a row of poles along the canal with rachet teeth such that a bobber inserted on one can move freely upward, but cannot move downward. This is the mechanical analog of some sort of peak detector. A wave train down the canal will push all the bobbers to the same height, that is, the crest amplitude. If a board is placed a few wavelengths downstream from the source, a standing wave will be set up. To measure this, assume that all bobbers have been held at the canal bottom until steady state has been achieved and then released. Next drain the canal and notice that the bobbers will be in the scalloped-shape pattern of the standing wave. ▲

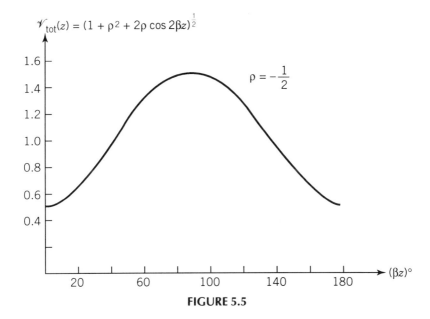

**FIGURE 5.5**

*Example 5.3* An illuminating lab demonstration to illustrate standing wave behavior is to place a fluorescent bulb near an open two-wire line. The tube glows in some proportion to the strength of the electric field. Therefore, one observes bright and dark regions along the tube, and the center-to-center distance between the dark regions is half a wavelength of the signal on the line. ▲

## 5.2 COMPLEX REPRESENTATION (PHASORS)

For sinusoidal excitation, the solutions to the wave equations:

$$\frac{\partial \mathscr{V}}{\partial z} = -L \frac{\partial \mathscr{I}}{\partial t}$$

$$\frac{\partial \mathscr{I}}{\partial z} = -C \frac{\partial \mathscr{V}}{\partial t}$$

are

$$\mathscr{V}^+ = A \cos(\omega t - \beta z + \varphi^+) \tag{5.2a}$$

$$\mathscr{V}^- = B \cos(\omega t + \beta z + \varphi^-) \tag{5.2b}$$

where $A, B, \varphi^+, \varphi^-$ are arbitrary constants. We will study TLs using ideas similar to those of ac circuit theory, and the first step is to utilize complex numbers to represent our traveling waves. We write

$$\mathscr{V}^+ = \operatorname{Re}\{Ae^{j\omega t}e^{j(-\beta z+\varphi^+)}\} \tag{5.3a}$$

$$\mathscr{V}^- = \operatorname{Re}\{Be^{j\omega t}e^{j(\beta z+\varphi^-)}\} \tag{5.3b}$$

where Re means the "real part of" the portion in the braces. We define the phasor representations for $\mathscr{V}^+$ and $\mathscr{V}^-$ as

$$\mathscr{V}^+ \leftrightarrow Ae^{j(-\beta z+\varphi^+)} \equiv V^+(z) \tag{5.4a}$$

$$\mathscr{V}^- \leftrightarrow Be^{j(\beta z+\varphi^-)} \equiv V^-(z) \tag{5.4b}$$

Notice that these phasors $V^+(z)$, $V^-(z)$ are slight extensions to those of circuit theory, since the angles change with distance $z$. Recall from circuits that a voltage waveform was represented as

$$\mathscr{V}(t) = C\cos(\omega t + \varphi)$$
$$= \operatorname{Re}\{Ce^{j(\omega t+\varphi)}\} = \operatorname{Re}\{Ce^{j\varphi}e^{j\omega t}\}$$

suppress $e^{j\omega t}$; then

$$\mathscr{V}(t) \leftrightarrow Ce^{j\varphi} \equiv V$$

where $V$ is the complex representation for $\mathscr{V}(t)$. The phasor representation for the net voltage at any point is therefore the sum of $V^+(z)$ and $V^-(z)$:

$$V_{\text{tot}}(z) = V^+(z) + V^-(z) \tag{5.5}$$

Figure 5.6 shows this expression for four values of $z$. The projection of the total phasor on the horizontal axis is the magnitude of $\mathscr{V}(z, t)$ and this projection varies with $z$ as shown in Example 5.1.

With the definitions

$$\mathscr{V}(z, t) = \operatorname{Re}\{V(z)e^{j\omega t}\} \tag{5.6a}$$

$$\mathscr{I}(z, t) = \operatorname{Re}\{I(z)e^{j\omega t}\} \tag{5.6b}$$

our basic line equations are (including losses)

$$\frac{\partial}{\partial z}\operatorname{Re}\{V(z)e^{j\omega t}\} = -L\frac{\partial}{\partial t}\operatorname{Re}\{I(z)e^{j\omega t}\} - R\operatorname{Re}\{I(z)e^{j\omega t}\}$$

$$\frac{\partial}{\partial z}\operatorname{Re}\{I(z)e^{j\omega t}\} = -C\frac{\partial}{\partial t}\operatorname{Re}\{V(z)e^{j\omega t}\} - G\operatorname{Re}\{V(z)e^{j\omega t}\}$$

These are satisfied if the following are satisfied, since when two complex quanti-

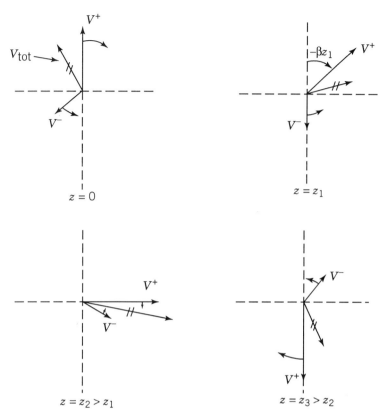

**FIGURE 5.6** Sketches of Eq. (5.5) for various values of $z$. Notice that $V^+(z)$ rotates clockwise, whereas $V^-(z)$ rotates counterclockwise with increasing $z$. Here, $|V^+| = 2|V^-|$, $\varphi^+ = \pi/2$, and $\varphi^- = 5\pi/4$.

ties are equal, both the real and imaginary parts are equal separately:

$$\frac{\partial}{\partial z}[V(z)e^{j\omega t}] = -j\omega L[I(z)e^{j\omega t}] - R[I(z)e^{j\omega t}]$$

$$\frac{\partial}{\partial z}[I(z)e^{j\omega t}] = -j\omega C[V(z)e^{j\omega t}] - G[V(z)e^{j\omega t}]$$

The common factor $e^{j\omega t}$ cancels from each term, giving the basic equation in the phasor domain:

$$\frac{\partial V}{\partial z} \to \frac{dV}{dz} = -(R + j\omega L)I \equiv -\mathbf{Z}I \quad (5.7a)$$

$$\frac{\partial I}{\partial z} \to \frac{dI}{dz} = -(G + j\omega C)V \equiv -\mathbf{Y}V \quad (5.7b)$$

where **Z** and **Y** are the impedance and admittance per unit length along the line. Notice that a slight notation problem occurs here. The "z" position variable is expressed with a lowercase letter, whereas the line impedance is expressed with a capital **Z**. In print, no problem exists, but for hand-written discussions, confusion might occur if one does not strictly follow the same notation scheme. However, by the nature of the calculations, such confusion rarely occurs.

***Example 5.4*** Discuss the term "complex amplitude" for phasors defined in traveling wave situations.

From Eq. (5.4a), we notice

$$V^+(z) = Ae^{-j\beta z}e^{j\varphi^+}$$

where $A$ is real [the amplitude of $\mathscr{V}^+(z,t)$]. We further write

$$V^+(z) = Ae^{j\varphi^+}e^{-j\beta z} \equiv \underline{A}e^{-j\beta z}$$

where $\underline{A}$ is complex, and the angle of primary concern is $\beta z$. The quantity $\underline{A}$ is sometimes called the "complex amplitude" of the phasor $V^+(z)$. Notice that this is in contrast to phasors of ordinary circuit theory wherein the amplitude is always real. ▲

## 5.3 GENERAL SOLUTION IN THE PHASOR DOMAIN

We start by writing Eqs. (5.7a) and (5.7b) as

$$\frac{dV}{dz} = -\mathbf{Z}I \qquad (5.8a)$$

$$\frac{dI}{dz} = -\mathbf{Y}V \qquad (5.8b)$$

where the partial derivatives are replaced by total derivatives since $V(z)$ and $I(z)$ do not depend on time. Eliminate $I$ from them to find

$$\frac{d^2V}{dz^2} - \mathbf{ZY}V = 0 \qquad (5.9)$$

which has solution:

$$V(z) = V^+e^{-\gamma z} + V^-e^{+\gamma z} \qquad (5.10)$$

where

$$\gamma^2 = \mathbf{ZY} = (R + j\omega L)(G + j\omega C) \qquad (5.11)$$

$$\gamma = \alpha + j\beta$$

## GENERAL SOLUTION IN THE PHASOR DOMAIN

$$\alpha = \{\tfrac{1}{2}[\sqrt{(R^2 + \omega^2 L^2)(G^2 + \omega^2 C^2)} + RG - \omega^2 LC]\}^{1/2} \tag{5.12}$$

$$\beta = \{\tfrac{1}{2}[\sqrt{(R^2 + \omega^2 L^2)(G^2 + \omega^2 C^2)} - RG + \omega^2 LC]\}^{1/2} \tag{5.13}$$

and $V^+, V^-$ are arbitrary complex constants determined by boundary conditions at both source and load planes. The solution for the current is found in a similar fashion and is

$$I(z) = I^+ e^{-\gamma z} + I^- e^{\gamma z} \tag{5.14}$$

where again we must relate $V^+, V^-, I^+$, and $I^-$ for a general TL. Since

$$-\mathbf{Z}I = \frac{dV}{dz} = -\gamma V^+ e^{-\gamma z} + \gamma V^- e^{\gamma z}$$

$$I = \frac{\gamma}{\mathbf{Z}}(V^+ e^{-\gamma z} - V^- e^{\gamma z})$$

define $\gamma/\mathbf{Z} = \sqrt{Y/\mathbf{Z}} = 1/\mathbf{Z}_0$, where $\mathbf{Z}_0$ is the complex characteristic impedance of the line. Then

$$I = \frac{1}{\mathbf{Z}_0}(V^+ e^{-\gamma z} - V^- e^{\gamma z}) \tag{5.15}$$

Using the linear independence of exponentials permits us to state

$$I^+ = \frac{V^+}{\mathbf{Z}_0}, \quad I^- = -\frac{V^-}{\mathbf{Z}_0} \tag{5.16}$$

after comparing Eqs. (5.14) and (5.15).

The basic geometry is shown in Figure 5.7, where we notice that the origin is located at the load plane. Recall in the time domain chapters that the origin was almost exclusively at the source plane. The reader must be careful to notice where the origin is located in various treatments, since certain terms have sign changes as the origin shifts. Thus, quick mixing of results from various sources can induce bothersome sign errors. At the load, we write

$$V(0) = V^+ + V^-$$

$$I(0) = \frac{1}{\mathbf{Z}_0}(V^+ - V^-)$$

and define the load reflection coefficient:

$$\Gamma_L = \frac{V^-(0)}{V^+(0)} \equiv \rho_L \underline{/\varphi_L} \tag{5.17}$$

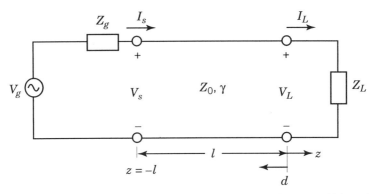

**FIGURE 5.7** The notation for a TL in the sinusoidal steady state. Although $Z_0$ is generally complex, we will almost always assume it to be pure real. We will often keep the real part of $\gamma$ since it is important to include this damping. Thus, we are not completely self-consistent in our analysis, but this is the standard way to treat low loss lines.

while using the bc $V(0) = I(0)Z_L$, we have

$$\Gamma_L = \frac{Z_L - Z_0}{Z_L + Z_0} \tag{5.18}$$

We can eliminate the arbitrary constants $V^+$, $V^-$ once and for all as follows. At the source plane, we have

$$V(-l) = V^+ e^{\gamma l} + V^- e^{-\gamma l} = V_s \tag{5.19a}$$

$$I(-l) = \frac{1}{Z_0}(V^+ e^{\gamma l} - V^- e^{-\gamma l}) = I_s \tag{5.19b}$$

Also

$$V_g = I(-l)Z_g + V_s, \quad [I(-l) \equiv I_s] \tag{5.19c}$$

Using Eq. (5.17) to eliminate $V^-$, we have

$$V_s = V^+(e^{\gamma l} + \Gamma_L e^{-\gamma l}) \tag{5.20a}$$

$$I_s = \frac{V^+}{Z_0}(e^{\gamma l} - \Gamma_L e^{-\gamma l}) \tag{5.20b}$$

Now using Eq. (5.19c) with the above gives

$$V^+ = \frac{V_g Z_0}{Z_0 + Z_g} \frac{e^{-\gamma l}}{1 - \Gamma_L \Gamma_s e^{-2\gamma l}} \tag{5.21}$$

where

$$\Gamma_s = \frac{Z_g - Z_0}{Z_g + Z_0} \qquad (5.22)$$

which agrees with the results of Example 2.8 developed in the general time-varying case. With $V^+$ expressed in terms of the known quantities in Eq. (5.21), we may write the general solutions as

$$V(z) = \frac{V_g}{1 + Z_g/Z_0} \frac{e^{-\gamma l}(e^{-\gamma z} + \Gamma_L e^{\gamma z})}{(1 - \Gamma_L \Gamma_s e^{-2\gamma l})} \qquad (5.23a)$$

$$I(z) = \frac{V_g}{Z_0 + Z_g} \frac{e^{-\gamma l}(e^{-\gamma z} - \Gamma_L e^{\gamma z})}{(1 - \Gamma_L \Gamma_s e^{-2\gamma l})} \qquad (5.23b)$$

which we recognize are the closed-form solutions for the infinite series of waves that comprise the sinusoidal steady state. Recall that the term

$$\frac{1}{1 - \Gamma_L \Gamma_s e^{-2\gamma l}}$$

is the closed-form expression for the infinite geometric series.

## 5.4 PROPERTIES OF THE GENERAL SOLUTION

One of the main properties to be determined in Figure 5.7 is the input impedance at the source plane, $\mathbf{Z}_{in} = \mathbf{Z}(-l) = V_s/I_s$. Notice that $\mathbf{Z}_{in}$ is the impedance looking toward the load. If one wants the impedance looking toward the generator, one uses $V_s/(-I_s)$. In other words, define the current to be that which points into the direction one is "looking." Using Eqs. (5.20a) and (5.20b), we find

$$\mathbf{Z}_{in}(-l) = \frac{V_s}{I_s} = \mathbf{Z}_0 \frac{1 + \Gamma_L e^{-2\gamma l}}{1 - \Gamma_L e^{-2\gamma l}} \qquad (5.24)$$

which for $\alpha = 0$ reduces to

$$\mathbf{Z}_{in}(-l) = \mathbf{Z}_0 \frac{Z_L + jZ_0 \tan \beta l}{Z_0 + jZ_l \tan \beta l} \qquad (5.24a)$$

The next quantity is the reflection coefficient at any plane along the line. We write $V(z)$ in the form

$$V(z) = V^+ e^{-\gamma z} + V^- e^{\gamma z}$$

and define

$$\Gamma(z) = \frac{V^- e^{\gamma z}}{V^+ e^{-\gamma z}} \tag{5.25}$$

to be the reflection coefficient at any plane $z$. Observe that $\Gamma(z)$ is indeed the reflected signal over the incident one. The above reduces to

$$\Gamma(z) = \frac{V^-}{V^+} e^{2\gamma z}$$

and $\Gamma(z=0) \equiv \Gamma_L = V^-/V^+$, so

$$\Gamma(z) = \Gamma_L e^{2\gamma z} \tag{5.25a}$$

and for any point on the line, $z = -l_1$,

$$\Gamma(-l_1) = \Gamma_L e^{-2\gamma l_1} \tag{5.26}$$

which is a very basic and important result. Using this last result in Eq. (5.24) yields

$$\mathbf{Z}_{in}(-l_1) = \mathbf{Z}_0 \frac{1 + \Gamma(-l_1)}{1 - \Gamma(-l_1)} \tag{5.27}$$

which will, in part, form the basis of the next chapter.

We now consider the envelope of the standing wave; the total voltage may be expressed as follows:

$$V(z) = V^+ e^{-\gamma z} + V^- e^{\gamma z}$$
$$= V^+ e^{-\gamma z}(1 + \Gamma_L e^{2\gamma z}) \tag{5.28}$$

Assume no loss, so $\alpha = 0$ and $\gamma = j\beta$:

$$V(z) = V^+ e^{-j\beta z}(1 + \Gamma_L e^{j2\beta z}) \tag{5.29}$$

Take the magnitude and manipulate as follows:

$$|V(z)| = |V^+| |e^{-j\beta z}| |1 + \Gamma_L e^{j2\beta z}|$$
$$= |V^+| |1 + \Gamma_L e^{j2\beta z}| \tag{5.29a}$$

since $|e^{-j\beta z}| = 1$. Now use $\Gamma_L = \rho_L e^{j\varphi_L}$:

$$|V(z)| = |V^+| |1 + \rho_L e^{j(2\beta z + \varphi_L)}|$$
$$= |V^+| \{(1 + \rho_L \cos\psi)^2 + (\rho_L \sin\psi)^2\}^{1/2} \tag{5.29b}$$

where $\psi = 2\beta z + \varphi_L$. Then

$$|V(z)| = |V^+|\{(1 + \rho_L^2 + 2\rho_L \cos\psi\}^{1/2} \tag{5.30}$$

which has its extrema at the positions:

*Maxima*

$$\cos\psi = +1; \quad 2\beta z_{max} + \varphi_L = \pm 2m\pi, \quad m = 0, 1, 2, \ldots \tag{5.31}$$

$$|V(z)|_{max} = |V^+|(1 + \rho_L)$$

*Minima*

$$\cos\psi = -1; \quad 2\beta z_{min} + \varphi_L = \pm(2m+1)\pi, \quad m = 0, 1, 2, \ldots \tag{5.32}$$

$$|V(z)|_{min} = |V^+|(1 - \rho_L)$$

The VSWR is defined as

$$\text{VSWR} = S = \frac{|V(z)|_{max}}{|V(z)|_{min}} = \frac{1 + \rho_L}{1 - \rho_L} \tag{5.33}$$

which is the quantity quite often measured directly in the laboratory. Using Eq. (5.25a), we can write Eq. (5.29a) as

$$|V(z)| = |V^+||1 + \Gamma(z)| \tag{5.34}$$

and if we compare this with Eqs. (5.31) and (5.32), we see

$$\Gamma(z) = \rho_L \quad \text{at a voltage maximum}$$

$$\Gamma(z) = -\rho_L \quad \text{at a voltage minimum}$$

Now use Eq. (5.27) and let $(-l_1)$ correspond to the positions of $z_{max}$ and $z_{min}$. Then

$$Z_{in} = Z_0 \frac{1 + \rho_L}{1 - \rho_L} = Z_0 S \quad \text{at a voltage maximum} \tag{5.33a}$$

$$Z_{in} = Z_0 \frac{1 - \rho_L}{1 + \rho_L} = \frac{Z_0}{S} \quad \text{at a voltage minimum} \tag{5.33b}$$

which will prove useful later.

Notice that $|V(z)|$ as expressed in Eq. (5.30) is the same as the peak amplitude of $\mathscr{V}_{\text{tot}}(z,t)$ as determined in Example 5.1, that is,

$$|\mathscr{V}_{\text{tot}}(z,t)| = \{A^2 + B^2 + 2AB\cos(2\beta z + \xi)\}^{1/2}$$

$$= A\left\{1 + \frac{B}{A} + 2\left(\frac{B}{A}\right)\cos(2\beta z + \xi)\right\}^{1/2}$$

Compare this with Eq. (5.30) and note

$$\rho_L = \frac{B}{A}$$

$$\varphi_L = \xi = \varphi^- - \varphi^+$$

$$|V^+| = A$$

which agrees with our intuition concerning the traveling waves.

## 5.5 POWER FLOW ALONG THE LINE

The time-averaged power flow into the TL and load of Figure 5.7 are, respectively,

$$P(z=-l) = P_{\text{in}} = \tfrac{1}{2}\text{Re}\{V_s I_s^*\} \qquad (5.34a)$$

$$P(z=0) = P_L = \tfrac{1}{2}\text{Re}\{V_L I_L^*\} \qquad (5.34b)$$

which is the definition from circuit theory; see Example 5.5. For a lossless line, these two quantities must be equal, but when line loss occurs, we should develop an expression for this loss. For then, $P_{\text{in}} = P_{\text{loss}} + P_L$, where $P_{\text{loss}}$ is that power dissipated into heat on the TL. Start by defining the power flowing toward the load at any plane along the line:

$$P(z) = \tfrac{1}{2}\text{Re}\{V(z)I(z)^*\} \qquad (5.35)$$

If we use

$$V(z) = V^+(e^{-\gamma z} + \Gamma_L e^{\gamma z}) \qquad (5.36a)$$

$$I(z) = \frac{V^+}{Z_0}(e^{-\gamma z} - \Gamma_L e^{\gamma z}) \qquad (5.36b)$$

then $P(z)$ becomes

$$P(z) = \tfrac{1}{2}\text{Re}\left\{V^+(e^{-\gamma z} + \Gamma_L e^{\gamma z})\left(\frac{V^+}{Z_0}\right)^*(e^{-\gamma^* z} - \Gamma_L^* e^{\gamma^* z})\right\}$$

POWER FLOW ALONG THE LINE    203

and simplify this using the following definitions and properties of complex numbers. For a complex number $C = x + jy$, we have

$$x = \text{Re}\{C\} \tag{5.37a}$$

$$y = \mathscr{I}_m\{C\} \tag{5.37b}$$

$$C^* = x - jy \tag{5.37c}$$

$$C + C^* = 2x \tag{5.37d}$$

$$x = \frac{C + C^*}{2} \tag{5.37e}$$

$$C - C^* = j2y = j2\mathscr{I}_m\{C\} \tag{5.37f}$$

$$y = \frac{C - C^*}{2j} \tag{5.37g}$$

$$CC^* = |C|^2 \tag{5.37h}$$

$$|C|^2 = \sqrt{x^2 + y^2} \tag{5.37i}$$

and let

$$\gamma = \alpha + j\beta$$

$$Y_0 = \frac{1}{Z_0} = G_0 + jB_0$$

$$Y_0^* = G_0 - jB_0 \tag{5.37j}$$

Then we find

$$P(z) = \tfrac{1}{2}\text{Re}\{|V^+|^2 Y_0^*(e^{-2\alpha z} - \Gamma_L^* e^{-j2\beta z} + \Gamma_L e^{j2\beta z} - |\Gamma_L|^2 e^{2\alpha z})\}$$

Now notice that the second term is the complex conjugate of the third one, so using Eq. (5.37f), we see

$$P(z) = \tfrac{1}{2}\text{Re}\{|V^+|^2 Y_0^*(e^{-2\alpha z} - |\Gamma_L|^2 e^{2\alpha z} + j2\mathscr{I}_m\{\Gamma_L e^{j2\beta z}\})\}$$

which simplifies to

$$P(z) = \tfrac{1}{2}G_0|V^+|^2(e^{-2\alpha z} - |\Gamma_L|^2 e^{2\alpha z}) + B_0|V^+|^2 \mathscr{I}_m\{\Gamma_L e^{j2\beta z}\} \tag{5.38}$$

Next evaluate this at the source and load planes:

$$P(z=-l) = P_{in} = \tfrac{1}{2}G_0|V^+|^2(e^{2\alpha l} - |\Gamma_L|^2 e^{-2\alpha l}) + B_0|V^+|^2 \mathscr{I}_m\{\Gamma_L e^{-j2\beta l}\} \quad (5.39)$$

$$P(z=0) = P_L = \tfrac{1}{2}G_0|V^+|^2(1-|\Gamma_L|^2) + B_0|V^+|^2 \mathscr{I}_m\{\Gamma_L\} \quad (5.40)$$

Then the power lost in the line is

$$P_{loss} = P_{in} - P_L = \tfrac{1}{2}G_0|V^+|^2[(e^{2\alpha l}-1) + |\Gamma_L|^2(1-e^{-2\alpha l})]$$
$$+ B_0|V^+|^2[\mathscr{I}_m\{\Gamma_L e^{-j2\beta l}\} - \mathscr{I}_m\{\Gamma_L\}] \quad (5.41)$$

In most cases, we do not use the above formulas since low loss lines are normally employed. The first approximation is

$$Z_0 = \sqrt{\frac{R+j\omega L}{G+j\omega C}} \doteq \sqrt{\frac{L}{C}} \quad (R \sim G \sim 0) \quad (5.42)$$

so the impedance $Z_0$ and admittance $Y_0 = 1/Z_0$ are pure real. Then $B_0 = 0$ and the above formulas reduce to

$$P_{in} = \tfrac{1}{2}G_0|V^+|^2(e^{2\alpha l} - |\Gamma_L|^2 e^{-2\alpha l}) \quad (5.43)$$

$$P_L = \tfrac{1}{2}G_0|V^+|^2(1-|\Gamma_L|^2) \quad (5.44)$$

$$P_{loss} = \tfrac{1}{2}G_0|V^+|^2[(e^{2\alpha l}-1) + |\Gamma_L|^2(1-e^{-2\alpha l})] \quad (5.45)$$

which are the normally used forms. When the losses are negligible ($\alpha \to 0$), we find

$$P_{in} = P_L = \tfrac{1}{2}G_0|V^+|^2(1-|\Gamma_L|^2) \quad (5.46)$$

which can be interpreted as the combination of leftward and rightward moving power. See ref. 2. for more discussion of power flow.

To show this, recall Eq. (5.21):

$$|V^+| = \left|\frac{V_g Z_0}{Z_0 + Z_g}\right| \frac{|e^{-\gamma l}|}{|1 - \Gamma_L \Gamma_s e^{-2\gamma l}|}$$

and if $\Gamma_s = 0$ (generator matched to line) along with $\alpha = 0$, we have

$$|V^+| = \left|\frac{V_g}{2}\right| \frac{|e^{-j\beta l}|}{1} = \left|\frac{V_g}{2}\right|$$

Then we write

$$P_{in} = \tfrac{1}{2}G_0 \left|\frac{V_g}{2}\right|^2 (1-|\Gamma_L|^2) \quad (5.47)$$

which we notice is independent of the line length. One may interpret the power leaving the generator as

$$P_{incident} = \tfrac{1}{2} G_0 \frac{|V_g|^2}{4} = \frac{G_0}{8}|V_g|^2 \qquad (5.48)$$

and that reflected by the load and returning toward the source as

$$P_{reflected} = +|\Gamma_L|^2 P_{incident}$$

Thus, we define a power reflection coefficient by

$$|\Gamma_L|^2$$

which is the magnitude of the ratio of reflected to incident power. The incident power is also called the available power from the generator (see Example 5.6).

**Example 5.5** Develop Eq. (5.34) using Figure 5.8.
From circuit theory, the power into the network is

$$P_{ave} = \tfrac{1}{2} \operatorname{Re}\{VI^*\} \qquad (1)$$

which is derived as follows. The average power is

$$P_{ave} = \frac{1}{T}\int_0^T v(t)i(t)\,dt \qquad (2)$$

where

$$v(t) = A\cos(\omega t + \varphi)$$
$$i(t) = B\cos(\omega t + \theta)$$

and

$$V = Ae^{j\varphi}$$
$$I = Be^{j\theta}$$

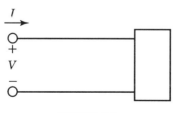

FIGURE 5.8

are corresponding phasors. Write

$$v(t) = \text{Re}\{Ve^{j\omega t}\}$$
$$i(t) = \text{Re}\{Ie^{j\omega t}\}$$

Now consider any complex number $C = x + jy$. The real part may be expressed as

$$\text{Re}\{C\} = x = \tfrac{1}{2}(C + C^*) \tag{3}$$

where $C^* = x - jy$ is the complex conjugate of $C$. This permits us to write

$$v(t) = \tfrac{1}{2}[V^*e^{-j\omega t} + Ve^{j\omega t}]$$
$$i(t) = \tfrac{1}{2}[I^*e^{-j\omega t} + Ie^{j\omega t}]$$

Substituting these into Eq. (2) yields

$$P_{ave} = \frac{1}{T}\int_0^T \tfrac{1}{4}[V^*I^*e^{-j2\omega t} + V^*I + VI^* + VIe^{j2\omega t}]\,dt$$

$$= \tfrac{1}{4}[V^*I + VI^*] + \frac{1}{4T}\int_0^T [V^*I^*e^{-j2\omega t} + VIe^{j2\omega t}]\,dt$$

$$= \tfrac{1}{4}[(VI^*)^* + VI^*] + 0$$

where the integral is identically zero (expand exponentials into $\cos 2\omega t$ and $j\sin 2\omega t$). We have also used $V^*I = (VI^*)^*$. Then from Eq. (3),

$$P_{ave} = \tfrac{1}{4}[2\,\text{Re}\{VI^*\}]$$
$$= \tfrac{1}{2}\text{Re}\{VI^*\}$$

which is Eq. (5.34). Expanding this further gives

$$P_{ave} = \tfrac{1}{2}\text{Re}\{Ae^{j\varphi}Be^{-j\theta}\}$$
$$= \tfrac{1}{2}AB\cos(\varphi - \theta)$$

where $(\varphi - \theta)$ is the power factor angle and $\cos(\varphi - \theta)$ is the power factor. A word of caution is in order. When we write the expressions for $v(t)$ and $i(t)$, the amplitudes may be peak values (assumed here) or rms values, that is, $A_{rms} = A_{peak}/\sqrt{2}$. If rms values are used, then the result for power becomes

$$P_{ave} = \text{Re}\left\{\frac{V}{\sqrt{2}}\frac{I^*}{\sqrt{2}}\right\}$$
$$= \text{Re}\{V_{rms}I^*_{rms}\}$$

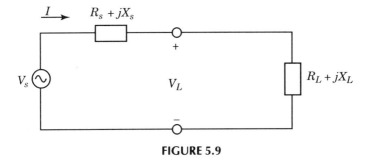

**FIGURE 5.9**

However, one often finds both peak and rms given the same symbol, that is, $V$, so factors of 2, 4, and in some situations, 8 can crop up. ▲

**Example 5.6** Develop the expression for the available power from a generator using lumped circuit methods.

In this discussion, all phasors are for peak values. We start by writing the average power into the load see Figure. 5.9:

$$P_{ave}(\text{in } \mathbf{Z}_L) = \tfrac{1}{2}\text{Re}\{V_L I^*\} = \tfrac{1}{2}\text{Re}\{|I|^2 \mathbf{Z}_L\}$$
$$= \tfrac{1}{2}|I|^2 R_L$$

Now writing this in terms of $R_L$ and $X_L$ gives

$$P_{ave}(R_L, X_L) = \tfrac{1}{2} R_L \frac{|V_s|^2}{[(R_L + R_s)^2 + (X_L + X_s)^2]}$$

Maximize this by inspection ($R_L = R_s$, $X_L = -X_s$):

$$P_{ave}(\text{max}) = \frac{\tfrac{1}{2}|V_s|^2 R_s}{(2R_s)^2 + 0} = \frac{1}{8}\frac{|V_s|^2}{R_s}$$

If $V_s$ is in rms, the result is

$$P_{ave}(\text{max}) = \frac{1}{4}\frac{|V_s|^2}{R_s} \equiv P_{avail}$$ ▲

***Example 5.7*** Develop Eq. (5.45) using simpler arguments.

Consider power flowing only to the right, that is, $\Gamma_L = 0$. Then

$$V(z) = V^+ e^{-j\beta z}$$

$$I(z) = \frac{1}{R_0} V^+ e^{-j\beta z}$$

$$P_{ave} = \tfrac{1}{2}\text{Re}\{V(z)I(z)^*\} = \tfrac{1}{2}\text{Re}\left\{V^+ e^{-j\beta z} \frac{1}{R_0}(V^+)^* e^{+j\beta z}\right\}$$

$$= \tfrac{1}{2}\text{Re}\left\{\frac{|V^+|^2}{R_0}\right\} = \frac{|V^+|^2}{2R_0} \tag{1}$$

but

$$|V^+| = \left|\frac{V_g Z_0}{Z_0 + Z_g}\right|$$

since $\Gamma_L = 0$, $\alpha = 0$. Then if $Z_g = Z_0 = R_0$,

$$P_{ave} = \frac{1}{2R_0}\left|\frac{V_g}{2}\right|^2 = \frac{G_0 |V_g|^2}{8} \quad \text{where } G_0 = \frac{1}{R_0}$$

which is $P_{incident}$.

Suppose now power flows only to the left as shown in Figure. 5.10. Then

$$V(z) = V^- e^{j\beta z}$$

$$I(z) = -\frac{1}{R_0} V^- e^{j\beta z}$$

$$P_{ave} = \tfrac{1}{2}\text{Re}\left\{V^- e^{j\beta z}\left(-\frac{1}{R_0}\right)(V^-)^* e^{-j\beta z}\right\}$$

$$= -\tfrac{1}{2}\text{Re}\left\{|V^-|^2 \frac{1}{R_0}\right\} = -\frac{|V^-|^2}{2R_0} \tag{2}$$

If we choose to define $V^- \equiv \Gamma_L V^+$, then

$$P_{ave} = -\frac{1}{2R_0}|\Gamma_L|^2 |V^+|^2$$

and assume that $|V^+|^2$ is as before. Thus, the above is $P_{reflected}$.

The main points in this example are to show that waves moving to the right have power dictated by Eq. (1), whereas those moving to the left are expressed by

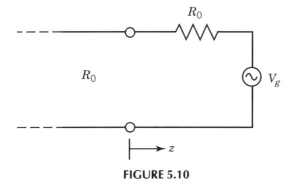

**FIGURE 5.10**

Eq. (2). In both cases, each generator sees a matched line, so we can think of "power waves." ▲

## 5.6 LOW LOSS CONDITIONS AND ATTENUATION FORMULAS

Recall the general expression for the propagation constant:

$$\gamma = \sqrt{(R+j\omega L)(G+j\omega C)} \equiv \alpha + j\beta \tag{5.49}$$

and the characteristic impedance $Z_0$:

$$Z_0 = \sqrt{\frac{R+j\omega L}{G+j\omega C}} \tag{5.50}$$

The attenuation constant becomes, in the low loss limit ($R \to 0, G \to 0$),

$$\alpha \doteq \frac{1}{2}\left(R\sqrt{\frac{C}{L}} + G\sqrt{\frac{L}{C}}\right) \tag{5.51}$$

as detailed in Example 5.8. The impedance $Z_0$ is

$$Z_0 = \sqrt{\frac{L}{C}}\sqrt{\frac{(j\omega + R/L)}{(j\omega + G/C)}}$$

which, in general, is always complex. In the special case $(R/L) = (G/C)$, we have $Z_0 = \sqrt{L/C}$, and this case is called the distortionless condition. For this special case, we quickly notice

$$\gamma = R\sqrt{\frac{C}{L}} + j\omega\sqrt{LC}$$

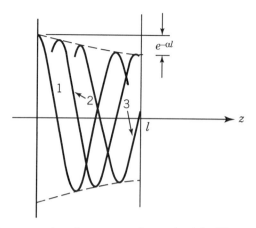

**FIGURE 5.11** The attenuation of a wave moving to the right. The sequence of curves lie between exponential envelopes that define the attenuation per unit length.

so

$$\alpha = R\sqrt{\frac{C}{L}}, \qquad \beta = \omega\sqrt{LC} \qquad (5.52)$$

and $\beta$ is the same as that for zero losses. Since $\beta$ varies linearly with $\omega$, the wave velocity $v_p = \omega/\beta = 1/\sqrt{LC}$ is the same at all frequencies, the condition for distortionless transmission. The distortionless case is only of academic interest and generally one strives for extremely low loss. Unless specifically stated, $\mathbf{Z}_0$ will always be treated as purely real, since $\omega L \gg R$ and $\omega C \gg G$ are usually satisfied.

Consider an infinitely long TL with small loss as shown in Figure 5.11. The sketch is of the voltage distribution at three successive instants. By the time a crest has moved the distance $l$, it has decreased by the factor $e^{-\alpha l}$. We define the attenuation over this interval by

$$\text{Attenuation} \equiv \ln\left(\frac{V_1}{V_2}\right) = \alpha l \qquad (5.53)$$

to which we attach the nomenclature of (Np). Thus, $\alpha$ has the nomenclature of Np/m. Since a neper is the natural logarithm of a voltage ratio, it has, of course, no "units." However, by convention, we say "units" of nepers. In the laboratory, one does not measure voltage at microwave frequencies, but instead measures power. Assume that an ideal power meter senses average power at two points $z_3$ and $z_4$. Then define the attenuation in this case as

$$\text{Attenuation} \equiv 10\log_{10}\left[\frac{P(z_3)}{P(z_4)}\right] \text{dB} \qquad (5.54)$$

Since the input resistance is the same for both power measurements (the same, or identical meters), the power sensed is proportional to the square of the voltages at

LOW LOSS CONDITIONS AND ATTENUATION FORMULAS    211

the points. Then

$$\text{Attenuation} = 10 \log_{10}\left(\frac{\mathscr{V}_3}{\mathscr{V}_4}\right)^2 = 20 \log_{10}\left(\left|\frac{\mathscr{V}_3}{\mathscr{V}_4}\right|\right) \text{dB} \quad (5.55a)$$

and the attenuation in nepers between these points is

$$\text{Attenuation} = \ln\left(\frac{\mathscr{V}_3}{\mathscr{V}_4}\right) \text{Np} \quad (5.55b)$$

Now using $\log_{10} X = 0.434 \ln X$, we can relate nepers to decibels as

$$20 \log_{10}\left(\frac{\mathscr{V}_3}{\mathscr{V}_4}\right) \ (\text{dB}) = 20(0.4343) \ln\left(\frac{\mathscr{V}_3}{\mathscr{V}_4}\right)$$

$$= 8.686 \ (\text{Np})$$

Then we write, using dB for decibels and Np for nepers,

$$\text{Attenuation} \ (\text{in dB}) = 8.686 \cdot (\text{Attenuation in Np}) \quad (5.56)$$

1 Np corresponds to 8.686 dB, or 1 dB corresponds to 0.1151 Np. Most microwave power meters are calibrated in dB, as this is the "unit" in the lab. The Neper on the other hand is used for the most part in textbook calculations (easier equations). Since Eq. (5.54) is the definition of dB, one must be careful when relating power values and corresponding voltages. For example, suppose two different power meters are used and have different internal resistances. Then $P_3 = \mathscr{V}_3^2/2R_a$, $P_4 = \mathscr{V}_4^2/2R_b$, where $R_a$ and $R_b$ are the corresponding internal values. Then the expression for attenuation becomes

$$\text{Attenuation} = 10 \log_{10}\left(\frac{P_3}{P_4}\right) = 10 \log_{10}\left[\left(\frac{\mathscr{V}_3}{\mathscr{V}_4}\right)^2 \frac{R_b}{R_a}\right]$$

$$= 10 \log_{10}\left(\frac{\mathscr{V}_3}{\mathscr{V}_4}\right)^2 + 10 \log_{10}\left(\frac{R_b}{R_a}\right)$$

and the conversion to nepers is now ambiguous.

**Example 5.8** Show the algebra needed to arrive at Eq. (5.51).
Starting with Eq. (5.49)

$$\gamma = \sqrt{RG + j\omega RC + j\omega LG - \omega^2 LC}$$

neglect the product $RG$ with respect to $-\omega^2 LC$:

$$\gamma \doteq \left\{-\omega^2 LC\left[1 + \frac{j}{-\omega LC}(RC + LG)\right]\right\}^{1/2}$$

$$\doteq j\omega\sqrt{LC}\left[1 - \frac{j}{\omega LC}(RC + LC)\right]^{1/2}$$

Use the binomial expansion:

$$\gamma \doteq j\omega\sqrt{LC}\left[1 - \frac{j}{2\omega LC}(RC + LG)\right]$$

$$= j\omega\sqrt{LC} + \frac{1}{2}\left(R\sqrt{\frac{C}{L}} + G\sqrt{\frac{L}{C}}\right)$$

so we notice

$$\alpha = \text{Re}\{\gamma\} = \frac{1}{2}\left(R\sqrt{\frac{C}{L}} + G\sqrt{\frac{L}{C}}\right)$$

$$\beta = \mathscr{I}_m\{\gamma\} = \omega\sqrt{LC} \qquad \blacktriangle$$

***Example 5.9*** Show the attenuation in nepers between two measurement points by using the expressions for average power measured at the two planes.

The average power expression is

$$P_{\text{ave}} = \frac{1}{2}\text{Re}\left\{V\left(\frac{V}{R}\right)^*\right\} = \frac{1}{2R}|V|^2$$

Then

$$V = V^+ e^{-\gamma z} = V^+ e^{-\alpha z} e^{-j\beta z}$$

$$|V|^2 = |V^+|^2 e^{-2\alpha z}$$

so

$$P_2 = P(z_1) = \frac{1}{2R}|V^+|^2 e^{-2\alpha z_1}$$

$$P_1 = P(z_2) = \frac{1}{2R}|V^+|^2 e^{-2\alpha z_2}$$

and finally,

$$\frac{P_2}{P_1} = e^{-2\alpha z_1 + 2\alpha z_2}$$

$$\frac{V_2^2}{V_1^2} = e^{2\alpha(z_2 - z_1)}$$

$$\frac{V_2}{V_1} = e^{\alpha(z_2 - z_1)}$$

$$\ln\frac{V_2}{V_1} = \alpha(z_2 - z_1) \equiv \text{Attenuation} \qquad \blacktriangle$$

## SUPPLEMENTARY EXAMPLES

**5.1** Assume that a coaxial cable, 5 m long, has the following distributed parameters at 100 MHz:

$$R = 0.3\,\Omega/\text{m}$$
$$G = 12\,\mu\mho/\text{m}$$
$$L = 0.2\,\mu\text{H}/\text{m}$$
$$C = 120\,\text{pF}/\text{m}$$

Determine the following quantities $\alpha$, $\beta$, $Z_0$, $\lambda$, $v_p$, and the electrical length at 100 MHz.

Start with Eq. (5.11):

$$\gamma^2 = (R + j\omega L)(G + j\omega C)$$
$$\gamma = [0.3 + j2\pi(10^8)(0.2 \times 10^{-6})][12 \times 10^{-6} + j2\pi(10^8)(12 \times 10^{-11})]$$
$$\gamma^2 = -9.475 + j2.413 \times 10^{-2}$$
$$\gamma = 3.919 \times 10^{-3} + j3.078$$

Thus,

$$\alpha = \text{Re}\{\gamma\} = 3.919 \times 10^{-3}\,\text{Np/m}$$
$$\beta = \mathscr{I}_m\{\gamma\} = 3.078\,\text{rad/m}$$

$$Z_0 = \sqrt{\frac{R + j\omega L}{G + j\omega C}} = \sqrt{1.667 \times 10^3 - j3.714} = 40.82 - j0.0455\,\Omega$$

$$\lambda = \frac{2\pi}{\beta} = 2.04\,\text{m}$$

$$v_p = \frac{\omega}{\beta} = \frac{2\pi(10^8)}{3.078} = 2.04 \times 10^8\,\text{m/s}.$$

The electrical length is the "length" of the cable in units of the wavelength at the frequency of·interest.

Thus,

$$\text{Electrical length} = \frac{5\,\text{m}}{2.04\,\text{m}} = 2.449\,\text{wavelengths}$$

# SINUSOIDAL STEADY STATE

and this is generally expressed in either radians or degrees. Then

$$\text{Electrical length} = \frac{l}{\lambda}\left(2\pi \frac{\text{rad}}{\text{wavelength}}\right) = 2\pi\left(\frac{l}{\lambda}\right) \text{rad}$$

$$= \frac{2\pi}{\lambda} l = \beta l = (3.078)(5) = 15.39 \text{ rad} \quad (881.8°)$$

Therefore, the value of $\beta l$ is the "electrical length" that is often called just the length of the line. This length therefore changes with frequency.

If one does not have a pocket calculator or software package with complex manipulations, use Eqs. (5.12) and (5.13) for $\alpha$ and $\beta$. For the square root of complex numbers, use either of the following.

For a complex number,

$$C = x \pm jy$$

$$\sqrt{C} = \pm\left[\sqrt{\frac{r+x}{2}} \pm j\sqrt{\frac{r-x}{2}}\right], \quad \begin{array}{l} x \text{ positive or negative} \\ y \text{ must be positive} \\ r = \sqrt{x^2 + y^2} \end{array}$$

or if

$$C = A\underline{/\theta}$$

$$A = \sqrt{x^2 + y^2}, \quad \tan\theta = \frac{y}{x}$$

then

$$\sqrt{C} = \sqrt{A}\underline{/\frac{\theta}{2}}$$

**5.2** Suppose that a cable 10 m long has a capacity of 600 pF. Determine the distributed inductance $L$ and $Z_0$. Assume that the velocity is $200 \times 10^6$ m/s and the cable may be considered lossless.

The distributed capacity is

$$C = \frac{600 \text{ pF}}{10 \text{ m}} = 60 \text{ pF/m}$$

since

$$v^2 = \frac{1}{LC}$$

$$L = \frac{1}{v^2 C} = \frac{1}{(2 \times 10^8)^2 (60 \times 10^{-12})} = 0.417 \, \mu\text{H/m}$$

then
$$Z_0 = \sqrt{\frac{L}{C}} = 83.3\,\Omega$$

**5.3** Consider two counterstreaming voltage waves that form $\mathscr{V}(z,t)$:

$$\mathscr{V}(z,t) = |V^+|e^{-\alpha z}\cos(\omega t - \beta z + \varphi^+) + |V^-|e^{\alpha z}\cos(\omega t + \beta z + \varphi^-)$$

These waves that include damping are the generalizations of those in Eqs. (5.1a) and (5.1b). Write $\mathscr{V}(z,t)$ in compact phasor notation:

$$\mathscr{V}(z,t) = \text{Re}\{|V^+|e^{-\alpha z}e^{j(\omega t - \beta z + \varphi^+)} + |V^-|e^{\alpha z}e^{j(\omega t + \beta z + \varphi^-)}\}$$
$$= \text{Re}\{[|V^+|e^{j\varphi^+}e^{-\alpha z}e^{-j\beta z} + |V^-|e^{j\varphi^-}e^{\alpha z}e^{j\beta z}]e^{j\omega t}\}$$
$$= \text{Re}\{[V^+e^{-(\alpha+j\beta)z} + V^-e^{(\alpha+j\beta)z}]e^{j\omega t}\}$$

where

$$V^+ = |V^+|e^{j\varphi^+}$$
$$V^- = |V^-|e^{j\varphi^-}$$

$$\mathscr{V}(z,t) = \text{Re}\{[V^+e^{-\gamma z} + V^-e^{\gamma z}]e^{j\omega t}\}$$
$$= \text{Re}\{V(z)e^{j\omega t}\}$$

suppress Re and $e^{j\omega t}$. Then

$$\mathscr{V}(z,t) \leftrightarrow V(z)$$

**5.4** For a lossless line, the magnitudes of $V(z)$ and $I(z)$ can be sketched vs. position along the TL by using the following graphical technique. Start with Eq. (5.29a):

$$|V| = |V^+||1 + \Gamma_L e^{j2\beta z}| \tag{1}$$

Since $I(z) = 1/Z_0[V^+e^{-\gamma z} - V^-e^{+\gamma z}]$, it follows that

$$|I(z)| = \frac{|V^+|}{Z_0}|1 - \Gamma_L e^{j2\beta z}| \tag{2}$$

These can be written as

$$|V(z)| = |V^+||1 + \Gamma(z)|$$
$$|I(z)| = \frac{|V^+|}{Z_0}|1 - \Gamma(z)|$$

and introduce the complex $\Gamma$ plane as shown in Figure 5.12.

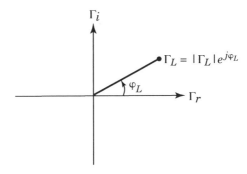

**FIGURE 5.12**

As we move along the line, $\Gamma(z)$ varies as [see Eqs. (5.25a) and (5.26)]

$$\Gamma(z = -l) = \Gamma_L e^{-2\gamma l}$$
$$= \Gamma_L e^{-j2\beta l} \quad (\alpha = 0)$$

While we move away from the load, the magnitude of $\Gamma(z)$ remains at $|\Gamma_L|$, but the phase decreases from $\varphi_L$ by $2\beta l$ radians. See sketches in Figure 5.13. Now a vector $|1 + \Gamma(z)|$ can be constructed as shown in Figure 5.14. Thus by plotting $|1 + \Gamma(z)|$ vs. $l$ in Figure 5.15, we obtain $|V(z)|$ to within the constant $|V^+|$:

For the current, we just need to plot $|1 - \Gamma(z)|$ as shown in Figure 5.16 which gives (current standing wave envelope dotted) Figure 5.17. Thus, $I(z)$ is plotted to within the constant $|V^+|/\mathbf{Z}_0$. Notice that $V(z)$ is at a minimum when $\Gamma(z)$ has angle $\pi$. Also, $V(z)$ has a maximum when $I(z)$ is at its minimum, and vice versa.

It is important to observe that one 360° rotation corresponds to the angle of $\Gamma(z)$ changing by

$$2\beta z = 2\pi$$

or

$$\beta z = \pi$$

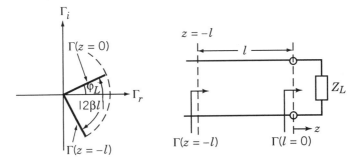

**FIGURE 5.13**

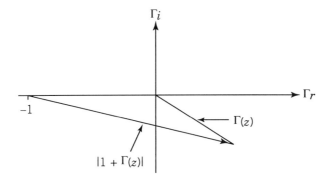

**FIGURE 5.14**

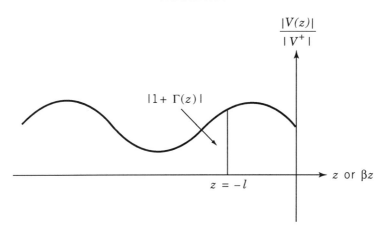

**FIGURE 5.15**

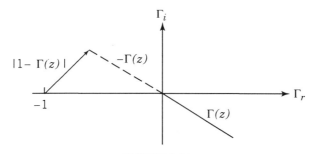

**FIGURE 5.16**

or

$$\frac{2\pi}{\lambda}(z = l) = \pi$$

$$\therefore \quad l = \frac{\lambda}{2}$$

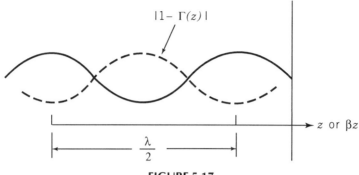

**FIGURE 5.17**

so the reflection coefficient is periodic with the distance of $l = \lambda/2$. In other words, when one moves $\lambda/2$ or $180°$ ($\pi$ radians) along the TL, the reflection has changed by $360°$. Alternatively, when one has moved the distance along the line corresponding to one wavelength, the standing wave pattern has repeated itself twice, and the reflection coefficient has rotated two complete revolutions in the $\Gamma$ plane.

**5.5** For a 50-$\Omega$ line, VSWR $= S = 4$ is known, and the minimum of the voltage standing wave pattern is $\lambda/4$ from the load. What is $\mathbf{Z}_L$?

From Eq. (5.33), we can determine $\rho_L$:

$$\rho_L = \frac{S-1}{S+1} = \frac{3}{5}$$

But from Eq. (5.18), $\mathbf{Z}_L$ is determined from $\Gamma_L$:

$$\mathbf{Z}_L = \mathbf{Z}_0 \frac{1+\Gamma_L}{1-\Gamma_L}$$

and since $\Gamma_L = \rho_L e^{j\varphi_L}$, we need $\varphi_L$. From the previous problem, we know $\Gamma(z)$ has angle $\pi$ at the minimum points and angle 0 at the maxima. We also know that the maximum and minimum are separated $\lambda/4$ (since successive minima are $\lambda/2$ apart). We are given the fact that the minimum occurs $\lambda/4$ from the load plane; thus, for this special case, the load is at a maximum of $|V(z)|$. Then

$$\underline{/\Gamma(z)} = (\varphi_L - 2\beta z) = 0$$

and $z = 0$, so $\varphi_L = 0$. Then

$$\mathbf{Z}_L = \mathbf{Z}_0 \frac{1+0.6}{1-0.6} = 50(4) = 200\ \Omega$$

**5.6** For a 50-Ω line, 3.4 m long, terminated in 100 Ω, filled with dielectric $\mu_r = 1$, $\varepsilon_r = 2.25$, and operated at 200 MHz, find $\mathbf{Z}_{in}$.

From Eq. (5.24a),

$$\mathbf{Z}_{in} = \mathbf{Z}_0 \frac{\mathbf{Z}_L + j\mathbf{Z}_0 \tan \beta l}{\mathbf{Z}_0 + j\mathbf{Z}_L \tan \beta l}$$

we need to obtain $\beta$:

$$\beta = \frac{\omega}{v} \quad \text{and} \quad v = \frac{c}{\sqrt{\varepsilon_r}}, \quad \text{where } c = \text{speed of light in free space}$$

$$= 3 \times 10^8 \text{ m/s}$$

$$\therefore v = \frac{3 \times 10^8}{\sqrt{2.25}} = 2 \times 10^8 \text{ m/s}$$

$$\beta = \frac{(2\pi)(2 \times 10^8)}{(2 \times 10^8)} = 2\pi \text{ rad/m}$$

$$\beta l = (2\pi)(3.4) = 21.36 \text{ rad} \Rightarrow 144°$$

$$\tan \beta l = -0.7265$$

$$\therefore \mathbf{Z}_{in} = 50 \frac{100 + j50(-0.7265)}{50 + j100(-0.7265)}$$

$$= 49.106 + j35.027 \, \Omega$$

**5.7** For the general TL case in Figure 5.18, find the following:

(a) $\Gamma_L, \Gamma_{in}$
(b) $V_L, V_S$
(c) $I_L, I_s$
(d) Sketch a phasor diagram showing $V_L, V_s, V_g$.

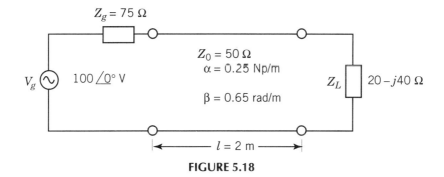

FIGURE 5.18

**220**   SINUSOIDAL STEADY STATE

(a) Equation (5.18):

$$\Gamma_L = \frac{Z_L - Z_0}{Z_L + Z_0} = \frac{-30 - j40}{70 - j40} = 0.62 \underline{/-97°}$$

$$= -0.0769 - j0.6154$$

Equation (5.25a):

$$\Gamma_{in} = \Gamma(z = -2) = \Gamma_L e^{2\gamma(-2)} = \Gamma_L e^{-4\gamma}$$

$$e^{-4\gamma} = e^{-4(0.25 + j0.65)} = e^{-1} e^{-j2.6} = 0.368(1 \underline{/-148.97°})$$

$$\therefore \ \Gamma_{in} = (0.62 \underline{/-97°})(0.368 \underline{/-148.97°}) = 0.228 \underline{/-246°}$$

$$= -0.092 + j0.209$$

(b) Equation (5.23a):

$$V(z) = \frac{V_g}{1 + Z_g/Z_0} \frac{e^{-\gamma l}(e^{-\gamma z} + \Gamma_L e^{\gamma z})}{(1 - \Gamma_L \Gamma_s e^{-2\gamma l})}$$

$$e^{-\gamma l} = e^{-(0.25 + j0.65)(2)} = 0.607 \underline{/-37.2°}$$

$$\Gamma_s = \frac{Z_g - Z_0}{Z_g + Z_0} = 0.2$$

$$V_L = V(0) = \frac{100}{2.5} \frac{e^{-\gamma l}(1 + \Gamma_L)}{(1 - 0.2 \Gamma_L e^{-2\gamma l})} = 25.75 \underline{/-71.3°}$$

$$V_s = V(z = -2) = 40 \frac{e^{-2\gamma}(e^{2\gamma} + \Gamma_L e^{-2\gamma})}{1.045 + j0.00666}$$

$$= 35.64 \underline{/12.6°}$$

(c) $$I_L = \frac{V_L}{Z_L} = \frac{8.258 - j24.39}{20 - j40} = 0.57 - j0.079 = 0.576 \underline{/-7.9°}$$

$$I_s = \frac{V_g - V_s}{Z_g} = 0.8696 - j0.103 = 0.876 \underline{/-6.8°}$$

Notice that $V_g$ and $V_s$ are at the same point (plane) in space, as in a phasor diagram in lumped circuits. In Figure 5.19, however, $V_L$ is at the load plane. Thus, the phasors in distributed circuits are located at different points (planes) in space.

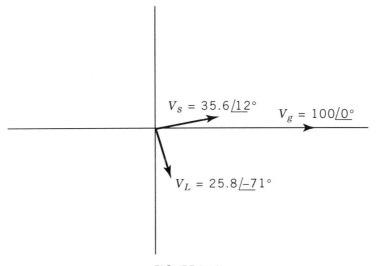

**FIGURE 5.19**

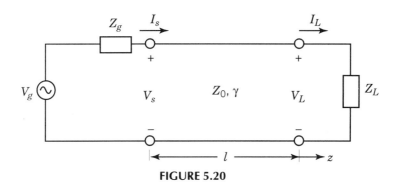

**FIGURE 5.20**

**5.8** For the general line shown in Figure 5.20, find the relationship between the load ($V_L$, $I_L$) and source ($V_S$, $I_s$) variables.

Start with the general expression for voltage:

$$V(z) = V^+ e^{-\gamma z} + V^- e^{\gamma z}$$

$$V_L = V(0) = V^+ + V^- \triangleq TV^+$$

Thus,

$$V^+(1 + \Gamma_L) = TV^+$$

or $T = (2\mathbf{Z}_L)/(\mathbf{Z}_L + \mathbf{Z}_0)$. Then write

$$V^+ = \frac{V_L}{T}$$

Now the manipulation is as follows:

$$V(z=-l) = V_s = \frac{V_L}{T}[e^{\gamma l} + \Gamma_L e^{-\gamma l}]$$

$$I_s = \frac{V_L}{TZ_0}[e^{\gamma l} - \Gamma_L e^{-\gamma l}]$$

or

$$V_s = \frac{V_L}{2Z_L}[(\mathbf{Z}_L + \mathbf{Z}_0)e^{\gamma l} + (\mathbf{Z}_L - \mathbf{Z}_0)e^{-\gamma l}]$$

$$I_s = \frac{V_L}{2Z_L Z_0}[(\mathbf{Z}_L + \mathbf{Z}_0)e^{\gamma l} - (\mathbf{Z}_L - \mathbf{Z}_0)e^{-\gamma l}]$$

or

$$I_L = \frac{V_L}{Z_L}$$

$$V_s = V_L\left[\cosh \gamma l + \frac{\mathbf{Z}_0}{\mathbf{Z}_L}V_L \sinh \gamma l\right]$$

$$I_s = V_L\left[\frac{1}{\mathbf{Z}_0}\sinh \gamma l + I_L \sinh \gamma l\right]$$

so finally,

$$\begin{bmatrix} V_s \\ I_s \end{bmatrix} = \begin{bmatrix} \cosh \gamma l & \mathbf{Z}_0 \sinh \gamma l \\ \frac{1}{\mathbf{Z}_0}\sinh \gamma l & \cosh \gamma l \end{bmatrix} \begin{bmatrix} V_L \\ I_L \end{bmatrix}$$

Now invert

$$\begin{bmatrix} V_L \\ I_L \end{bmatrix} = \begin{bmatrix} \cosh \gamma l & -\mathbf{Z}_0 \sinh \gamma l \\ -\frac{1}{\mathbf{Z}_0}\sinh \gamma l & \cosh \gamma l \end{bmatrix} \begin{bmatrix} V_s \\ I_s \end{bmatrix}$$

The ABCD parameters of a two-port are just those in the first matrix above, that is,

$$\begin{bmatrix} V_1 \\ I_1 \end{bmatrix} = \begin{bmatrix} A & B \\ C & D \end{bmatrix} \begin{bmatrix} V_2 \\ I_2 \end{bmatrix}$$

so

$$A = \cosh \gamma l, \qquad B = \mathbf{Z}_0 \sinh \gamma l$$

$$C = \frac{1}{\mathbf{Z}_0}\sinh \gamma l, \qquad D = \cosh \gamma l$$

as shown in Figure 5.21.

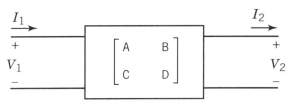

**FIGURE 5.21**

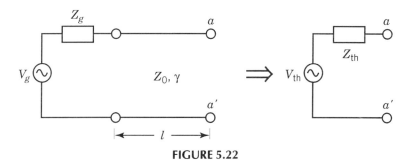

**FIGURE 5.22**

**5.9** Find the Thévenin equivalent seen by the load in Figure 5.22. From Eq. (5.23a), we have

$$V(z) = \frac{V_g}{1 + Z_g/Z_0} \frac{e^{-\gamma l}(e^{-\gamma z} + \Gamma_L e^{\gamma z})}{(1 - \Gamma_L \Gamma_s e^{-2\gamma l})}$$

here $z = 0$, $\Gamma_L = 1$, and

$$V(0) = \frac{2V_g Z_0}{Z_0 + Z_g} \frac{e^{-\gamma l}}{(1 - \Gamma_s e^{-2\gamma l})}$$

so

$$V_{th} = V_{oc} = V(0)$$

Next apply a short to the line and determine $I(0)$. Use Eq. (5.23b):

$$I(z) = \frac{V_g}{Z_0 + Z_g} \frac{e^{-\gamma l}(e^{-\gamma z} - \Gamma_L e^{\gamma z})}{(1 - \Gamma_L \Gamma_s e^{-2\gamma l})}$$

Now

$$I(0) = \frac{V_g}{Z_0 + Z_g} \frac{2e^{-\gamma l}}{(1 + \Gamma_s e^{-2\gamma l})}$$

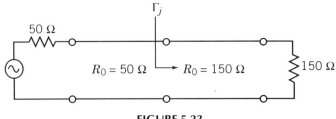

**FIGURE 5.23**

By definition,

$$Z_{th} = \frac{V_{oc}}{I_{sc}} = \frac{V(0)}{I(0)}$$

so

$$Z_{th} = \frac{1 + \Gamma_s e^{-2\gamma l}}{1 - \Gamma_s e^{-2\gamma l}}(Z_0)$$

**5.10** For the two lines connected as shown in Figure 5.23, what is $S$ at the junction, and what is it at the source plane?

$$\Gamma_J = \frac{150 - 50}{150 + 50} = \frac{1}{2}$$

$$S = \frac{1 + |\Gamma|}{1 - |\Gamma|} = 3$$

The VSWR ($S$), by the definition, is not a function of position along the line. For lossless lines, the ratio of maximum to minimum is invariant, no matter which maximum or minimum pair is chosen. For lossy lines, the extrema vary with position, so the VSWR is ambiguous. For this problem, assume that lossless conditions apply, then the second part of the question is meaningless.

**5.11** A TL has $L = 0.1\,\mu\text{H}$ and $C = 250\,\text{pF}$.

(a) Find $Z_0$.
(b) If $\varepsilon_r = 2.25$, find the wavelength at 100 MHz ($\mu_r = 1$).

(a) $Z_0 = \sqrt{\dfrac{1 \times 10^{-7}}{250 \times 10^{-12}}} = 20\,\Omega$

(b) $\lambda = \dfrac{v}{f} = \dfrac{c/\sqrt{\varepsilon_r}}{1 \times 10^8} = \dfrac{3 \times 10^8\,\text{m/s}}{\sqrt{2.25}(1 \times 10^8)} = 2\,\text{m}$

# SUPPLEMENTARY EXAMPLES 225

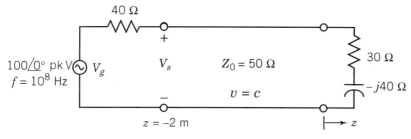

**FIGURE 5.24**

**5.12** For the lossless line shown, in Figure 5.24, find (a) VSWR, (b) $Z_{in}$, (c) the voltage at the source plane $V_S$, (d) the average power absorbed in the load, and (e) the load current.

(a) 
$$\Gamma_L = \frac{30 - j40 - 50}{30 - j40 + 50} = -j0.5$$

$$S = \frac{1 + |\Gamma|}{1 - |\Gamma|} = 3$$

(b) 
$$Z_{in}(-l) = Z_0 \frac{Z_L + jZ_0 \tan \beta l}{Z_0 + jZ_L \tan \beta l}, \qquad \beta l = \frac{2\pi(10^8)}{3 \times 10^8}(2)$$

$$= \frac{4\pi}{3}$$

$$Z_{in} = 50 \frac{(30 - j40) + j50(1.732)}{50 + j(30 - j40)(1.732)}, \qquad \tan \beta l = 1.732$$

$$Z_{in} = 17.72 + j11.814 \, \Omega$$

(c) From Figure 5.25,

$$V_s = V_g \frac{Z_{in}}{Z_{in} + R_g}$$

$$V_s = 100 \frac{17.72 + j11.814}{57.72 + j11.814}$$

$$= 33.486 + j13.614$$

$$= 36.15 \underline{/22.1^\circ}$$

(d) The power to load is the same as that into $Z_{in}$ since the line is lossless:

$$P_{ave} = \tfrac{1}{2} \operatorname{Re}\{V_s I_s^*\} = \tfrac{1}{2} \operatorname{Re}\{V_s Y_s^* V_s^*\} = \tfrac{1}{2} |V_s|^2 \operatorname{Re}\{Y_{in}^*\}$$

$$= \frac{(36.15)^2}{2}(0.03907) = 25.53 \text{ W}$$

226  SINUSOIDAL STEADY STATE

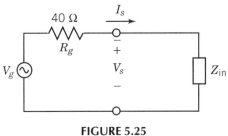

**FIGURE 5.25**

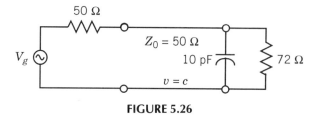

**FIGURE 5.26**

(e) $$I(0) = \frac{100}{50+40} \frac{e^{-j4\pi/3}(1-\Gamma_L)}{1-\Gamma_L\Gamma_s e^{-j8\pi/3}}, \quad \begin{array}{l}\Gamma_L = -j0.5 \\ \Gamma_s = -\frac{1}{9}\end{array}$$

$$= (-0.971 + j0.679)\text{A} = 1.18 \underline{/145°}\text{ A}$$

**5.13** For the load given in Figure 5.26, calculate $\Gamma_L$, $S$ and determine where the first voltage maximum from the load is located:

$$Y_L = \frac{1}{Z_L} = \frac{1}{72} + j2\pi(50 \times 10^6)(10 \times 10^{-12})$$

$$\therefore Z_L = 68.5 - j15.49\,\Omega$$

$$\Gamma_L = \frac{Z_L - Z_0}{Z_L + Z_0} = 0.1703 - j0.1085 = 0.202\underline{/-32.5°}$$

$$S = \frac{1+|\Gamma|}{1-|\Gamma|} = 1.506$$

From Eq. (5.31),

$$2\beta z_{\max} + \varphi_L = \pm 2m\pi, \quad m = 0, 1, 2, \ldots$$

Choose $m$ and the sign of right side such that the smallest negative value of

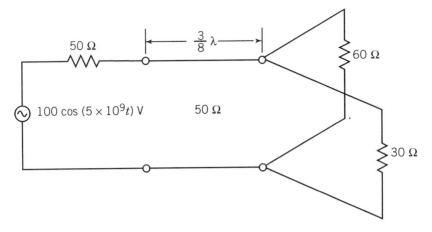

**FIGURE 5.27**

$z_{max}$ occurs. Let $m = -1$:

$$2\left(\frac{\pi}{3}\right)z_{max} = -2\pi + 32.5°$$

$$\therefore \quad z_{max} = -2.73 \text{ m}$$

$$\lambda = \frac{c}{f} = \frac{3 \times 10^8}{50 \times 10^6} = 6 \text{ m} \quad \text{or} \quad z_{max} = -0.455\lambda \text{ from load}$$

**5.14** Assume that the line in Figure 5.27 is lossless and determine the power delivered to each load:

$$\beta l = \left(\frac{2\pi}{\lambda}\right)\left(\frac{3\lambda}{8}\right) = \frac{3\pi}{4} = 135°, \quad e^{-j\beta l} = 0.707(-1-j)$$

Use the Thévenin equivalent as seen by the load plane in Figure 5.28 (see Supplementary Example 5.9):

$$\Gamma_S = 0, \quad V(0) = -70.7(1+j)$$

$$Z_{th} = Z_0 = 50 \Omega$$

$$I = \frac{-70.7(1+j)}{50+20} = -1.01(1+j)$$

Use current division:

$$I_{60} = I\frac{30}{60+30} = -0.3367 - j0.3367$$

$$I_{30} = I - I_{60} = -0.6733 - j0.6733$$

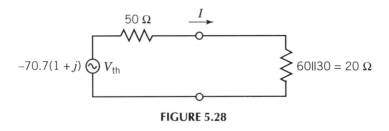

FIGURE 5.28

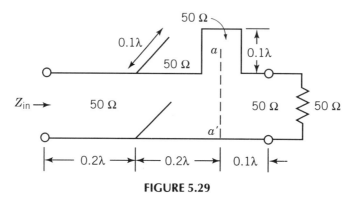

FIGURE 5.29

Then
$$P_{60\Omega} = \tfrac{1}{2}|I_{60}|^2(60) = \tfrac{1}{2}|0.4762|^2(60) = 6.8 \text{ W}$$
$$P_{30\Omega} = \tfrac{1}{2}|0.9522|^2(30) = 13.6 \text{ W}$$

**5.15** Find the input impedance $Z_{in}$ in Figure 5.29. The line inserted in series with the main line is called a series stub and is assumed to add a contribution only at the plane of its centerline.

First, we find the input impedance for both shorted and open segments of line see Figure 5.30.

$$Z_{in}(\text{short}) = jZ_0 \tan\beta l \qquad\qquad Z_{in}(\text{open}) = -jZ_0 \cot\beta l$$

Here, $\beta l = \left(\dfrac{2\pi}{\lambda}\right)\left(\dfrac{\lambda}{10}\right) = 36° \qquad Z_{in}(\text{open}) = -j68.82\,\Omega$

$$Z_{in}(\text{short}) = j36.33\,\Omega$$

FIGURE 5.30

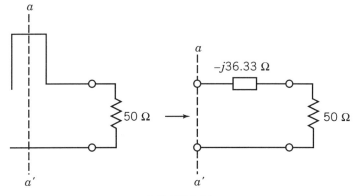

**FIGURE 5.31**

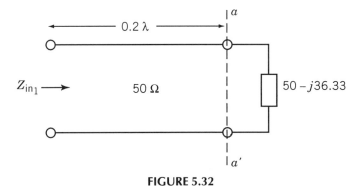

**FIGURE 5.32**

This allows the reduction to the form, see Figure 5.31:

Then from Figure 5.32,

$$\mathbf{Z}_{in_1} = 50 \frac{(50 - j36.33) + j50(0.7265)}{50 + j(50 - j36.33)(0.7265)} = 26.69 - j12.69 \, \Omega$$

Finally, we have the result depicted in Figures 5.33 and 5.34

$$\tan \beta l = \tan \left( \frac{2\pi}{\lambda} \frac{2\lambda}{10} \right)$$

$$= \tan(0.4\pi)$$

$$= 0.7265$$

$$\mathbf{Z}_{in} = 50 \frac{(17.18 - j16.34) + j50(0.7265)}{50 + j(17.18 - j16.34)(0.7265)}$$

$$\therefore \quad \mathbf{Z}_{in} = 16.47 + j12.82 \, \Omega$$

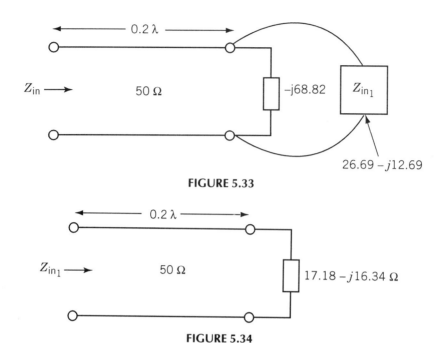

**FIGURE 5.33**

**FIGURE 5.34**

**5.16** For the line shown in Figure 5.35 with VSWR = 3, find $X_L$.

$$\rho_L = \frac{S-1}{S+1} = \frac{3-1}{3+1} = 0.5$$

$$|\Gamma_L|^2 = \left|\frac{(150+jX_L)-75}{(150+jX_L)+75}\right|^2 = (0.5)^2$$

$$\left(\frac{150+jX_L-75}{150+jX_L+75}\right)\left(\frac{150-jX_L-75}{150-jX_L+75}\right) = 0.25$$

or

$$X_L^2 = 9375$$

$$\therefore \quad X_L = \pm 96.82 \,\Omega$$

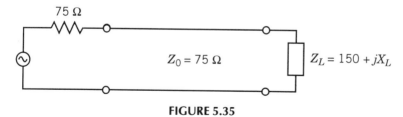

**FIGURE 5.35**

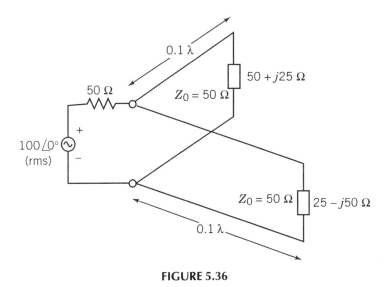

**FIGURE 5.36**

**5.17** Two TLs in parallel are connected to a generator as shown in Figure 5.36.

(a) What is the available power from the generator?

$$P_\text{avail} = \frac{|V_g|^2}{4Z_g} = \frac{10^4}{4(50)} = 50 \text{ W}$$

(b) What percent of this power is absorbed by the two loads in Figure 5.37?

$$\beta l = 36°, \quad \tan \beta l = 0.7265$$

$$\mathbf{Z}_{\text{in}_1} = 50 \frac{(50 + j25) + j50(0.7265)}{50 + j(50 + j25)(0.7265)}, \quad \mathbf{Z}_{\text{in}_2} = 12.27 - j10.5$$

$$\therefore \mathbf{Z}_{\text{in}_1} = 81.86 + j2.92$$

Then

$$\mathbf{Z}_{\text{in}_1} \| \mathbf{Z}_{\text{in}_2} = 11.63 - j7.82$$

From Figure 5.38 the current is

$$I = \frac{100 \underline{/0°}}{61.63 - j7.82} = 1.62 \underline{/7.2°}$$

$$P_\text{load} = |I|^2 \text{ Re}\{\mathbf{Z}_L\} = |1.62|^2 (11.63) = 30 \text{ W}$$

$$\therefore \quad \frac{P_\text{load}}{P_\text{avail}} = 60\%$$

**232** SINUSOIDAL STEADY STATE

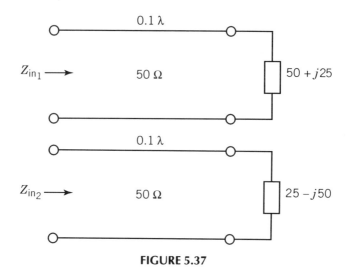

**FIGURE 5.37**

**FIGURE 5.38**

**5.18** Determine the line characteristic $R_0$ shown in Figure 5.39, such that the VSWR is a minimum for a general load $R_L + jX_L$.
By inspection of the expression for $S$ (the VSWR), we see it is smallest when $|\Gamma_L|$ is smallest:

$$|\Gamma_L| = \Gamma_L \Gamma_L^* = \left(\frac{R_L + jX_L - R_0}{R_L + jX_L + R_0}\right)\left(\frac{R_L - jX_L - R_0}{R_L - jX_L + R_0}\right) = \frac{(R - R_0)^2 + X_L^2}{(R + R_0)^2 + X_L^2}$$

$$\frac{\partial |\Gamma_L|^2}{\partial R} = 0 \Rightarrow R_0^2 = R_L^2 + X_L^2$$

or

$$R_0 = \sqrt{R_L^2 + X_L^2}$$

Then

$$|\Gamma| = \frac{X_L}{R_0 + R_L}$$

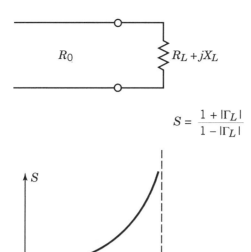

**FIGURE 5.39**

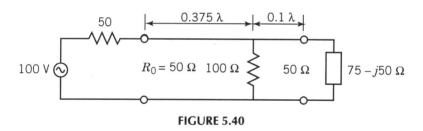

**FIGURE 5.40**

**5.19** For the network in Figure 5.40, sketch the magnitude of $V(z)$ over both sections of line:

First, obtain the equivalent load for the 0.375 $\lambda$ line. The input impedance to the right of the 100-$\Omega$ shunt resistor is (see Fig. 5.41)

$$Z_{in} = 27.5 - j25.27 \, \Omega$$

This combined with the 100-$\Omega$ resistor gives

$$Z_{in} \parallel 100 \, \Omega = 24.53 - j14.96 \, \Omega$$

Now ascertain the distribution on the line. From Eq. (5.21), we find

$$V^+ = \frac{V_g Z_0}{Z_0 + Z_g} \frac{e^{-\gamma l}}{1 - \Gamma_L \Gamma_s e^{-2\gamma l}} = 50 \text{ V}$$

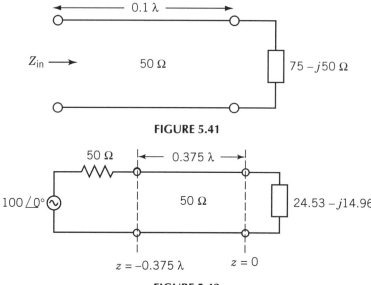

FIGURE 5.41

FIGURE 5.42

Then evaluating at the load and source ends see Figure 5.42 gives

$$|V(z)| = |V^+|\{1 + \rho_L^2 + 2\rho_L \cos\psi\}^{1/2}, \quad \psi = 2\beta z + \varphi_L$$

$$\Gamma_L = 0.3886 \underline{/-138.2°}$$

$$|V(0)| = 37.8 \text{ V}$$

$$\left|V\left(\frac{-3\lambda}{8}\right)\right| = 64.6 \text{ V}$$

The extremes of $|V(z)|$ are

$$|V(z)|_{\max} = 50(1 + 0.3886) = 69.4$$

$$|V(z)|_{\min} = 50(1 - 0.3886) = 30.6$$

which permits a quick sketch as shown in Figure 5.43.

For the other segment, first Thévenize the 0.375-$\lambda$ line just to the left of the 100-$\Omega$ resistor (Fig. 5.44):
Combine with the 100-$\Omega$ resistor (Fig. 5.45):
Then obtain the voltage distribution at both ends (Fig. 5.46):

$$|V(0)| = 36.6 \text{ V}$$

$$|V(-0.1\lambda)| = 26.8 \text{ V}$$

Combining both sketches yields Figure 5.47.
Notice the break at the plane of the shunting element in Figure 5.48.

SUPPLEMENTARY EXAMPLES **235**

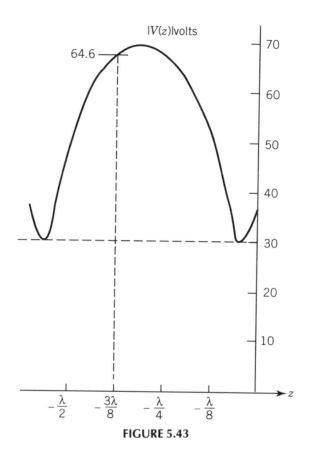

**FIGURE 5.43**

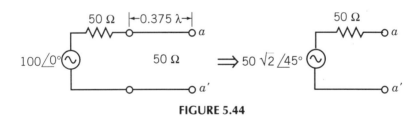

**FIGURE 5.44**

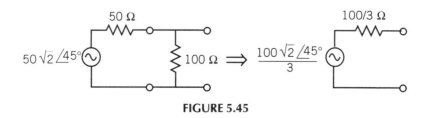

**FIGURE 5.45**

**236** SINUSOIDAL STEADY STATE

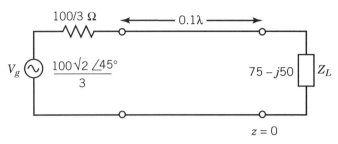

**FIGURE 5.46**

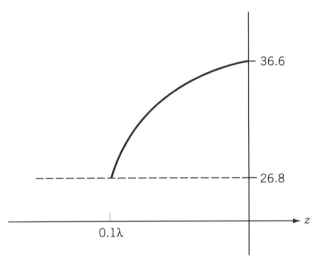

**FIGURE 5.47**

**5.20** For the special case wherein both the load and generator are matched to the line, one can quickly calculate the reflection, absorption, and transmission of an element in shunt with the line as shown in Figure 5.49. The power terms are

$$P_1 = \frac{1}{2R_0} V_1^+ (V_1^+)^*$$

$$P_2 = \frac{1}{2R_0} V_1^- (V_1^-)^* = \frac{\Gamma\Gamma^*}{2R_0} V_1^+ (V_1^+)^*$$

Let the power reflection coefficient be $\Gamma\Gamma^* = R$:

$$V_{aa'} = V_1^+ + V_2^+ = V_1^+ (1 + \Gamma)$$

$$\therefore P_3 = \frac{1}{2R_0} \text{Re}\{V_{aa'} V_{aa'}^*\} = \frac{V_1^+ (V_1^+)^* (1 + \Gamma)(1 + \Gamma)^*}{2R_0}$$

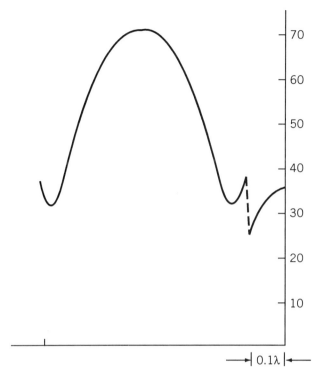

**FIGURE 5.48**

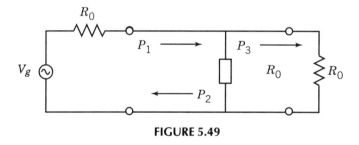

**FIGURE 5.49**

Then the power transfer coefficient is

$$T = (1 + \Gamma)(1 + \Gamma)^*$$

and the power absorbed by the shunt element is

$$A = P_1 - P_2 - P_3 = P_1\left(1 - \frac{P_2}{P_1} - \frac{P_3}{P_1}\right)$$
$$= P_1(1 - R - T)$$
$$= P_1(-1)[2\,\text{Re}\{\Gamma\} - 2|\Gamma|^2]$$

**5.21** For the lumped network in Figure 5.50, find the available power and develop the equations for plotting contours of constant power to the load as both $R_L$ and $X_L$ vary.

The loop current is

$$I = \frac{V_g}{(R_g + R_L) + j(X_g + X_L)}$$

$$|I|^2 = \frac{|V_g|^2}{[(R_g + R_L)^2 + (X_g + X_L)^2]}$$

Then the power and maximum power (available power) are

$$P = \tfrac{1}{2}|I|^2 R_L$$

$$P_{max} = P_{avail} = \frac{|V_g|^2}{8R_g}$$

Form the ratio:

$$\frac{P}{P_{max}} = \frac{4R_g R_L}{(R_g + R_L)^2 + (X_g + X_L)^2}$$

Let

$$R' = \frac{R_L}{R_g}, \qquad X' = \frac{(X_g + X_L)}{R_g}$$

Then

$$\frac{P}{P_{max}} = \frac{4R'}{(1 + R')^2 + (X')^2} \equiv r$$

or

$$-(R')^2 - 2R'\left(1 - \frac{2}{r}\right) + (X')^2 + 1 = 0$$

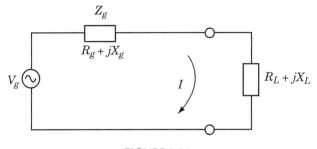

FIGURE 5.50

Complete the square:

$$\left\{R' + \left(1 - \frac{2}{r}\right)\right\}^2 + (X')^2 = 2\frac{(r-1)^2}{r^2}$$

which we recognize as a family of circles with centers and radii of

Centers
$$\left(\frac{2}{r} - 1\right)$$

Radii
$$\frac{\sqrt{2}}{r}(r - 1)$$

as shown in Figure 5.51.

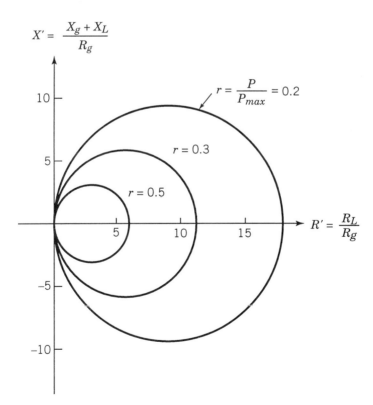

**FIGURE 5.51**

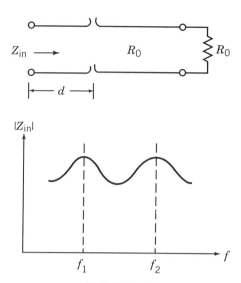

**FIGURE 5.52**

**5.22** Discuss the location of a reflecting point on a TL to the periodic variation of the line's input impedance with frequency.

If the input impedance $Z_{in}$ is plotted vs. frequency for a line with a small discontinuity and matched at the far end as shown in Figure 5.52, the trace will display an almost periodic variation as indicated:
The cause is due to the distance to the discontinuity becoming integral multiples of the wavelength. Therefore, $|Z_{in}|$ must be repeated, since it is periodic (via the tangent functions of which it is composed). The physical distance $d$ to the reflecting point is determined as follows:

$$d = k\lambda_1 = (k + \tfrac{1}{2})\lambda_2 \quad \text{for some } k$$

(recall that $Z_{in}$ is periodic for every half wavelength). Eliminate $k$, using

$$v_1 = f_1\lambda_1, \qquad v_2 = f_2\lambda_2$$

Then

$$d = \frac{v_1 v_2}{2(v_1 f_2 - v_2 f_1)}$$

If $v_1 = v_2$ (as it should), $v_1 = v_2 = v$, and

$$d = \frac{v}{2(f_2 - f_1)}$$

**5.23** Rearrange the basic expression for the net voltage on a line and show that it may be interpreted as a combination of a single traveling wave to the right and a standing wave:

$$V(z) = V^+ e^{-j\beta z} + V^- e^{+j\beta z}$$

Assume the standard situation of a load. Then

$$V(z) = V^+ e^{-j\beta z} + \rho_L e^{j\varphi_L} V^+ e^{j\beta z} \qquad \left(\Gamma_L = \frac{V^-}{V^+} = \rho_L e^{j\varphi_L}\right)$$

Now add and subtract the term $V^+ \rho_L e^{-j\beta z}$:

$$V(z) = V^+(1 - \rho_L) e^{-j\beta z} + V^+ \rho_L e^{j\varphi_L/2}(e^{-j\beta z - j\varphi_L/2} + e^{j\varphi_L/2 + j\beta z})$$

Let $V^+ = |V^+| e^{j\theta}$. Some arranging shows

$$\mathscr{V}(z,t) = \operatorname{Re}\left\{|V^+|\left[(1-\rho_L)e^{j(\omega t - \beta z - \theta)} + \rho_L e^{j(\omega t + \theta + \varphi_L/2)} 2\cos\left(\beta z + \frac{\varphi_L}{2}\right)\right]\right\}$$

$$= \underbrace{|V^+|(1 - \rho_L)\cos(\omega t - \beta z - \theta)}_{\text{Traveling to the right} \rightsquigarrow}$$

$$+ \underbrace{2|V^+|\rho_L \cos\left(\beta z + \frac{\varphi_L}{2}\right)}_{\text{Standing}} \underbrace{\cos\left(\omega t + \theta + \frac{\varphi_L}{2}\right)}_{\text{No "}\beta z\text{"}\atop \text{term here}}$$

The first term is recognized as a wave moving to the right. The second is a standing wave (the argument of the cosine does not contain a $\beta z$ term), which is the result of two counterstreaming waves. This indicates the variety of "representations" that are possible when interpreting the sinusoidal steady-state response.

**5.24** For the rather unique situations shown in Figure 5.53, find the input impedance. One is cautioned about such connections in the real world, as radiation may start to creep in rendering our equations even more approximate than we might expect.

(a) $\qquad\qquad c = \lambda$

## 242 SINUSOIDAL STEADY STATE

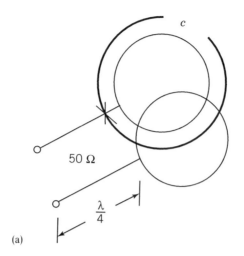

(a)

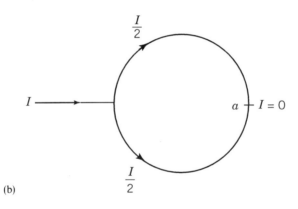

(b)

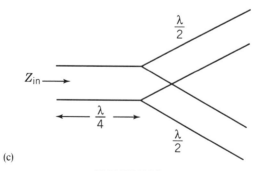

(c)

**FIGURE 5.53a–c**

SUPPLEMENTARY EXAMPLES  243

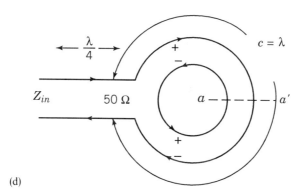

(d)

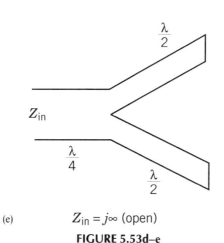

(e)  $Z_{in} = j\infty$ (open)

**FIGURE 5.53d–e**

By symmetry, the current at the junction splits equally:
Then $I = 0$ at point $a$ (hence, an open):
Thus,

$$\therefore \mathbf{Z}_{in} = 0 \quad \text{(short)}$$

**(b)** $\qquad c = \lambda.$

Extend the normal line currents onto the outer loop. Sketch next the corresponding currents in the inner loop. Assume power moving away from the junction; then voltage polarity is as shown. This means a short appears at plane $a - a'$.

## PROBLEMS

**5.1** Show that the general TL problem may be summarized as follows in Figure P5.1.

$$V_1 = \frac{V_s}{2}\left[1 - \frac{\Gamma_s - \Gamma_L e^{-j2\theta}}{1 - \Gamma_s \Gamma_L e^{-j2\theta}}\right]$$

$$V_2 = \frac{V_s}{2}\left[\frac{(1-\Gamma_s)(1+\Gamma_L)e^{-j\theta}}{1 - \Gamma_s \Gamma_L e^{-j2\theta}}\right]$$

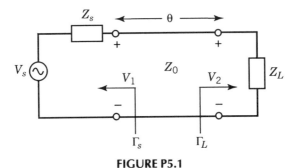

**FIGURE P5.1**

**5.2** Plot the constant power circle to the load in Figure P5.2 for $r = 0.5$, as discussed in Supplementary Example 5.21:

Determine a load that can absorb this value.

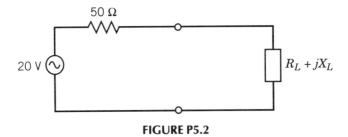

**FIGURE P5.2**

**5.3** For two counterstreaming waves of the form:

$$\mathscr{V}^+(z,t) = 10\cos(\omega t - \beta z + 20°)$$
$$\mathscr{V}^-(z,t) = 7\cos(\omega t + \beta z + 60°)$$

**(a)** Calculate the average power flow in each direction ($R_0 = 50\,\Omega$).

(b) Sketch the voltage standing wave pattern.
(c) What is the VSWR?
(d) What are the complex amplitudes of each wave?

**5.4** For a TL with parameters

$$R = 0.2\,\Omega/m, \qquad G = 20\,\mu\mho/m$$
$$L = 1\,\mu H/m, \qquad C = 150\,pF/m$$

Find $Z_0$, $\alpha$, $\beta$, $\gamma$, $v_p$, the electrical length $\theta$ for $l = 50$ ft.

**5.5** For a line with $Z_0 = 35\,\Omega$, $S = 2.2$, and the maximum in the voltage standing wave pattern $\lambda/16$ from the load, find the load impedance.

**5.6** Develop the lumped equivalent network at the plane $a - a'$, in Figure P.5.6 Assume $\omega = 2\pi \times 10^{10}$ rad/s:

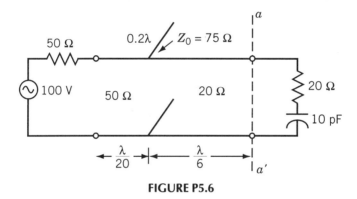

**FIGURE P5.6**

**5.7** What is the available power at the plane $a - a'$ in the previous problem?

**5.8** A TL with parameters

$$R = 1\,\Omega/m, \qquad G = 25\,m\mho/m$$
$$L = 10\,\mu H/m, \qquad C = 200\,pF/m$$

is 100 feet long. A generator with open circuit voltage 45 V (peak) and internal impedance 75 Ω at 5 MHz is driving the line that is terminated in 75 Ω. What power is absorbed in the load. Explain all assumptions.

**5.9** Using Figure P5.9, find the reflection coefficients $\Gamma_1$, $\Gamma_2$, $\Gamma_3$.

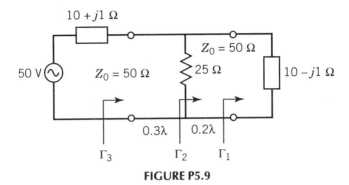

**FIGURE P5.9**

**5.10** Find the input impedances at the planes where $\Gamma_2$ and $\Gamma_3$ are defined in the previous problem.

## REFERENCES

1. Shive, J. N., *Similarities in Wave Behavior*, New Jersey. Bell Telephone Laboratories, Inc., 5th printing, May 1973. (An educational film, upon which the publication is based may be available.)
2. Vernon, R. J. and Seshadri, S. R., "Reflection Coefficient and Reflected Power on a Lossy Transmission Line," *IEEE Proc.*, **57**, Jan. 1969, pp. 101–102.

CHAPTER SIX

# The Smith Chart

## 6.1 MOTIVATION

The previous chapter dealt with the analytical steps used to solve TL problems. After developing many example solutions, it became clear that the calculations of $\Gamma_{in}$ and $Z_{in}$ were among the most repetitive steps. The Smith chart (a graphical calculator) was developed in the 1930s to alleviate some of the calculation drudgery. With reasonable care, the chart could be used to obtain results accurate to a few percent. With the advent of pocket calculators and PCs, the need for accuracy from the chart has diminished; however, the need to understand the chart is still the same. Unlike the slide rule that has taken its place among obsolete tools, the Smith chart will probably have enduring usefulness due to the unique visual insights it provides.

With pocket calculators, PCs, and all the available commercial software, it may come as a surprise that the percentage of pages containing Smith charts in the technical literature has not declined. For example, in the *IEEE Transactions on Microwave Theory and Techniques*, the percentage from the years 1963 to 1993 is around 2.5 to 3% when averaged over a few consecutive years. In many microwave design short courses offered several times a year throughout the country, a substantial amount of time is given in the early session to the chart. About 29% of all overheads used in the courses employ the chart. In other words, the CAD results developed are most easily displayed on the Smith chart. As a final example in the text by Gonzalez (the design of microwave transistor amplifiers; see Ref. 1), approximately 30% of the pages contain a chart. The reason for this state of affairs is that the chart is the "roadmap" for many TL discussions, and as such, considerable attention is given it in this chapter.

The Smith chart is shown in Figure 6.1, and it has been rumored (in dark hallways) that it can be used (only by experienced and deft-fingered individuals) to actually prepare a snack or a full meal. All one needs, of course, is a microwave oven!

## IMPEDANCE OR ADMITTANCE COORDINATES

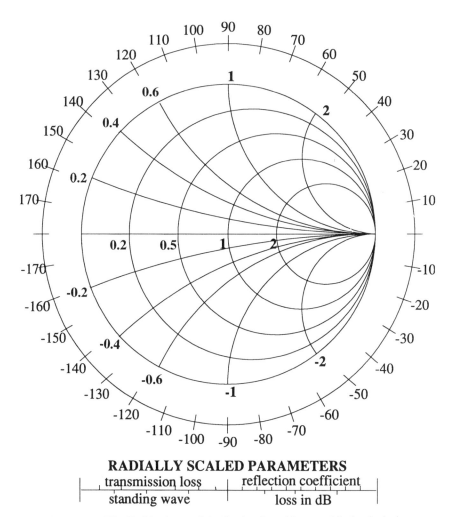

**FIGURE 6.1** The Smith chart originally developed for graphical calculations.

### 6.2 BASIS FOR THE SMITH CHART

Essentially, the chart is the complex impedance plane mapped onto the complex reflection coefficient plane. It gives reflection coefficient and impedance at a glance, and allows frequency variation to be seen clearly. As a matter of fact, the HP 8510 series network analyzers have the Smith chart display as one of the primary screen options. With reference to Figure 6.2, we can define both a reflection coefficient and input impedance $[\Gamma(z), Z_{in}(z)]$ at any point along the

## BASIS FOR THE SMITH CHART

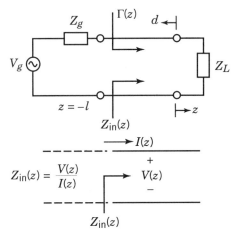

**FIGURE 6.2** The notation for the reflection coefficient $\Gamma(z)$ and input impedance $Z_{IN}(z)$ at any point along the line.

line. Recall (assume $\alpha = 0$ unless otherwise stated) the line voltage,

$$V(z) = V^+ e^{-j\beta z} + V^- e^{j\beta z} \tag{6.1}$$

and define the reflection coefficient as

$$\Gamma(z) = \frac{V^- e^{j\beta z}}{V^+ e^{-j\beta z}} = \frac{V^-}{V^+} e^{+j2\beta z} \tag{6.2}$$

At the load, we have

$$\Gamma(z=0) = \Gamma_L = \frac{V^-}{V^+}$$

so

$$\Gamma(z) = \Gamma_L e^{j2\beta z}$$

Now at some point on the line $z = -l$, we have

$$\Gamma(-l) = \Gamma_L e^{-j2\beta l} \tag{6.3}$$

which is one of the equations on which the chart is based. Physically, it says the reflection at some distance $l$ from the load has the same magnitude as that at the load, but the phase is changed by subtracting $2\beta l$ radians.

Recall from Eq. (5.27) that

$$Z_{in}(-l) = Z_0 \frac{1 + \Gamma(-l)}{1 - \Gamma(-l)} \tag{6.4}$$

which relates input impedance to reflection coefficient. Rewrite this as

$$\Gamma(-l) = \frac{Z_{in}(-l) - Z_0}{Z_{in}(-l) + Z_0} \tag{6.4a}$$

which is the basic relationship for the chart. The first thing to note is that the connection between $\Gamma(-l)$ and $Z_{in}(-l)$ is tied up with $Z_0$ of the particular line. To eliminate the $Z_0$ dependence, we define

$$\bar{Z}_{in}(-l) = \frac{Z_{in}(-l)}{Z_0} \tag{6.5}$$

where $\bar{Z}_{in}(-l)$ is a normalized impedance. Then we have

$$\Gamma(-l) = \frac{\bar{Z}_{in}(-l) - 1}{\bar{Z}_{in}(-l) + 1} \tag{6.6}$$

from which the chart is constructed. Consider the left side, $\Gamma(-l)$, first. The reflection coefficient is a complex number and can have a magnitude from 0 to 1. If we define a reflection coefficient plane as shown in Figure 6.3a, all possible values of $\Gamma(-l)$ must lie inside the unit circle. Thus, only the passive loads for $Z_L$ are considered for now. Now write (let $z$ be any point on the line)

$$\Gamma(z) = \Gamma_r + j\Gamma_i \tag{6.7}$$

$$\bar{Z}_{in}(z) = r + jx \tag{6.8}$$

and Eq. (6.6) becomes

$$\Gamma_r + j\Gamma_i = \frac{r + jx - 1}{r + jx + 1} \tag{6.9}$$

Now equate real and imaginary parts (see Supplementary Example 6.2) to find

$$r = \frac{1 - \Gamma_r^2 - \Gamma_i^2}{(1 - \Gamma_r)^2 + \Gamma_i^2} \tag{6.10a}$$

$$x = \frac{2\Gamma_i}{(1 - \Gamma_r)^2 + \Gamma_i^2} \tag{6.10b}$$

or

$$\left(\Gamma_r - \frac{r}{1+r}\right)^2 + \Gamma_i^2 = \left(\frac{1}{1+r}\right)^2 \tag{6.10c}$$

$$(\Gamma_r - 1)^2 + \left(\Gamma_i - \frac{1}{x}\right)^2 = \left(\frac{1}{x}\right)^2 \tag{6.10d}$$

## BASIS FOR THE SMITH CHART

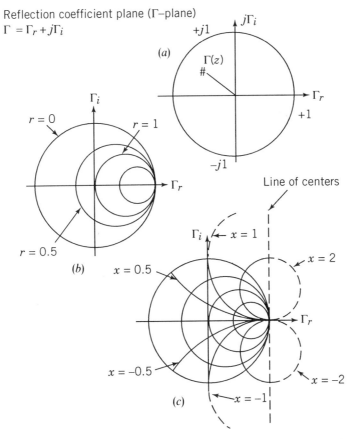

**FIGURE 6.3** The unit circle in the complex $\Gamma$-plane, along with constant $r$ circles and constant $x$ circles. In the top sketch, all values of $\Gamma(z)$ lie inside the unit circle.

We notice these are equations of circles in the $\Gamma$ plane. The first is a family with centers along the $\Gamma_r$ axis, and a few are shown in Figure 6.3b. The second represents a family with centers along the line $\Gamma_r = 1$, see Figure 6.3c. Only the portions of the circles inside the $|\Gamma| = 1$ circle are relevant and are drawn in solid arcs. Those portions outside the unit circle are shown as dotted. Notice that all possible values for $r$ and $x$ are permissible ($0 \leqslant r \leqslant \infty$), ($-\infty \leqslant x \leqslant \infty$) in the above equations; thus, the entire normalized impedance plane ($\bar{Z}_{in}$ plane) is plotted inside the circle. Figure 6.4 shows both sets of circles plotted inside the $|\Gamma| = 1$ circle, which is the Smith chart. Formally, Eq. (6.9) is a mapping between two complex variables, $\bar{Z}_{in} = r + jx$ and $\Gamma = \Gamma_r + j\Gamma_i$. Figure 6.4 shows that this mapping transforms all points in the $\bar{Z}_{in}$ plane into the unit circle of the $\Gamma$ plane. The $x$ axis in the $\bar{Z}_{in}$ plane is mapped into the $|\Gamma| = 1$ circle. The $r$ axis is mapped

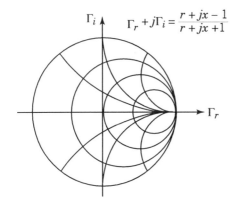

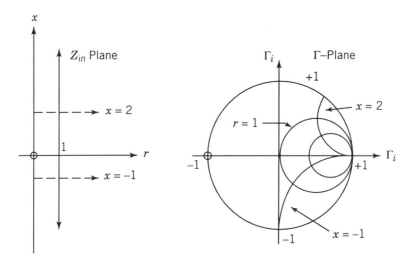

**FIGURE 6.4** Contours of constant $r$ and $x$ in the mapping given in Eq. (6.9). The lower portion shows the $r = 1$ vertical line in the $Z_{IN}$ plane and the circle into which it maps in the $\Gamma$ plane. Similar sketches for $x = 2$ and $-1$ are also shown.

into the interval $-1 \leq \Gamma_r \leq +1$. The $r = $ constant vertical lines map into circles with centers on the $\Gamma_r$ axis; horizontal lines ($x = $ constant) map into arcs of circles. The transformation is conformal (preserves angles), so the corners of a rectangle in the $\bar{Z}_{in}$ plane remain at 90° when the rectangle is mapped into the $\Gamma$ plane. Impedance vectors $r + jx$ (straight lines) map into curved lines as shown in Figure 6.5.

The Smith chart is therefore the mapping of the complex normalized impedance plane onto the reflection coefficient plane. It graphically solves Eq. (6.9). In Figure 6.5b, we sketch a reflection coefficient in the $\Gamma$ plane; the polar angle is

$\varphi$ that is measured from the horizontal. This line terminates at the intersection of the $r_1$ and $x_1$ circles. These values satisfy Eq. (6.9), that is,

$$\Gamma_{r_1} + j\Gamma_{i_1} = \frac{r_1 + jx_1 - 1}{r_1 + jx_1 + 1}$$

Thus if one plots a reflection coefficient, its end point corresponds to $\bar{Z}_{in}$ at that point, which can be read directly off the chart. Alternatively, plotting a $\bar{Z}_{in}$ point permits the sketching of the reflection coefficient; draw a line from the center of the chart to the plotted point. The length is the magnitude of $\Gamma$, and the angle $\varphi$ is shown in Figure 6.5b.

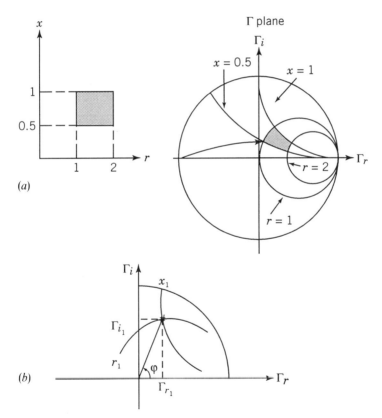

**FIGURE 6.5** (a) A region of the complex $Z_{IN}$ plane and its shape and position in the complex $\Gamma$ plane. Also shown is a vector $(r + jx)$ and its curved appearance on the chart. (b) Below is a detail of the two sets of coordinates $(r, x)$, $(\Gamma_r, \Gamma_i)$ for a given "point", which may be interpreted as either a point for the reflection coefficient, or for a normalized impedance.

## 6.3 BASIC CHART CALCULATIONS

The primary calculation with the chart is the determination of the input impedance $Z_{in}(-l)$, given $Z_L$ and the line $Z_0$ and length. It is convenient to view the chart as just a unit circle with a plastic overlay that contains the normalized impedance "circles." Then a point plotted on either sheet is immediately visible on the other. We move on either sheet depending on the particular application and read values from both. The idea is to enter, move, and read in the easiest possible sequence. For example, to find $Z_{in}(-l)$ given the load $Z_L$, line impedance $Z_0$, and length, the steps are:

1. Calculate $\bar{Z}_L = Z_L/Z_0$ and plot.
2. Draw $|\Gamma|$ (line from center to the $\bar{Z}_L$ point just plotted).
3. Rotate this line clockwise (moving toward the generator) by $2\beta l$ degrees.
4. After this rotation, the line segment is terminated on the point corresponding to $\bar{Z}_{in}(-l)$.
5. Find $Z_{in}(-l)$ by "denormalizing": $Z_{in} = \bar{Z}_{in}Z_0$.

Notice we "enter" via the normalized $\bar{Z}_{in}$ grid, then immediately drop down to the lower surface ($\Gamma$ plane) to perform the rotation that connects the values of $\Gamma(z)$ at two points. After the rotation on this plane, we read the grid of the "upper" or overlay plane, that is, the $\bar{Z}_{in}$ plane.

If $Z_{in}$ at the source is known, the value at the load is found as follows:

1. Normalize (i.e., $\bar{Z}_{in} = Z_{in}/Z_0$) and plot $\bar{Z}_{in}$.
2. Rotate counterclockwise (toward load) by $2\beta l$ degrees.
3. Read $\bar{Z}_L$; then "denormalize" to find $Z_L = \bar{Z}_L Z_0$.

The corresponding analytical steps to find $Z_{in}$ given $Z_L$ and line length $l$ are:

1. Find $\bar{Z}_L = \dfrac{Z_L}{Z_0}$

2. Find $\Gamma_L = \dfrac{\bar{Z}_L - 1}{\bar{Z}_L + 1}$

3. Find $\Gamma_{in} = \Gamma_L e^{-j2\beta l}$

4. Find $\bar{Z}_{in} = \dfrac{1 + \Gamma_{in}}{1 - \Gamma_{in}}$

5. Find $Z_{in} = \bar{Z}_{in} Z_0$

The above graphical and analytical steps are really the heart of all Smith chart operations. In summary, the chart relates $\bar{Z}_L$, $\bar{Z}_{in}(-l)$, and $l$. Thus, given two of the three, the third can be determined quickly. Due to the periodicity, however, $\beta l$ can only be determined to within $n\lambda/2$, which is not of real concern (only crops up on exam problems).

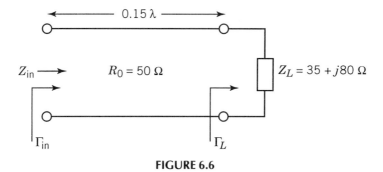

**FIGURE 6.6**

***Example 6.1*** Find $Z_{in}$ using the Smith chart and Figure 6.6.

The sequence of steps is as follows:

1. $\bar{Z}_L = \dfrac{35 + j80}{50} = 0.7 + j1.6$

   plot this as shown below in Figure 6.7.
2. Draw a line from the center through $\bar{Z}_L$ and extend to the periphery. We immediately have the angle of $\Gamma_L$; it is $\cong 57.2°$. The magnitude of $\Gamma_L$ can be found using the upper-most scale on the lower right as shown. Using mechanical dividers on the scale, we measure

$$|\Gamma_L| = 0.691$$

The last digit is, of course, interpolated. We should always agree to the second decimal point, if not, one of us has made an error. Now we have $\bar{Z}_L$ and $\Gamma = 0.691 \angle 57.2°$. Let us check ourselves by using the equation on step 2 of the previous section:

$$\Gamma_L = \frac{\bar{Z}_L - 1}{\bar{Z}_L + 1} = \frac{0.7 + j1.6 - 1}{0.7 + j1.6 + 1} = 0.376 + j0.5872$$

$$= 0.6973 \angle 57.36°$$

The errors, in percent, are

$$|\Gamma_i|; \quad \frac{0.6973 - 0.691}{0.6973}(100) = 0.9 \simeq 1\%.$$

$$\angle \Gamma_L; \quad \frac{57.36 - 57.2}{57.36}(100) = 0.279 \sim 0.3\%$$

which is about the range one should expect.

## IMPEDANCE OR ADMITTANCE COORDINATES

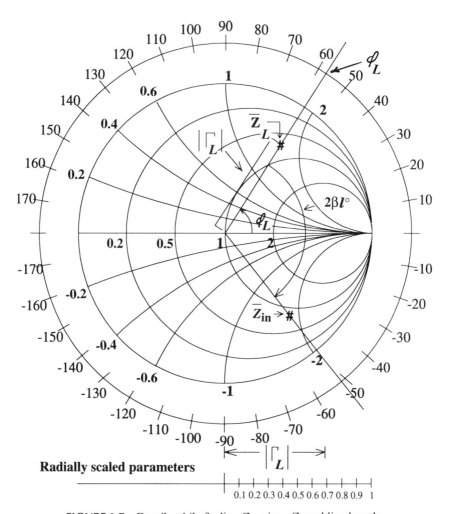

**FIGURE 6.7** Details while finding $Z_{IN}$ given $Z_L$ and line length.

3. The line is $0.15\lambda$; then

$$\beta l = \frac{2\pi}{\lambda}\left(\frac{15\lambda}{10}\right) = 0.3\pi = 54°$$

Less formally, just say $l = 0.15\lambda$ and since $\lambda \Leftrightarrow$ (corresponds to 360°), $0.15\lambda = 0.15(360°) = 54°$. Next rotate clockwise $2\beta l°$ (using dividers to maintain the length of Γ) (108°) and stop at

$$\frac{\begin{array}{r} 57.2° \\ -108.0° \end{array}}{-50.8°}$$

Now $\Gamma_{in} = 0.691 \angle -50.8°$.

4. Now read the $\bar{Z}_{in}$ grid and find

$$\bar{Z}_{in} = 0.86 - j1.78$$

5. Remove the normalization:

$$Z_{in} = \bar{Z}_{in}R_0 = \bar{Z}_{in}(50) = 43 - j89 \, \Omega$$

The analytical result gives

$$Z_{in} = 50\frac{(35 + j80) + j50 \tan 54°}{50 + j(35 + j80) \tan 54°}$$

$$= 42.68 - j89.58$$

which shows the accuracy one can expect in some cases. However, when $|\Gamma_L|$ is large, it becomes apparent that considerable error occurs near $\angle \Gamma_L \simeq 0°$, since that portion of the chart is so compact. ▲

**Example 6.2** The measured input impedance is $Z_{in} = 600 - j30 \, \Omega$. Find the load impedance $Z_L$. See Figure 6.8.

Find

$$\bar{Z}_{in} = \frac{600 - j30}{50} = 12 - j0.6$$

Now rotate $2\beta l = 400°$, so move only $(400° - 360°) = 40°$ on the angle Γ scale; see

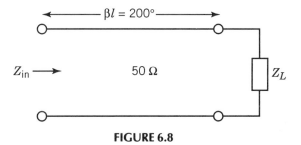

**FIGURE 6.8**

## IMPEDANCE OR ADMITTANCE COORDINATES

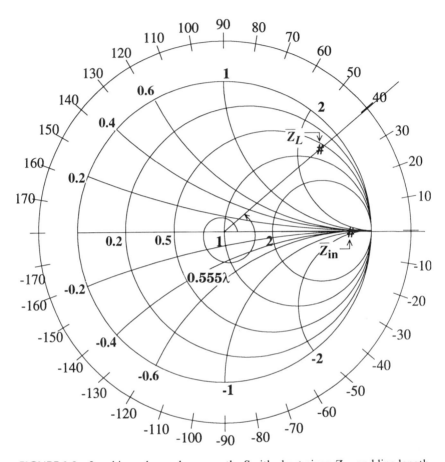

**FIGURE 6.9** Load impedance shown on the Smith chart given $Z_{IN}$ and line length.

Figure 6.9. Read

$$\bar{Z}_L = 0.69 + j2.62$$

Remove the normalization

$$Z_L = \bar{Z}_L(50) = 34.5 + j131 \, \Omega$$

You have noticed the dual use of the symbol $\lambda$ so far in the chapter. When $\lambda$ is given in units of meters, then $0.15\lambda$ means a certain physical length. However, in other cases, $\lambda$ represents the angular equivalence of 360°; then $0.15\lambda$ corresponds to 54°.

BASIC CHART CALCULATIONS 259

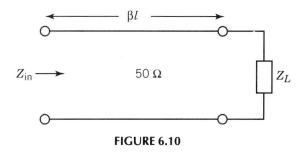

**FIGURE 6.10**

## IMPEDANCE OR ADMITTANCE COORDINATES

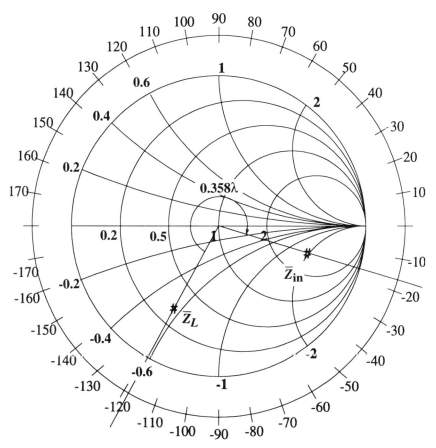

**FIGURE 6.11** Finding line length given $Z_{IN}$ and $Z_L$.

**Example 6.3** For the TL shown in Figure 6.10, $Z_L = 15 - j27.5$, $Z_{in} = 160 - j90$. Find the length of the line.
Calculate

$$\bar{Z}_L = \frac{15 - j27.5}{50} = 0.3 - j0.55$$

$$\bar{Z}_{in} = \frac{160 - j90}{50} = 3.2 - j1.8$$

The movement in Figure 6.11 is from 0.415 to 0.273 on the WTG (wavelengths toward generator) scale. Since we cross the zero of the scale, we have

$$\begin{array}{rr} 0.500 & 0.085 \\ -0.415 & +0.273 \\ \hline 0.085 & 0.358 \end{array}$$

Then

$$\beta l = 0.358\lambda + n\frac{\lambda}{2}, \quad n = 0, 1, 2, \ldots$$

since the input impedance is periodic with $\lambda/2$. ▲

## 6.4 PROPERTIES OF THE SMITH CHART

The properties of the chart are probably uncountable since we have a complex plane unit circle with a rectangular half-plane mapped onto it. Many geometric constructions similar to those performed in high school geometry are possible—for example, one may determine the sine or cosine of an angle (Ref. 2). There has also appeared an article concerned with Fermat's theorem on algebra and a possible interpretation of it with voltage vectors on the chart (Ref. 3). In addition, an article concerning the TL equations and the Lorentz transformations has been published (Ref. 4), and if one is so inclined (having a lot of spare time to waste), one could propose a method to chart a trip to some far-off galaxy using Smith chart constructions and relating them to space and time intervals. Here, we will be satisfied with just the simple constructions used to design microwave networks.

First, inspect the scales near the periphery of the chart. The scale closest to the $|\Gamma| = 1$ circle is that for the angle of $\Gamma$ ($\angle \Gamma$). Note that this scale is outside of the unit circle by a small amount, and its origin is on the right side as is customary for polar coordinates. The angle measure increases to $\pm 180°$ in the counterclockwise and clockwise fashions, respectively. The next ring with tick marks has both of the "wavelengths" scales. Both of these have their common origin on the left

PROPERTIES OF THE SMITH CHART   261

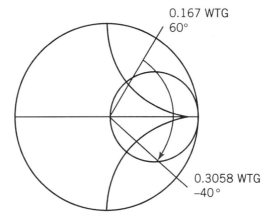

$$\frac{50°}{360°} = 0.1389$$

−40° (corresponds to) 0.3058 on WTG scale

**FIGURE 6.12** Moving from load to source plane using the wavelengths toward generator (WTG) scale.

side of the chart (adjacent to the $\angle \Gamma = \pm 180°$ point). The outer scale is the "wavelengths toward generator," whereas the inner one is the "wavelengths toward load." Henceforth, we will refer to these as WTG and WTL scales, respectively. With reference to Figure 6.12, we explain their use. Assume that the load has a reflection coefficient angle of $+60°$ as shown. The angle of $\Gamma_{in}$ will be $(60° - 2\beta l°)$, and for the example, this amounts to $-40°$. Notice that we have moved a physical distance of $l$ meters along the line. The fractions of wavelength corresponding to this movement are

$$\frac{50°}{360°} = 0.1389$$

Notice that the value on the WTG scale along the radial direction of $\Gamma_L$ is 0.167. If

**262** THE SMITH CHART

we add the fractions of wavelength traveled (0.1389), we obtain 0.3059. This is the value on the WTG scale that is opposite the $-40°$ on the $\angle \Gamma$ scale at the radial position corresponding to the source plane. Thus, if we move $l$ meters on the TL, it corresponds to a certain fraction of wavelengths. This fraction is just $l/\lambda$. We know, however, that $\Gamma(z)$ repeats every half wavelength, so if we construct a scale about the chart, it should show 0.5 wavelengths traversed for each complete revolution. Notice that the "wavelengths" scales are constructed in this manner. Therefore, when moving along the line by one half wavelength, one completes a full revolution on the chart. This at first can be confusing, because on the one hand, we have moved a half wavelength while on the other (with respect to the "geometry" of the chart), we have rotated about the chart once and thus rotated 360°, which corresponds to one complete wavelength. The point to remember is that for each degree traveled "linearly" along the line, one "rotates" twice that number on the chart. This is due to the fact that the reflection coefficient is periodic along the line with a space period of one

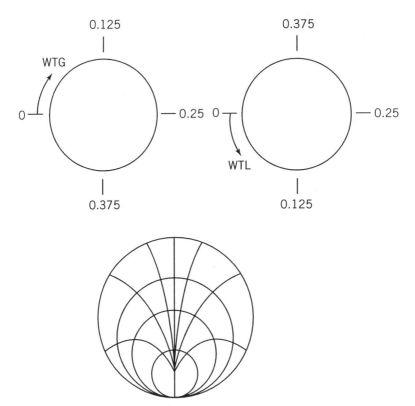

**FIGURE 6.13** The two "wavelengths" scales. That toward the generator is on the left, whereas that toward the load is on the right. The lower sketch showing the chart with the real axis vertical is now largely obsolete.

half wavelength. But when plotted in polar form, it must be periodic for one complete revolution.

Figure 6.13 shows an empty chart with just the wavelengths scales indicated so that one can see clearly their relationship. The direction one travels on the line determines which scale to use. The choice for the zero position for the wavelengths scales being on the left side has been questioned (Ref. 6), as it may be more logical to place it on the right so as to line up with that for $\angle \Gamma$. A more objectional practice may be the use of the chart while oriented such that the real axis ($r$ axis) is vertical. This practice (see Fig. 6.13) is now fairly obsolete. Reference 7 gives a historical review of the various charts used in TL calculations. Reference 5 discusses the question of whether the Smith chart will cease to be used, as well as developing it from equations. Many of the remaining references included in the (Bibliography) are to indicate the wide range of topics and viewpoints concerning graphical techniques for the TL equations. The book by P. H. Smith himself (Ref. 8) indicates that indeed, complete volumes can be devoted to the subject.

The remaining properties we will find useful are listed below:

1. Once the normalized load impedance is plotted, one can infer all values of impedance along the line. This is because the locus for $\bar{Z}_{in}$ on a lossless line is a circle shown in Figure 6.14. Thus, all values for $\bar{Z}_{in}$ lie on the particular circle, which we call the $\bar{Z}_{in}$ circle.

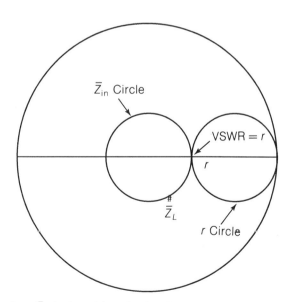

**FIGURE 6.14** Once $\bar{Z}_L$ is plotted for a lossless line, all possible values of $\bar{Z}_{IN}(-l)$ along the line are those on the $\bar{Z}_{IN}$ circle. The $r$ circle that is tangent on the right side is numerically equal to the VSWR on the line.

**264** THE SMITH CHART

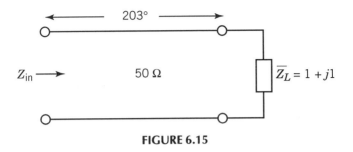

**FIGURE 6.15**

## IMPEDANCE OR ADMITTANCE COORDINATES

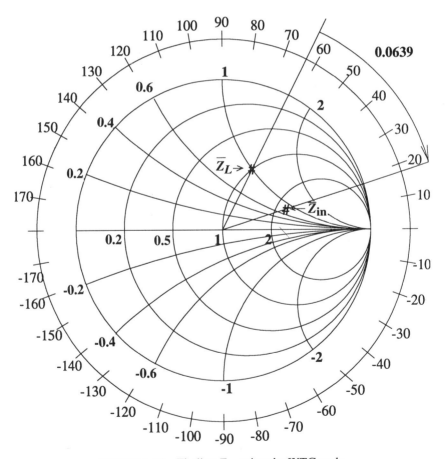

**FIGURE 6.16** Finding $Z_{IN}$ using the WTG scale.

2. The circle mentioned above permits one to determine the VSWR ($S$) on the line. It is numerically equal to the constant $r$ circle that is tangent to the $\bar{Z}_{in}$ circle, as seen in Figure 6.14.
3. The $\bar{Z}_{in}$ circle intercepts with the $\Gamma_r$ axis occur at the extrema of the standing wave pattern. This is seen in the expression for $|V(z)|$, since the maximum and minimum values occur for $\angle \Gamma = 0°$ and $180°$, respectively. Since the current standing wave is shifted $\lambda/4$ from that for voltage, $I(z)_{max}$ occurs where $V(z)_{min}$, and $I(z)_{min}$ occurs for $V(z)_{max}$.
4. The input admittance $\bar{Y}_{in} = 1/\bar{Z}_{in}$ is determined by inverting the vector for $\bar{Z}_{in}$ through the center of the chart.
5. It turns out that circles in the rectangular normalized impedance plane map into circles on the Smith chart, and vice-versa.
6. Other properties will become evident in later sections of the chapter.

***Example 6.4*** Find $Z_{in}$ using the Smith chart and Figure 6.15.
The length in wavelengths is

$$\frac{203}{360} = 0.56389$$

This distance is greater than 0.5 units, so in essence, one is rotating about the chart one complete revolution, and then the additional distance of $0.56389 - 0.5 = 0.06389$. Now plot $\bar{Z}_L$ and note that it is along the 0.162 point on the WTG scale. See Figure 6.16. Now add 0.06389 to the above:

$$0.162 + 0.06389 = 0.22589$$

which says we rotate the radial line to this stopping point. Keeping the dividers at the same separation as was found for $\bar{Z}_L$, we read $\bar{Z}_{in}$, which is $2.25 + j0.77$. Now remove the normalization:

$$Z_{in} = 50\bar{Z}_{in} = 112.5 + j38.5 \, \Omega \qquad \blacktriangle$$

## 6.5 THE ADMITTANCE CHART

So far, we have shown that the Smith chart is the mapping of the right half of the normalized impedance plane ($r, jx$) onto the voltage reflection coefficient $\Gamma^V$ plane. Similar steps show that the right half of a complex admittance plane ($g, jb$) maps onto the current reflection coefficient $\Gamma^I$ plane. Therefore, the Smith chart is either the voltage or current reflection coefficient plane with the impedance or admittance plane mapped on it. For example, suppose we have plotted $\bar{Z}_{in} = 1 + j2$ that has $\Gamma^V = 0.7 \angle 45°$. By reflecting this through the origin as shown in Figure 6.17, we obtain $\bar{Y}_{in} = 0.2 - j0.4$ that has $\Gamma^I = 0.7 \angle -135°$. We

## IMPEDANCE OR ADMITTANCE COORDINATES

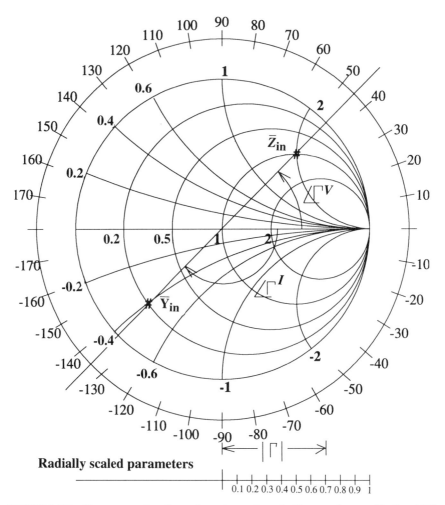

**FIGURE 6.17** The conversion from impedance to admittance along with the 180° change from voltage to current reflection coefficient. Here, $\Gamma^V = 0.7\underline{/45°}$, $\Gamma^I = -\Gamma^V = 0.7\underline{/-135°}$.

can summarize the interrelationships between treating a given problem on either an impedance or admittance basis by referring to Figure 6.18. On the left, impedances are used, whereas admittances are used on the right. The same piece of paper that is, "grid," is used on both sides. Once $\bar{Y}_L$ is plotted, all the constructions and peripheral scales are the same, except now the grid yields

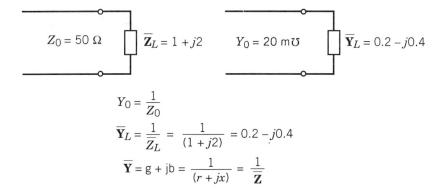

$$Y_0 = \frac{1}{Z_0}$$

$$\bar{Y}_L = \frac{1}{\bar{Z}_L} = \frac{1}{(1+j2)} = 0.2 - j0.4$$

$$\bar{Y} = g + jb = \frac{1}{(r+jx)} = \frac{1}{\bar{Z}}$$

Same grid

$$|\Gamma^V| = |\Gamma^I|$$

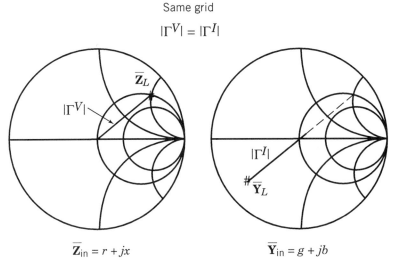

$\bar{Z}_{in} = r + jx$ $\qquad\qquad$ $\bar{Y}_{in} = g + jb$

**FIGURE 6.18** The conversion from impedance to admittance format for a specific problem. The normalized value $\bar{Z}_L$ and $\bar{Y}_L$ are reciprocals, which are obtained by reflection through the center of the chart.

values for $\bar{Y}_{in} = g + jb$.

$$Y_0 = \frac{1}{Z_0}$$

$$\bar{Y}_L = \frac{1}{\bar{Z}_L} = \frac{1}{1+j2} = 0.2 - j0.4$$

$$\bar{Y} = g + jb = \frac{1}{r+jx} = \frac{1}{\bar{Z}}$$

SAME GRID

$$|\Gamma^V| = |\Gamma^I|$$

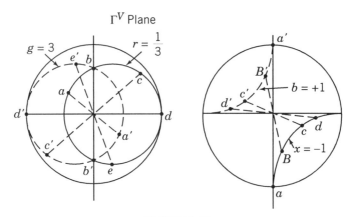

**FIGURE 6.19**

***Example 6.5*** Demonstrate the conversion from impedance to admittance viewpoint both analytically and graphically.

First, note that the corresponding reflection coefficients are

$$\Gamma^I = \frac{Y_L - Y_0}{Y_L + Y_0} = \frac{1/Z_L - 1/Z_0}{1/Z_L - 1/Z_0} = -\frac{Z_L - Z_0}{Z_L + Z_0} = -\Gamma^V$$

Then

$$\bar{Z} = \frac{Z}{Z_0}, \qquad \bar{Y} = \frac{Y}{Y_0} = \frac{G + jB}{Y_0} = g + jb = \frac{1}{\bar{Z}}$$

$$\bar{Y} = \frac{1}{r + jx} = \frac{r - jx}{r^2 + x^2} = \frac{r}{|\bar{Z}|^2} - j\frac{x}{|\bar{Z}|^2} = g + jb$$

$$\therefore g = \frac{r}{|\bar{Z}|^2}, \qquad b = -\frac{x}{|\bar{Z}|^2}$$

Note that $b > 0$ implies capacitive susceptance, whereas $b < 0$ implies inductive susceptance.

Graphically, the admittance chart can be obtained by inverting the impedance chart point by point. Inverting means reflection through the origin (chart center). Thus, in Figure 6.19, which leads to the the admittance chart on the $\Gamma^I$ plane. Notice that in Figure 6.20a we have also reflected the axes.

Now as in Figure 6.20b rotate by $180°$: ▲

***Example 6.6*** Find $Z_{in}$ for the TL in Figure 6.21 using calculations on the admittance chart ($Y$ chart).

## THE ADMITTANCE CHART   269

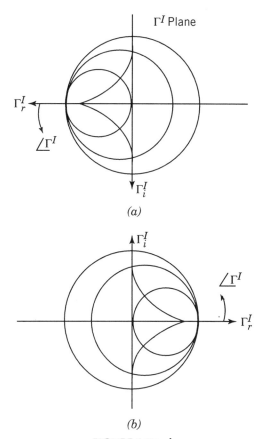

**FIGURE 6.20a, b**

Rotate

$$\frac{70}{360} = 0.194$$

$$0.476 \quad (\text{at } \bar{Y}_L)$$

$$\frac{+0.194}{0.670}$$

$$0.170 \quad \text{on WTG}$$

Read

$$\bar{Y}_{in} = 0.35 + j1.78$$

$$\bar{Z}_{in} = \frac{1}{\bar{Y}_{in}} = 0.106 - j0.54$$

$$\mathbf{Z}_{in} = 5.3 - j27\,\Omega \qquad \blacktriangle$$

**270** THE SMITH CHART

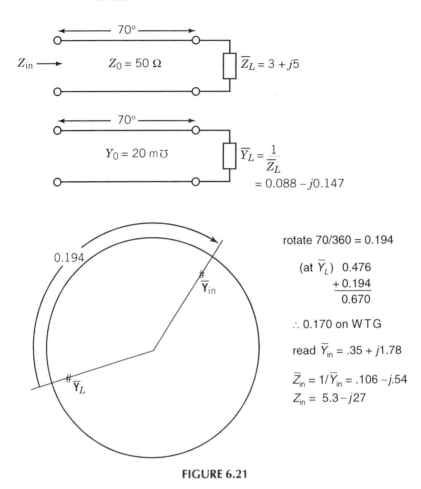

**FIGURE 6.21**

## 6.6 COMBINED Z–Y CHART

The use of the Smith chart as either an impedance or admittance diagram was standard chart practice until about 1970. Then there appeared charts with both $\bar{Z}$ and $\bar{Y}$ coordinates printed on the same sheet of paper. The $\bar{Z}$ coordinates (the old "standard" chart) normally appear in red ink, whereas the $\bar{Y}$ coordinates are in green. Sometimes, this dual-grid format is informally referred to as the Smith and Jones chart. The utility of this configuration will be demonstrated in the next problem. In essence, it allows one to use both $\bar{Z}$ and $\bar{Y}$ charts "simultaneously" on the same sheet of paper.

***Example 6.7*** The Smith and Jones (S and J) chart is useful when ladder networks are studied.

(a) For the network in Figure 6.22, with given normalized reactances, sketch the input impedance and admittance on separate charts.
(b) Now sketch these on the S and J chart.

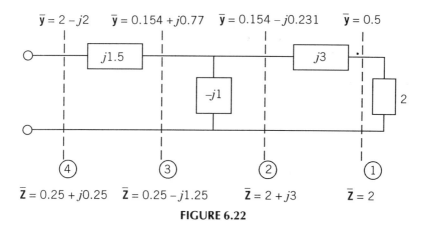

**FIGURE 6.22**

**IMPEDANCE COORDINATES**

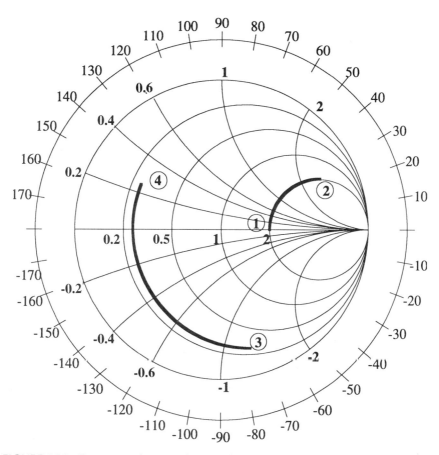

**FIGURE 6.23** Reactance changes when moving between planes ① and ② as well as ③ and ④.

**272** THE SMITH CHART

Notice that when moving between reference planes separated by a series element, we move along constant resistance $r$ circles on the impedance plane. On the other hand, when on the admittance plane, those reference planes separated by shunt elements are connected by movement along constant conductance $g$ circles. These arcs are the mapping of the straight-line impedance and admittance vectors in the rectangular immitance planes. Thus, the combined chart can show this "vector addition" quite easily, which is one of its strong points. Take care to notice that these arcs are not those of earlier sections, where we were constructing circles of constant $|\Gamma|$. Also notice that the network is completely lumped, and no normal TL length effects are considered. As we move from the load to the input to the circuit, we continuously jump from the **Z** to $Y$ planes, so as to use the property that impedances in series add, while admittances in parallel do likewise.

### ADMITTANCE COORDINATES

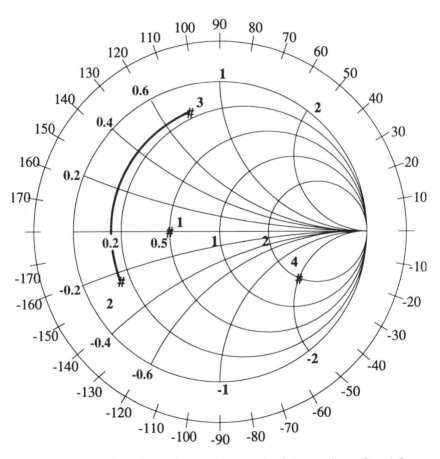

**FIGURE 6.24** Susceptance changes when passing between planes ② and ③.

## NORMALIZED IMPEDANCE & ADMITTANCE COORDINATES

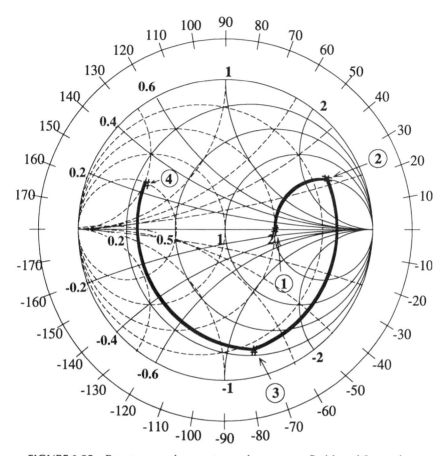

**FIGURE 6.25** Reactance and susceptance changes on a Smith and Jones chart.

▲

## 6.7 SLOTTED LINES

A slotted line is a measurement instrument that permits a probe to sample the standing wave of voltage and thus determine VSWR and the load impedance $Z_L$. The apparatus may be thought of as a coaxial cable with a longitudinal slot in the shield that allows the probe to move along the line. This device is outdated in a practical sense, but may still be part of a microwave laboratory course. One good reason for including it is that it permits one to directly measure the standing wave pattern and thus get a "feel" for the waves. The principle of the apparatus is shown in Figure 6.26. The load is always some distance from the end-of-travel

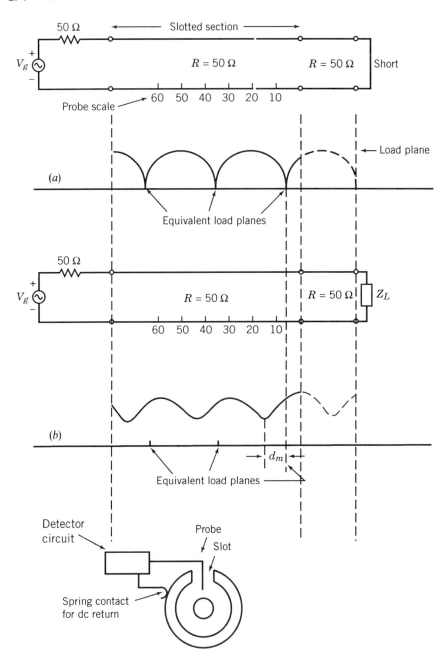

**FIGURE 6.26** A 50-Ω slotted line showing the standing wave patterns for both the calibration step (shorted) and the unknown load $\mathbf{Z}_L$. The lower sketch indicates the coupling probe and detector housing. The minimum of the pattern for the unknown load is $\mathbf{d}_m$ centimeters from an equivalent load plane.

SLOTTED LINES   275

point for the probe, so the standing wave value at the load is necessarily inferred. The low loss air line is treated as lossless, so the standing wave pattern is repeated exactly every half wavelength in the slotted region.

The measurement procedure starts with terminating the load plane with a short. We know that the voltage at this plane must be zero; thus, the minimum (here, zero) of the standing wave pattern is at the load plane. With reference to Figure 6.26a, the points on the slotted section where $|V(z)|$ is zero are planes that are equivalent to the load plane. These planes remain equivalent load planes regardless of the actual load, for they are integral multiples of $\lambda/2$ (in meters) from the actual load plane. This is the calibration step for the apparatus. Next, the unkown load is attached, and a pattern such as that in Figure 6.26b is produced. The load value $Z_L$ is next determined using the Smith chart. First, the VSWR is recorded by the probe since VSWR $= |V_{max}|/|V_{min}|$. From property 2 in Section 6.4, knowing $S$ allows us to sketch the $\bar{Z}_{in}$ circle as shown in Figure 6.27. We know $\bar{Z}_L$ is somewhere on this circle, and we use another property of the standing wave pattern to pin it down. This property is the fact that at a voltage minimum, one is on the left-hand portion of the real axis of the chart. Thus, when the probe is at the minimum position, one is simultaneously crossing the left side of the real axis on the chart (see Fig. 6.27a). We can go to the nearest load plane by moving $d_m/\lambda$ on the WTL scale, since the minimum is $d_m$ centimeters from one of the equivalent load planes. Performing this rotation yields $\bar{Z}_L$ as shown in Figure 6.27b.

The standing wave pattern $|V(z)|$ is sensed by a detector mounted in the probe housing (Fig. 6.28). The detector may be a baretter or crystal diode, both of which are termed "square law" devices. This term comes from the fact that these devices give dc or low-frequency amplitude responses that are proportional to the square of the rf voltage incident on them. Since the rf power is proportional to the square of the voltage, the devices give a response proportional to the incident rf power. A baretter is a thin metal film, the resistance of which changes predictably with heating due to the absorption of microwave power. This resistance change may be used in a bridge, so the final response is proportional to the microwave power.

The probe (regardless of detector type) couples primarily to the $\mathscr{E}$-field, so the response is directly proportional to the time average of $\mathscr{V}(z,t)$ at any position. From Example 5.1 and the end of Section 5.4, we know that the time average of $\mathscr{V}(z,t)$ is the same as $|V(z)|$. Thus, the probe gives a response directly proportional to $|V(z)|$. The probe must couple very lightly to the fields, so as not to disturb them very much. At the minimum positions, the probe tends to have the minimum disturbance. Baretters have rise times in the hundreds of microseconds range, whereas crystal diodes have hundreds of nanosecond values (including housing capacity, etc.). Diodes will be discussed somewhat, since they have many other uses in microwave systems. A general network for a diode used in the square law mode is given in Figure 6.29. The transfer characteristic is

$$v_0(t) = [v_i(t)]^2 \qquad (6.11)$$

## 276 THE SMITH CHART

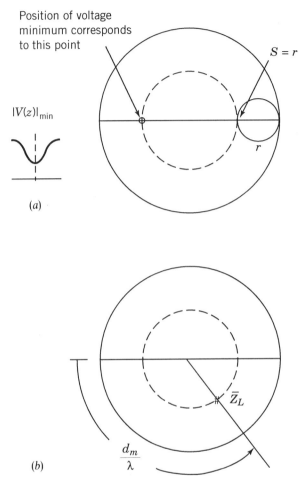

**FIGURE 6.27** From the first chart, knowing the VSWR (S) allows one to sketch the $\bar{Z}_{IN}$ circle on which $\bar{Z}_L$ lies. A reference point on this circle is its crossing of the real axis on the left side of the chart. At that point, one is physically at a minimum of the standing wave pattern. In the second chart, the load $\bar{Z}_L$ is found. Here, we rotate $\mathbf{d}_m/\lambda$ on the WTL scale, starting at the reference point.

which may be developed as follows (Ref. 10). The diode equation is

$$i_D(v_D) = I_R(e^{qv_D/nkT} - 1) \tag{6.12}$$

where $n$ is the correction factor to correct for nonideal behavior; it ranges from about 1 to 2. Expand the exponential in series form:

$$i_D(v_D) = I_R\left[\frac{qv_D}{nkT} + \left(\frac{qv_D}{2nkT}\right)^2 + \cdots\right]$$

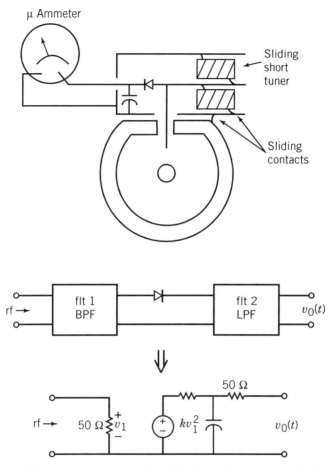

**FIGURE 6.28** Above we show a more detailed schematic of the slotted line apparatus. The sliding short tuner causes maximum absorption of rf energy by the diode, as well as providing a dc return path. If a dc current is not permitted to exist, the proper nonlinear behavior of the diode is not achieved. The lower sketch gives a general configuration for the diode. There it is between two filters, and a simplified equivalent circuit at low frequencies is shown.

The second term has a dc component, therefore, $I_{dc} \propto v_D^2 \propto P_{in}$, so the dc response is proportional to the input power. The dynamic range for the square law behavior can be up to 25 dB. Most often, however, the signal generator is amplitude-modulated at 1 kHz, and a high-gain narrow-band amplifier boosts the signal for further processing. Notice that the 1-kHz signal is also part of the second term.

In Figure 6.28, the diode has a dc and low-frequency return path, which allows for proper nonlinear behavior. In Figure 6.29, the assumed waveforms are sketched to show that the diode looks "active" in the dc or 1-kHz circuit. In the rf

**278**  THE SMITH CHART

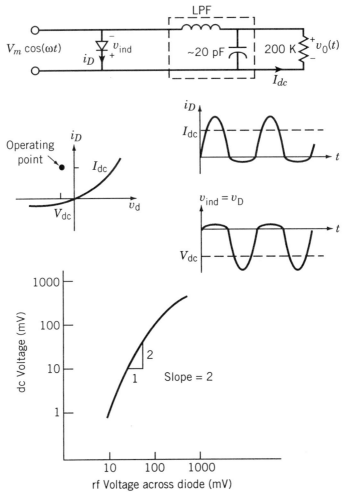

**FIGURE 6.29** An equivalent circuit for the square law detector circuit and the idealized waveforms for the diode. The diode appears passive at rf, but active at the down-converted frequency. The "operating point" is therefore in the second quadrant; the instantaneous $i_D - v_D$ trajectory is extremely complex and depends on details of the actual circuit. The dc voltage across about 10 MΩ vs. the rf voltage across the diode is sketched to indicate the slope = 2 region (square law).

portion of the circuit, the diode is passive, as it absorbs some rf power. Not all is lost as heat, as some is down-converted and transferred to the low-frequency loads. Thus, the operating point (due to self-bias here) is in the second quadrant of the static curve.

Although this apparatus is good for teaching purposes, the modern network analyzer is now the workhorse for impedance/admittance measurements. Notice

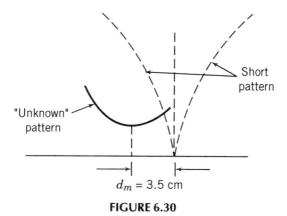

**FIGURE 6.30**

## IMPEDANCE OR ADMITTANCE COORDINATES

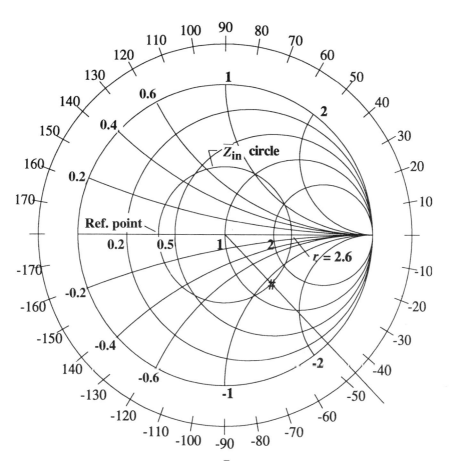

**FIGURE 6.31** Constructing the $\bar{Z}_{IN}$ circle, and finding the VSWR.

**280** THE SMITH CHART

that the slotted line is not well suited for swept measurements, which is one of the strong points for the analyzer.

***Example 6.8*** For the slotted line information given in Figure 6.30, determine $Z_L$. The operating frequency is 800 MHz, $S = 2.6$, and the "short" and "unknown" standing wave patterns are sketched. Assume a 50-$\Omega$ system.

1. First sketch the $\bar{Z}_{in}$ circle as shown in Figure 6.31. This circle cuts the real axis on the right side at $r = S = 2.6$.
2. With $\bar{Z}_L$ trapped somewhere on this circle, we can pin it down by moving 3.5 cm on the line toward the load. This is motion toward the load that corresponds to rotation toward load on the charts. Using the WTL scale, we start at the reference point (at voltage minimum) and rotate $3.5/18.75 = 0.18666$. (Note that the wavelength is $c/f$.)
3. Now read $\bar{Z}_L = 1.39 - j1.12$.
4. Unnormalize $Z_L = 69.5 - j56 \,\Omega$. ▲

## 6.8 LOSSY LINES AND THE RADIAL SCALES

Here, we treat low loss lines and assume that the characteristic impedance $Z_0$ is purely real. In this case, the $V^+$ and $V^-$ waves show damping by the factor $e^{-\alpha l}$. Thus, a given crest loses amplitude by this factor after moving $l$ meters in either direction on the line. Figure 6.32 shows the state of affairs. The reflection coefficient at various planes changes now its magnitude, as well as phase. We know from Chapter 5 that

$$\Gamma(-l) = \Gamma(0)e^{-2\gamma l} = \Gamma(0)e^{-2\alpha l}e^{-j2\beta l} \qquad (6.13)$$

which says the magnitude of $\Gamma(z)$ varies as a logarithmic spiral as shown. The standing wave ratio is now a function of position. Most calculations and procedures on the chart discussed so far remain the same, with the reduction factor $e^{-2\alpha l}$ in Eq. (6.13) properly included. The next example describes a method for determining $\alpha$.

When one discusses losses on transmission lines, one immediately becomes aware of a morass of terms, definitions, and subtle assumptions and approximations. We introduce just enough terms and definitions to discuss the radial scales at the bottom of Smith charts. Start with the notion of a matched line (both generator and load) and the attenuated forward wave; see Figure 6.33. Here, we sketch the signal at a particular instant. At the two planes separated $d$ meters, we know (only a forward wave here)

$$|V_2| = |V_1|e^{-\alpha d} \qquad (6.14)$$

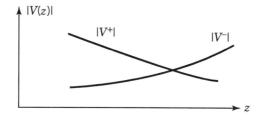

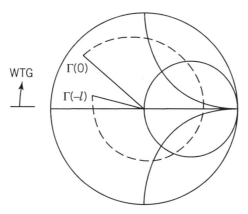

**FIGURE 6.32** The damping of the $V^+$ and $V^-$ phasor amplitudes by the factor $e^{-\alpha l}$. The reflection coefficient decreases as a logarithmic spiral as one moves toward the generator.

and for this special case, the power values passing the planes are

$$P_2 = P_1 e^{-2\alpha d} \tag{6.15}$$

that is,

$$P_2 \propto |V_2|^2, \qquad P_1 \propto |V_1|^2$$

Define the transmission loss or TL (dB) as

$$\text{TL (dB)} \triangleq -10 \log_{10}\left(\frac{P_2}{P_1}\right) \tag{6.16}$$

$$= -10 \log_{10} e^{-2\alpha d}$$

$$= 20(\alpha d) \log_{10} e$$

$$= 8.686(\alpha d) \tag{6.16a}$$

Notice that TL is a one-way loss due to line attenation; no reflection conditions

**282** THE SMITH CHART

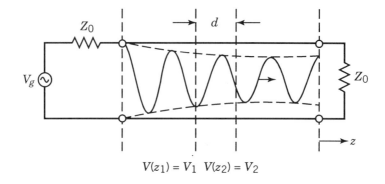

(a)

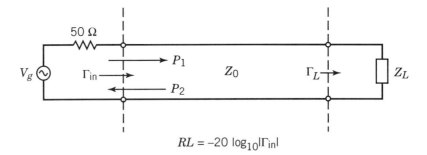

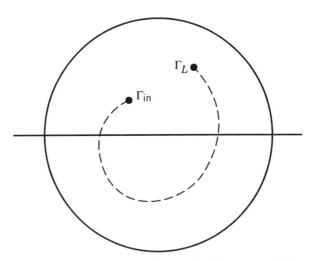

(b)

**FIGURE 6.33** (a) A line matched on both ends showing the exponential decay $\sim e^{-\alpha l}$ for the forward wave. (b) Return loss ($RL$) at the input plane; $P_2$ is reduced from $P_1$ by power absorbed in the load, and that lost in the line.

exist. The next term is the reflection loss that is

$$\text{Reflection loss (dB)} \triangleq -10\log_{10}(1-|\Gamma_L|^2) \qquad (6.17)$$

Notice that the factor $(1-|\Gamma_L|^2)$ is just the fraction of the power incident on the load that is absorbed by that load. That is,

$$\frac{P \text{ absorbed in load}}{P \text{ incident on load}} = (1-|\Gamma_L|^2) \qquad (6.18)$$

Observe here that line attenuation is not in the picture; this "loss" (power not absorbed by load) is due to the load reflecting some of the incident power. The next quantity is the return loss ($RL$) that is the ratio of power in the reflected wave to that in the incident wave:

$$RL \text{ (dB)} \triangleq -10\log_{10}\frac{P \text{ reflected wave}}{P \text{ incident wave}} \qquad (6.19)$$

$$= -10\log_{10}|\Gamma|^2$$

$$= 20\log_{10}\left(\frac{1}{|\Gamma|}\right) \qquad (6.19a)$$

Observe that RL is also defined in terms of the reflection coefficient, but is generally assumed to include line losses as follows. With reference to Figure 6.33, the input reflection coefficient is related to the incident and reflected power there by

$$\frac{P_2}{P_1} = |\Gamma_{in}|^2 \qquad (6.20)$$

However, $P_2$ is less than $P_1$ by the amount absorbed in the load, as well as that lost in the line. Thus, RL embraces both attenuation and the load's reflection. Note that the TL is for one-way travel of a wave, but RL includes loss incurred in both forward and reverse waves (two-way loss).

The last quantity is the standing wave loss factor (SWLF). This factor gives a very rough measure of line attenuation losses when the load does not match the line. Unlike the matched condition in Figure 6.33, Figure 6.34 gives the sketch of the total wave $|V(z)|$. For the given length of line, one could calculate the one-way TL as

$$\text{Loss}_1 = e^{-2\alpha_m d} \qquad (6.21)$$

where $\alpha_m$ is the normal attenuation factor (here, $m$ means the line is matched,

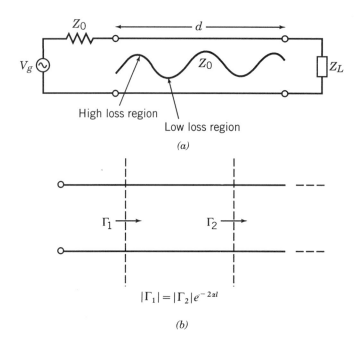

**FIGURE 6.34** (a) The situaton where $Z_L \neq Z_O$ and line losses are roughly estimated by the standing wave loss factor (SWLF). (b) The general expression for reflection coefficients $l$ meters apart.

$\Gamma_L = 0$, $S = 1$) if somehow the load were nonreflecting. For this hypothetical situation, the loss would be the difference between the power entering the line and that absorbed by the nonreflecting load. Now permit the load to reflect some power. In such a situation, the standing wave pattern has net voltage and current values at certain positions along the line that are now both larger and smaller than the values when no reflection was the case. Since ohmic losses are proportional to the magnitudes of $|V(z)|^2$ and $|I(z)|^2$, the losses at the maximum points are larger, and those at the minimum points are smaller than those in the reflectionless case. It turns out that the net loss is now larger (since the regions about minima are narrower than those about maxima) than in the first ($\Gamma_L = 0$) case. Call this loss$_2$ and define a corresponding $\alpha_r$. Here, $r$ means reflections are present, and $S > 1$. Now define the SWLF as

$$\text{SWLF} \triangleq \frac{\alpha_r}{\alpha_m}\bigg|_{P \text{ absorbed in load the same}} \qquad (6.22)$$

The analytical development is as follows (see Ref. 11). For a matched line (with the origin at the source), the power transmitted to some point $z$ is

$$P_T(z) = P_0 e^{-2\alpha z} \qquad (6.23)$$

where $P_0 = P(z = 0)$.

The power lost per unit length ($P_{LUL}$) is

$$-\frac{dP_T}{dz} = 2\alpha P_T \triangleq P_{LUL}$$

or

$$\alpha = \frac{P_{LUL}}{2P_T} \quad (6.24)$$

Write $P_{LUL}$ as (where now we assume a finite length and a reflected wave)

$$P_{LUL} = \tfrac{1}{2}(|I^+|^2 + |I^-|^2)R \quad W/m \quad (6.25)$$

where $R$ is the line's distributed resistance ($\Omega/m$).

This expression assumes that the losses in the forward and reverse waves are simply additive. The transmitted power is

$$P_T = \tfrac{1}{2}(|I^+|^2 - |I^-|^2)Z_0 \quad (6.26)$$

where again phase effects are neglected. Then using $I^- = -\Gamma I^+$ (where $\Gamma$ is the voltage reflection coefficient), we obtain

$$P_{LUL} = \tfrac{1}{2}|I^+|R(1+|\Gamma|^2) \quad (6.27)$$

$$P_T = \tfrac{1}{2}|I|^2 Z_0(1-|\Gamma|^2) \quad (6.28)$$

Now define

$$\alpha_r = \frac{P_{LUL}}{2P_T}\bigg|_{\Gamma \neq 0} = \frac{R}{2Z_0}\frac{1+|\Gamma|^2}{1-|\Gamma|^2} \quad (6.29)$$

The SWLF is $\alpha_r/\alpha_m$, where

$$\alpha_m = \frac{P_{LUL}}{2P_T}\bigg|_{\Gamma = 0} = \frac{R}{2Z_0} \quad (6.30)$$

$$\text{SWLF} = \frac{\alpha_r}{\alpha_m} = \frac{1+|\Gamma|^2}{1-|\Gamma|^2} \quad (6.31)$$

Some algebra shows that this may be written as

$$\text{SWLF} = \frac{S^2+1}{2S} \quad (6.31a)$$

The latter two equations are those found most often (see Refs. 8 and 11).

**286** THE SMITH CHART

Unfortunately, the term standing wave loss coefficient is often called the transmission loss coefficient [not to be confused with TL in Eq. (6.16)]. The topmost scale on the left of the "radially scaled parameters" on the Smith chart gives the values for Eq. (6.31) and is called the transmission loss (loss coefficient). The discussion in [Ref. 11] of this factor states explicitly that the term is applicable when phase effects are negligible (i.e., if $\alpha \to 0$). We can develop Eq. (6.31) in the following manner, following Ref. 12. With reference to Eqs. (5.43) through (5.45) repeated here for convenience.

$$P_{\text{in}} = \tfrac{1}{2} G_0 |V^+|^2 [e^{2\alpha l} - |\Gamma_L|^2 e^{-2\alpha l}] \tag{5.43}$$

$$P_{\text{load}} = \tfrac{1}{2} G_0 |V^+|[1 - |\Gamma_L|^2] \tag{5.44}$$

$$P_{\text{loss}} = P_{\text{in}} - P_{\text{load}}$$

$$= \tfrac{1}{2} G_0 |V^+|[(e^{2\alpha l} - 1) + |\Gamma_L|^2(1 - e^{-2\alpha l})] \tag{5.45}$$

where $P_{\text{in}}$, $P_{\text{load}}$, and $P_{\text{loss}}$ are the time-averaged power into the line and load, and that lost in the line, respectively.

We may write Eq. (6.22) as

$$\frac{\alpha_r}{\alpha_m} = \frac{(P_{\text{LUL}}/2P_{\text{load}})|_{\Gamma \neq 0}}{(P_{\text{LUL}}/2P_{\text{load}})|_{\Gamma = 0}}$$

$$= \frac{[(e^{2\alpha l} - 1) + |\Gamma_L|^2(1 - e^{-2\alpha l})]/[1 - |\Gamma_L|^2]}{[(e^{2\alpha l} - 1)]/(1)} \tag{6.32}$$

Now assume that $\alpha l \to 0$ and using L'Hospital's rule, we get

$$\frac{\alpha_r}{\alpha_m} = \frac{1 + |\Gamma_L|^2}{1 - |\Gamma_L|^2} \tag{6.33}$$

which yields the desired result.

With reference to the "radially scaled parameters" at the bottom of a Smith chart, we discuss each scale in turn. The four scales on the right fall under the general heading "reflection." When holding the chart to read the word reflection, we see that the leftmost scale is just that for $|\Gamma|$, the magnitude of the voltage reflection coefficient. The adjacent scale gives the power reflection coefficient that is $|\Gamma|^2$. The "loss in dB" scales are return and reflection losses, respectively; these are defined in Eqs. (6.19a) and (6.17). For example, if $|\Gamma| = 0.6$,

$$\text{RL} = 20 \log_{10} \frac{1}{0.6} = 4.44 \, \text{dB}$$

$$\text{Reflection loss} = -10 \log_{10}(1 - 0.36) = 1.94 \, \text{dB}$$

which are easily read from the scales. Now on the left side, the standing wave heading gives the VSWR both numerically and in dB notation:

$$\text{VSWR} = S = \frac{1+|\Gamma|}{1-|\Gamma|}$$

$$S(\text{dB}) = 20 \log_{10} S$$

Again, if $|\Gamma| = 0.6$, $S = 4$, $S$ (dB) $= 12.04$, which we verify using our dividers set at $|\Gamma| = 0.6$. Finally, we have the transmission loss heading. The rightmost scale is the SWLC of Eq. (6.33), under the loss coefficient subheading. For our example where $|\Gamma| = 0.6$, we find

$$\text{SWLC} = \frac{1+0.36}{1-0.36} = 2.125$$

(note that it is numerical, not a dB quantity). Finally, the 1-dB step scale is described. We know that (Fig. 6.34)

$$|\Gamma_1| = |\Gamma_2| e^{-2\alpha l}$$

or

$$\frac{|\Gamma_1|}{|\Gamma_2|} = e^{-2\alpha l}$$

Now define an attenuation by

$$A(\text{dB}) = -10 \log_{10} e^{-2\alpha l} = 20(\alpha l) \log_{10} e$$

$$= 8.686(\alpha l)$$

$$= -10 \log_{10}\left(\frac{|\Gamma_1|}{|\Gamma_2|}\right)$$

Notice that $A$(dB) is the same as the one-way TL of Eq. (6.16a). Let $\Gamma_2 = 1$ and change $A$ in 1-dB steps; calculate $\Gamma_1$ as shown below:

| $A$ (dB) | 1 | 2 | 3 | 4 | 5 | 6 |
|---|---|---|---|---|---|---|
| $\Gamma_1$ | 0.794 | 0.631 | 0.501 | 0.398 | 0.316 | 0.251 |

Observe that the tic marks on the 1-dB step scale correspond to $|\Gamma|$ decreasing as found in the calculation above. The next example demonstrates a use for this scale.

***Example 6.9*** Describe a method to obtain the line attenuation factor $\alpha$.

A basic scheme is to terminate the line in a short, then determine $|\Gamma_{\text{in}}|$. We know $|\Gamma_L| = 1$, as the load is a short. So

$$|\Gamma_{\text{in}}| = |\Gamma_L| e^{-2\alpha l} \tag{1}$$

288    THE SMITH CHART

or
$$|\Gamma_{in}| = |e^{-2\alpha l}| \tag{2}$$

Since $|\Gamma_{in}|$ is measured and $l$ known, we find $\alpha$ immediately:

$$\alpha = -\frac{1}{2l} \ln |\Gamma_{in}| \tag{3}$$

For example, suppose that the measured $\bar{Z}_{in}$ for a shorted line $3\lambda$ long is $0.2 - j0.2$; find $\alpha$ as follows. Plot $\bar{Z}_{in}$ and determine $|\Gamma_{in}| = 0.675$, so $\alpha = (1/6\lambda) \ln(0.675) = 0.066$ Np/wavelength.  ▲

***Example 6.10***   For a load $Z_L = 27.5 + j50$, find the VSWR in dB the return and reflection losses, and the fraction of the incident power reflected. Assume that the line is lossless.

The VSWR can be expressed as

$$S = \frac{1+|\Gamma|}{1-|\Gamma|} \quad \text{or} \quad |\Gamma| = \frac{S-1}{S+1}$$

$$S(\text{dB}) \triangleq 20 \log_{10}(S) = (8.686)\, 2 \tanh^{-1} |\Gamma|$$

That is, use

$$\tanh^{-1} x = \tfrac{1}{2} \ln \frac{1+x}{1-x}$$

Then
$$S(\text{dB}) = 5.95.$$

The return loss is

$$20 \log_{10} \frac{1}{|\Gamma|} = 4.52 = \text{RL} \quad (\text{dB})$$

Note that

$$S = \text{ctnh} \left[ \frac{1}{2} \frac{\text{RL(dB)}}{8.686} \right] = 3.93$$

which is indicated on Fig. 6.35.

The reflection loss is

$$10 \log_{10} \frac{1}{1-|\Gamma|^2} = 1.89\ \text{dB}$$

which is also called the load mismatch loss or the transmission loss on some occasions.

## LOSSY LINES AND THE RADIAL SCALES   289

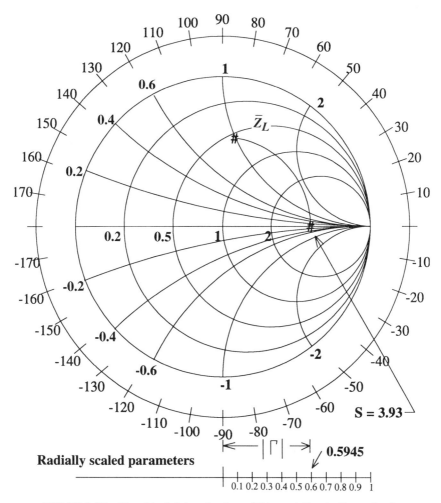

**FIGURE 6.35**  Graphical determination of $S$ by rotating to the real axis.

The absorbed power is

$$P_{incident}(1 - |\Gamma|^2) = P_{incident}(0.6466)$$

$$P_{reflected} = P_{incident} - P_{absored}$$

$$= P_{incident} - 0.6466 P_{incident}$$

$$\therefore \frac{P_{reflected}}{P_{incident}} = \frac{P_{incident}(1 - 0.6466)}{P_{incident}} = 0.353$$  ▲

## IMPEDANCE OR ADMITTANCE COORDINATES

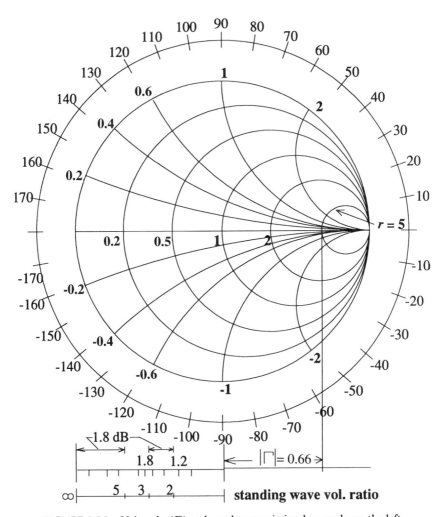

**FIGURE 6.36** Using the $|\Gamma|$ scale and transmission loss scale on the left.

*Example 6.11* Suppose that a shorted lossy line of given length has a measured $S$ of 5 at the input. Now an unknown load replaces the short and the new input VSWR is 2.0. What is the VSWR at the plane of this unknown load? See Figure 6.36 for constructions.

The first measurement shows that $S = \infty$ at the load plane decreases to $S = 5$ at the input. This corresponds to a separation on the $S$ scale as shown below. Using his distance on the 1-dB step scale, one must interpolate on a nonlinear scale. We bypass the interpolation here and measure $|\Gamma_{in}| = 0.66$ using dividers and the $|\Gamma|$ scale. Then using

$$A(\text{dB}) = -10\log_{10}(0.66) = +1.8$$

We see that this corresponds to 1.8 of the 1-dB scale. The next step is to start at the $S = 2$ value and read directly above on the 1-dB step scale. Observe that this is between the 4th and 5th tics; we interpolate the value to be about 4.8. Now we move back to the left on this scale 1.8 dB, since this is the same decrement (a property of the length of line, and not the load), regardless of the load. Moving to the point $4.8 - 1.8 = 3.0$ dB, we then move down to the $S$ scale and read $S \doteq 3.0$, the value sought. With calculators and PCs readily available, the scales on the left are seldom, if ever, used. Now only the $|\Gamma|$ and the return loss and reflection loss scales find occasional use.

Notice that the 1-dB steps are one-half the return loss steps; the first tic corresponds to the 2-dB value on the return loss scale. Thus, this scale is sometimes called the $\frac{1}{2}$ return loss scale.

## SUPPLEMENTARY EXAMPLES

**6.1** The load impedance is $175 - j130\,\Omega$ and $Z_0 = 50\,\Omega$.

(a) Plot $\bar{Z}_L$ and determine $\Gamma_L$. First form $\bar{Z}_L = (175 - j130)/50 = 3.5 - j2.6$. After plotting, read (see Fig. 6.37 for this example)

$$|\Gamma_L| = 0.69, \qquad \underline{/\Gamma_L} = -16.2°$$

(b) What is the VSWR? Rotate to the real axis on the right side and read $S = 5.5$. This circle with radius $|\Gamma_L|$ is generally called the VSWR circle for the load; we also refer to it as the $\bar{Z}_{\text{in}}$ circle.

(c) What is the VSWR in dB? Read the scale on the left: $S = 15$ dB.

(d) What is the standing wave loss coefficient (SWLC)? From the transmission loss heading, loss coefficient subheading, we read SWLC $= 2.88$.

(e) What is the load-normalized admittance $\bar{Y}_L$? Take $\bar{Z}_{\text{in}}$ diametrically through the chart center and read

$$\bar{Y}_L = 0.182 + j0.139$$

(f) What is the smallest real part of $Z_{\text{in}}$ ($R_{\text{in}}$) when $X_{\text{in}} = 0$? read $r_{\text{in}} = 0.179$ (on the left side of the $\bar{Z}_{\text{in}}$ circle). Thus,

$$R_{\text{in}} = 8.95\,\Omega \quad (\text{minimum})$$

(g) What is the largest real part of $Z_{\text{in}}$? Read $\bar{r}_{\text{in}} = 5.5$; then $R_{\text{in}} = 275\,\Omega$ (maximum).

## IMPEDANCE OR ADMITTANCE COORDINATES

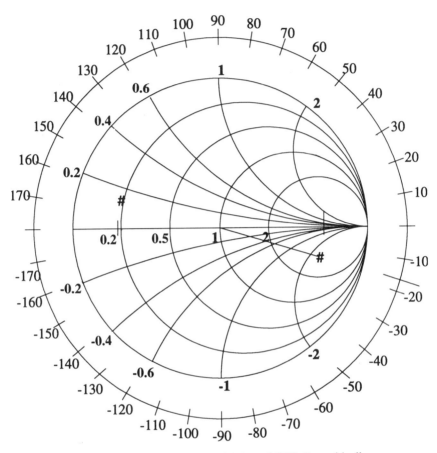

**FIGURE 6.37** Finding $\Gamma_L$, VSWR, and SWLC graphically.

**(h)** What is the distance in wavelengths to the first voltage minimum? To the first maximum?

The minimum occurs where $R_{in}$ is minimum; using the WTG scale, we find $V_{min}$ at

$$(0.5000 - 0.2728)\lambda = 0.2272\lambda$$

The maximum occurs where $R_{in}$ is maximum; this is always $0.25\lambda$ away from a minimum. Hence
$V_{max}$

$$(0.2272 + 0.2500)\lambda = 0.4772\lambda$$

(i) What are the return and reflection losses? From the scales, read

$$RL = 3.2 \, dB$$
$$\text{Reflection loss} = 2.8 \, dB$$

**6.2** To verify Eqs. (6.10a) and (6.10b), start with Eq. (6.9) and rearrange as

$$r + jx = \frac{1 + (\Gamma_r + j\Gamma_i)}{1 - (\Gamma_r + j\Gamma_i)}$$

$$= \frac{[(\Gamma_r + 1) + j\Gamma_i][(1 - \Gamma_r) + j(\Gamma_i)]}{(1 - \Gamma_r)^2 + \Gamma_i^2}$$

Equate real parts:

$$r = \frac{1 - \Gamma_r^2 - \Gamma_i^2}{(1 - \Gamma_r)^2 + \Gamma_i^2} \tag{6.20a}$$

Equate imaginary parts:

$$x = \frac{(\Gamma_r + 1)\Gamma_i + \Gamma_i(1 - \Gamma_r)}{(1 - \Gamma_r)^2 + \Gamma_i^2}$$

$$= \frac{2\Gamma_i}{(1 - \Gamma_r)^2 + \Gamma_i^2} \tag{6.10b}$$

which is the desired result.

**6.3** For the load in Figure 6.38, the VSWR = 4. Then a 3-dB pad (attenuator) is inserted as shown. What is $S$ on the line in the new configuration?

Start with $S = 4$ on the VSWR scale in Figure 6.39, and move up to the 1-dB step scale. Read approximately 2.2 on this unmarked scale. Now move 3-dB to the right (to decrease $|\Gamma|$) by the 3 dB of the pad and drop back down to the VSWR scale. The new $S = 1.86$.

**6.4** For the two lines in tandem, see Figure 6.40, find the input impedance at the junction and the input. Notice that at the junction of the 50-Ω and 30-Ω lines, the normalized values are different. The actual impedance is $10 + j12.5 \, \Omega$, and the normalized values are

$$\bar{Z}_{in}^{50} = 0.2 + j0.25, \qquad \bar{Z}_{in}^{30} = 0.33 + j0.4167$$

Thus, on a chart used with both normalizations displayed, the normalized contour has a "jump" as shown on the chart in Figure 6.41.

$$\bar{Z}_{in}^{50} = 0.2 + j0.25$$
$$\bar{Z}_{in}^{30} = 2.2 - j1.67$$
$$\therefore Z_{in} = 66 - j50.1 \, \Omega$$

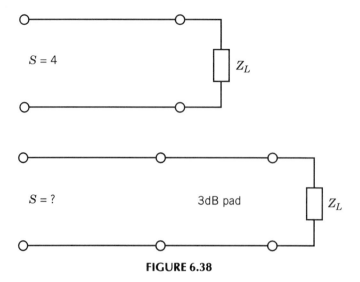

**FIGURE 6.38**

**6.5** An interesting property of the chart is the construction of "Q circles." Recall that $Q = |X|/R$ for a series network with $\mathbf{Z} = R + jX$. A constant $Q$ circle is one wherein the coordinates $r$ and $x$ are related by $X = kr$, where $k$ is the $Q$ of interest. In Figure 6.42, we have sketched the $Q = 1$ and five circles. The $Q = 1$ is quickly constructed, since its center is the $\Gamma = 1 \angle -90°$ point. Observe that the centers for the circles lie on a vertical line passing through the center of the chart (the $\Gamma_i$ axis).

**6.6** When lossy lines are analyzed, the characteristic impedance should be treated as complex, $\mathbf{Z}_0 = R_0 + jX_0$. Recall that we normally neglect $X_0$ in most practical calculations for slightly lossy lines. If we choose to include $X_0$ in the analysis, then the reflection coefficients must be in the form:

$$\Gamma_L = \frac{\mathbf{Z}_L - \mathbf{Z}_0^*}{\mathbf{Z}_L + \mathbf{Z}_0}, \quad \Gamma_s = \frac{\mathbf{Z}_s - \mathbf{Z}_0^*}{\mathbf{Z}_s + \mathbf{Z}_0} \tag{1}$$

which reduces to the normal form when $\mathbf{Z}_0$ is purely real (i.e., $\mathbf{Z}_0 = \mathbf{Z}_0^*$). To see why this is needed, start with the normal definition:

$$\Gamma_L = \frac{\mathbf{Z}_L - \mathbf{Z}_0}{\mathbf{Z}_L + \mathbf{Z}_0}, \quad \begin{array}{l} \mathbf{Z}_0 = R_0 + jX_0 \\ \mathbf{Z}_L = R_L + jX_L \end{array} \tag{2}$$

Then

$$|\Gamma_L|^2 = \Gamma_L \Gamma_L^* = \frac{(R_L - R_0)^2 + (X_L - X_0)^2}{(R_L + R_0)^2 + (X_L + X_0)^2}$$

$$= \frac{R_L^2 - 2R_L R_0 + R_0^2 + X_L^2 - 2X_L X_0 + X_0^2}{D},$$

$$D \stackrel{\Delta}{=} (R_L + R_0)^2 + (X_L + X_0)^2$$

## IMPEDANCE OR ADMITTANCE COORDINATES

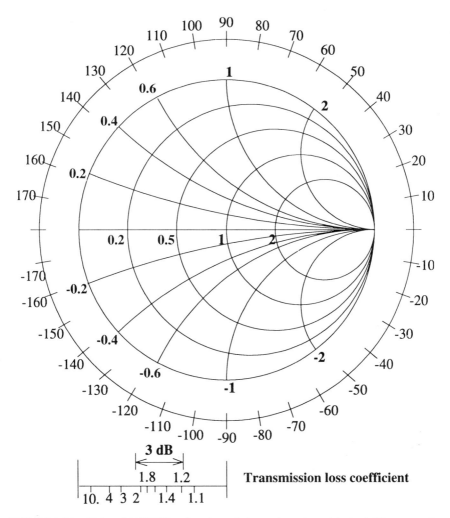

**FIGURE 6.39** Using the VSWR vol. ratio scale in conjunction with the 1 dB steps scale.

Now add and subtract $2R_L R_0$, $2X_L X_0$ and some algebra yields

$$|\Gamma_L|^2 = 1 - \frac{4(R_L R_0 + X_L X_0)}{D} \qquad (3)$$

Notice that $|\Gamma_L|^2 > 1$ if $(R_L R_0 + X_L X_0) < 0$. Since $R_L, R_s > 0$, the condition needed to keep $|\Gamma_L| \leqslant 1$ is $X_L X_0 < 0$, and $D > 4|R_L R_0 + X_L X_0|$.

**296** THE SMITH CHART

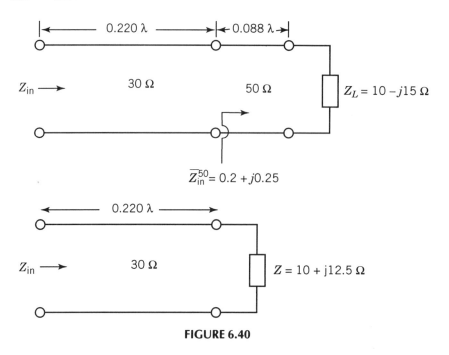

**FIGURE 6.40**

For example, let $Z_L = -j50$, $Z_0 = 40 + j30$. Then

$$\Gamma_L = \frac{Z_L - Z_0}{Z_L + Z_0} = -j2$$

$$\therefore |\Gamma_L| = 2$$

It turns out that $|\Gamma_L| \leq (1 + \sqrt{2})$ for all cases; see Ref. 2 in Chapter 5 and Ref. 9 here. If, however, we use

$$\Gamma_L = \frac{Z_L - Z_0^*}{Z_L + Z_0}$$

then similar steps lead to a modified form for Eq. (3):

$$|\Gamma_L|^2 = 1 - \frac{4R_L R_0}{D}$$

However, $4R_L R_0 > 0$ (passive loads), so $|\Gamma_L| \leq 1$ for all cases. For the same $Z_L$, $Z_0$ as above, we find $|\Gamma_L| = 1$.

## IMPEDANCE OR ADMITTANCE COORDINATES

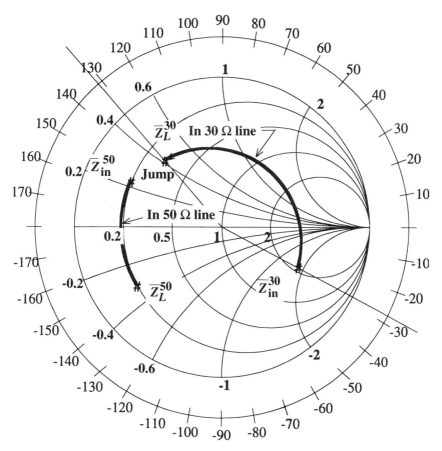

**FIGURE 6.41** The jump in normalized impedances when moving between lines of different characteristic impedances.

**6.7** For the lossy line in Figure 6.43 find the normalized input impedance. Also, what is the reduction factor for the change in $S$ from load to source plane?

Since

$$\beta = \frac{2\pi}{\lambda} = 0.4072, \qquad \lambda = 15.43 \text{ m}$$

So

$$\beta l = 2\pi \left( \frac{1.5}{15.43} \right) = 0.6108 \text{ rad} = 35°$$

## IMPEDANCE OR ADMITTANCE COORDINATES

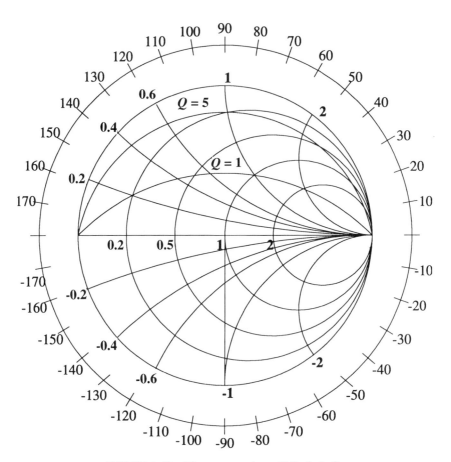

**FIGURE 6.42** The construction of "Q-circles".

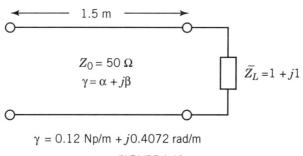

$\gamma = 0.12$ Np/m $+ j0.4072$ rad/m

**FIGURE 6.43**

## IMPEDANCE OR ADMITTANCE COORDINATES

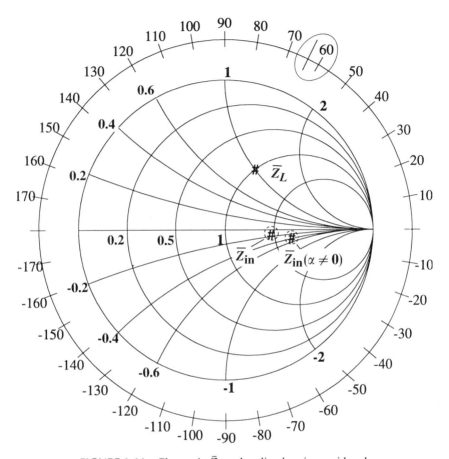

**FIGURE 6.44** Change in $\bar{Z}_{IN}$ when line loss is considered.

Plot $Z_L \to 0.162$ on the WTG scale, or $63.5°$. See Figure 6.44. After rotating $2\beta l^0 = 70°$ if $\alpha = 0$, $\bar{Z}_{in} = 2.52 - j0.31$, and $S = 2.55$, $|\Gamma_L| = 0.441$, then when $\alpha \neq 0$,

$$|\Gamma_{in}| = |\Gamma_L|e^{-2\alpha l} = (0.411)e^{-0.36}$$
$$= 0.30767$$
$$\therefore \bar{Z}_{in} = 1.86 - j0.15, \qquad S = 1.88$$

and the reduction factor is $1.88/2.59 = 72.670\%$.

## PROBLEMS

**6.1** Three complete rotations on a Smith chart correspond to how many wavelengths of motion along the line? How would you use a 50-Ω chart on a 75-Ω line? What does the term "50 Ω chart" mean?

**6.2** For $Z_0 = 75\,\Omega$ and $Z_L = 150 + j60\,\Omega$, find, on a SC, (a) the VSWR, (b) the distance in $\lambda$ to the first voltage maximum from the load, and (c) the angle of the current reflection coefficient.

**6.3** Plot the changes in input impedance or admittance on a Smith and Jones chart for the ladder network shown in Figure P6.3. Estimate the $Q$ circle that bounds the network.

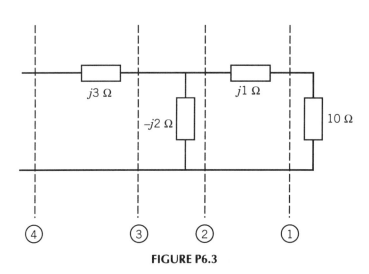

**FIGURE P6.3**

**6.4** With reference to Figure 6.26, the parameters of interest are $f = 750\,\text{MHz}$, $d_m = 13.5\,\text{cm}$. The voltage minimum for a short occurs at 8 cm on the scale.

(a) Sketch the standing wave pattern if the VSWR = 1.105.
(b) What is $Z_L$?
(c) Considering losses in the slotted section, what is the SWLF?

**6.5** Assume a slightly lossy line with $Z_0 = 50 - j3\,\Omega$. For a load $Z_L = 10 + j10\,\Omega$, what is $\Gamma_L$?

**6.6** For the lossy line in Figure P6.6, what is the reduction in $S$ from load to source?

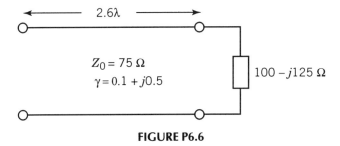

**FIGURE P6.6**

**6.7** What is the length of line needed in Figure P6.7 to make the $Z_{in}$ both real and as large as possible?

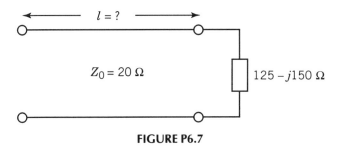

**FIGURE P6.7**

**6.8** Given the equation of a circle in the **Z** plane, what is the equation of that circle on the Smith chart. What is the easiest method of plotting such a circle in one plane given it in the other plane?

**6.9** Find $\Gamma_L$, $Z_L$, and $S$ in Figure P6.9.

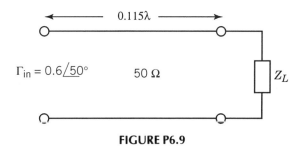

**FIGURE P6.9**

**6.10** Find $Z_{in}$ for Figure P6.10 both analytically and using the Smith chart. The frequency is 2 GHz, and all lines are air-filled. For a 50-$\Omega$ generator, what is the return loss?

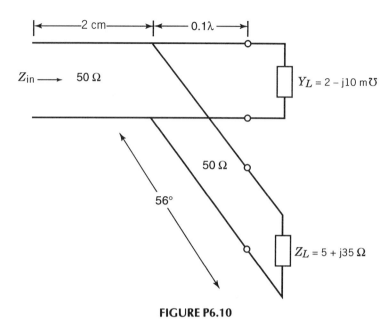

**FIGURE P6.10**

## REFERENCES

1. Gonzalez, G., *Microwave Transistor Amplifiers, Analysis and Design*, Prentice-Hall, Englewood Cliffs, NJ, 1984.
2. Senise, J. T., "Trigonometric Functions and the Smith Chart," *IEEE Trans. Microwave Theory Tech.*, *MTT-21*, Aug. 1973, pp. 569–570.
3. Clavier, P. A., "Pierre Fermat and the Smith Chart," *Proc. IEEE*, July 1966, p. 1004.
4. McCausland, I., "An Analogy Between the Lorentz Transformation and the Transmission-Line Equations," *Proc. IEEE*, 62, Oct. 1974, pp. 1398–1399.
5. White, J. F., "The Smith Chart: An Endangered Species?", *Microwave J.*, 22, Nov. 1979, pp. 49–54.
6. Schott, F. W., "On the Smith Chart," *IEEE Trans. Microwave Theory Tech.*, *MTT-26*, April 1978, p. 314.
7. Wheeler, H. A., "Reflection Charts Relating to Impedance Matching," *IEEE Trans. Microwave Theory Tech.*, *MTT-32*, Sept. 1984, pp. 1008–1021.
8. Smith, P. H., *Electronic Applications of the Smith Chart*, McGraw-Hill, New York, 1969.
9. Kretzschmar, J. and Schoonaert, D., "Smith Chart for Lossy Transmission Lines," *Proc. IEEE*, 57, Sept. 1969, pp. 1658–1660.
10. Lazarus, M. J. and Mak, L. K., "Diode RF Rectification: Predict It More Closely," *Microwaves*, 17, Feb. 1978, pp. 64–65.
11. Ragan, G. L. (ed.), *Microwave Transmission Circuits*, MIT Rad. Lab. series, no. 9, Boston Technical Publishers, Cambridge, MA, 1964, Section 2.6.

12. Chipman, R., *Transmission Lines*, Schaum's Outline Series, McGraw-Hill, New York, 1968, Section 9.10.

## BIBLIOGRAPHY

1. Bereskin, A. B., "Letters to the Editor," *Microwave J.*, *23*, March 1980, pp. 65–66 (expands on Ref. 5).
2. Blanchard, W. C., "New and More Accurate... Inside-Out Smith Chart," *Microwaves*, *6*, Nov. 1967, pp. 38–42.
3. Blume, H. C., "Input Impedance Display of One-Port Reflection-Type Amplifiers," *Microwave J.*, *10*, Nov. 1967, pp. 65–69.
4. Churchill, R. V., *Complex Variables and Applications*, 2nd ed., McGraw-Hill, New York, 1960, Section 34.
5. Dawirs, H. N., "Application of the Smith Chart to General Impedance Transformations," *Proc. IRE*, *45*, July 1957, pp. 954–956.
6. Graham, P. J. and Distler, R. J., "Use of the Smith Chart with Complex Characteristic Impedance," *IEEE Trans. Education*, *E-11*, June 1968, pp. 144–146.
7. LaPlante, P. and Borja, J., "An Automated Smith Charting Tool," *Microwave J.*, *30* May 1987, pp. 373–378.
8. Steere, R. M., "Novel Applications of the Smith Chart," *Microwave J.*, *3*, March 1960, pp. 97–100.
9. White, J. F., *Microwave Semiconductor Engineering*, Van Nostrand Reinhold, New York, 1982, Appendix J.

CHAPTER SEVEN

# Single Frequency Matching

## 7.1 BASICS, STUBS

The matching of one microwave component or network to another network is one of the most basic problems in microwave engineering. This chapter studies the special case where the bandwidth over which the match occurs is not part of the design process. Instead, one designs for the matching condition at a particular frequency $f_0$ and settles for the resulting degradation of the match as the frequency moves away from $f_0$. This is sometimes called single frequency or very narrow band matching. The first case we consider is given in Figure 7.1, where a single shunt "stub" (a section of TL) is used to match the load $\mathbf{Z}_L$ to the line with characteristic impedance $\mathbf{Z}_0$. The problem may be stated as follows. When the load is not equal to $\mathbf{Z}_0$, we have a reflection of some of the power traveling toward the load (a condition we seek to eliminate). The question is, is there a distance $d_1$ and length of shorted stub $d_2$ that will produce no reflected wave on the main line (to the left of the stub)? If this is possible, then all the power arriving from the generator is "absorbed" at this plane. Since the stub absorbs no power, it must all be consumed by the load. It turns out that such a "matched" condition is possible, and two distinct solutions (different distances $d_1$ with corresponding stub lengths) exist. From a traveling wave viewpoint, the reflected wave moving toward the generator produced by the stub's presence effectively cancels that part of the load's reflected wave that has been transmitted across the stub plane and is continuing toward the generator. This is shown schematically in Figure 7.1a, where wavelets $A$ and $B$ are those from the stub and load, respectively. They are equal in magnitude and $180°$ out of phase. Before continuing further, it is appropriate to review some lumped circuit ideas. From Figure 7.2a, we observe

$$I = \frac{V_G}{(R_G + R_L) + j(X_G + X_L)} \quad (7.1)$$

$$V_L = I Z_L$$

$$P_L = \frac{1}{2}\text{Re}\{V_L I^*\} = \frac{1}{2}\text{Re}\{|I|^2 Z_L\}$$

$$= \frac{1}{2}\text{Re}\left\{\frac{|V_G|^2(R_L + jX_L)}{(R_G + R_L)^2 + (X_G + X_L)^2}\right\} \quad (7.2)$$

$$= \frac{|V_G|^2}{2}\frac{R_L}{(R_G + R_L)^2 + (X_G + X_L)^2} \quad (7.3)$$

The maximum power the load can absorb is determined as follows. To maximize $P_L$, we note that the denominator is smallest if $X_L = -X_G$. Using this, we evaluate

$$\frac{dP_L}{dR_L} = \frac{d}{dR_L}\left[\frac{|V_G|^2}{2}\frac{R_L}{(R_G + R_L)^2}\right] = 0$$

which gives, of course, $R_L = R_G$. Then

$$P_L(\text{max}) = \frac{|V_G|^2 R_G}{2(2R_G)^2} = \frac{|V_G|^2}{8R_G} \quad (7.4)$$

which we recognize is the available power from the generator. Therefore, the maximum power is delivered to the load when the equivalent Thevenin generator has $Z_G = Z_L^*$:

$$P_L(\text{max}) = \frac{|V_G|^2}{8R_L} \quad (7.5)$$

When

$$R_L = R_G \quad (7.6a)$$

$$X_L = -X_G \quad (7.6b)$$

In Figure 7.2b, we show the lumped equivalent circuit at the plane of the stub. The load transformed the distance $d_1$ presents the admittance $Y_1$ at the stub plane, and the stub presents $Y_s$. Then the net impedance there is $Z = (Y_s + Y_1)^{-1}$. From Chapter 5, Supplementary Exercise 5.9, the source transformed to the stub plane has the representation:

$$V_{\text{th}} = V_G' = \frac{2V_G Z_0}{(Z_0 + Z_G)(1 - \Gamma_s \Gamma_L e^{-j2\beta l})} e^{-j\beta l} \quad (7.7)$$

$$Z_{\text{th}} = \frac{1 + \Gamma_s e^{-j2\beta l}}{1 - \Gamma_s e^{-j2\beta l}} Z_0 \quad (7.8)$$

**306** SINGLE FREQUENCY MATCHING

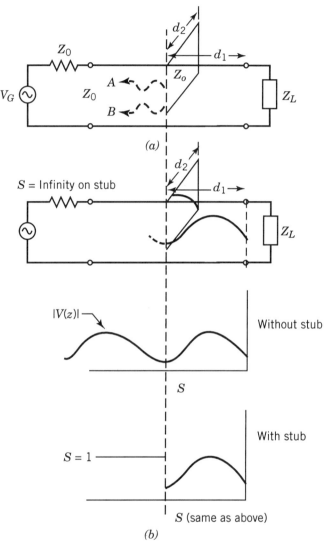

**FIGURE 7.1** The matching of a load $Z_L$ to a line with characteristic impedance $Z_0$, by the application of a shorted stub a distance $d_1$ from the load. The lower portions show $S$ both before and after the stub placement.

since $\Gamma_s = 0$,

$$V_{\text{th}} = V_G e^{-j\beta l} \tag{7.9}$$

and

$$\mathbf{Z}_{\text{th}} = \mathbf{Z}_0 \tag{7.10}$$

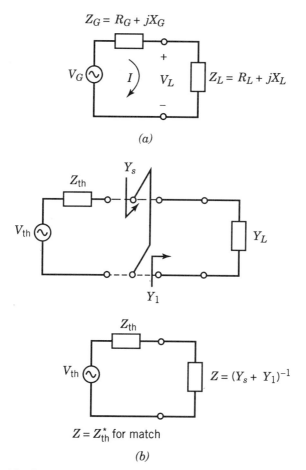

**FIGURE 7.2** (a) The general network to determine maximum power transfer from generator to load. (b) The reduction of a stub matched condition to a lumped network.

To have a match, we require

$$Z_0 = Z^* = \left(\frac{1}{Y_s + Y_1}\right)^* = \frac{1}{Y_0} \tag{7.11}$$

or

$$(Y_s + Y_1)^* = Y_0 \tag{7.11a}$$

This equation determines the lengths $d_1$ and $d_2$ in the following way. Since $Y_0$ is real (20 m℧ for $Z_0 = 50\,\Omega$), the left side is also purely real. Write this as

$$Y_s + Y_1 = Y_0^* = Y_0 \tag{7.12}$$

and let $Y_s = jB_s$, $Y_1 = G_1 + jB_1$. Then we have

$$jB_s + G_1 + jB_1 = Y_0 \tag{7.13}$$

Divide by $Y_0$:

$$jb_s + g_1 + jb_1 = 1 \tag{7.14}$$

Equating reals and imaginaries yields

$$g_1 = 1 \tag{7.15a}$$
$$b_s = -b_1 \tag{7.15b}$$

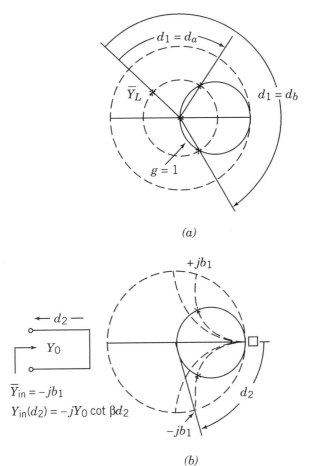

**FIGURE 7.3** (a) For the arbitrary load $\bar{Y}_L$, its VSWR circle is shown dotted. The intersections of this circle and the $g = 1$ circle are indicated by asterisks. The distances $d_1$ traveled from the load to the asterisks are denoted by $d_a$ and $d_b$. (b) For a shorted section of line, the input admittance is found on a SC as shown.

BASICS, STUBS    309

For the arbitrary load shown in Figure 7.3a, the $\bar{Y}_{in}$ circle (or VSWR circle) is shown dotted. We know that all values of $\bar{Y}_1$ (at distance $d_1$) lie somewhere on this circle. To satisfy Eq. (7.15a), we must move the distance $d_1$ until $g_1 = 1$, or we intersect the $g = 1$ circle on the SC. There are always two points of intersection (shown as asterisks); the shorter distance is denoted by $d_a$. Equation (7.15b) means the stub must present a susceptance that is the negative of that presented from the load. In Figure 7.3b, we show that the load presents $+jb_1$ (the imaginary circle through the asterisk), so the stub must present $-jb_1$. For a shorted line, the normalized admittance at the short is infinity (shown as a small square for visualization only), so the admittance at $d_2$ is

$$\bar{Y}_{stub} = \bar{Y}_s = -j \cot \beta d_2 \qquad (7.16)$$

Recall that the $\bar{Y}_{in}$ for a short is on the chart periphery. Figure 7.4 summarizes the conditions present at the plane of the stub when the condition of match is achieved.

***Example 7.1***  Using Figure 7.5, find the lengths $d_1$ and $d_2$ that match the load $Z_L$ to the $Z_0 = 50\,\Omega$ line. The stub also has $Z_0 = 50\,\Omega$.

First find $\bar{Y}_L = 1/\bar{Z}_L$ using the Smith chart of Figure 7.9:

$$\bar{Z}_L = \frac{100 + j100}{50} = 2 + j2$$

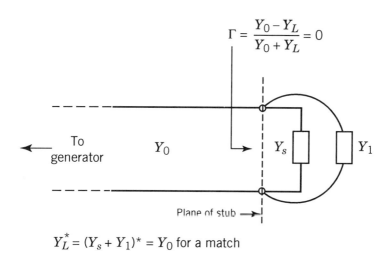

$Y_L^* = (Y_s + Y_1)^* = Y_0$ for a match

$Y_0$ Real

**FIGURE 7.4**  The conditions at the plane of the stub when the condition for a match (no reflection) is obtained. In all cases, we assume the main line to have a real characteristic admittance.

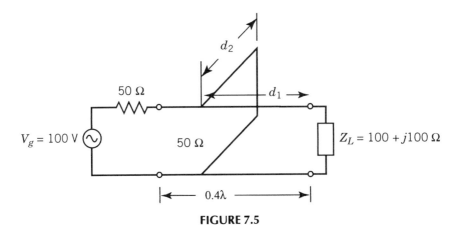

**FIGURE 7.5**

Reciprocating this by moving through the center of the chart yields

$$\bar{Y}_L = 0.25 - j0.25 \quad \text{(at 0.0415 WTL scale)}$$

Now bring $\bar{Y}_L$ up the line until we reach the $g = 1$ circle. This point has $\bar{Y}_1 = 1 + j1.6$ and is at 0.1785 on the WTG scale. Then the length $d_1$ is

$$d_1 = (0.0415 + 0.1785)\lambda = 0.22\lambda$$

The stub must present $-j1.6$ normalized susceptance, and its length is determined as shown. Here, we start at 0.25 WTG and stop at 0.3385. Thus,

$$d_2 = (0.3385 - 0.250)\lambda = 0.0885\lambda$$

The second solution (case B) is shown on the separate Smith chart of Figure 7.10. Here, we bypass the first crossing of the $g = 1$ circle and use the second intersection. The distance traversed is

$$d_1 = (0.0415 + 0.338)\lambda = 0.381\lambda$$

Now, the stub must present $+j1.6$ normalized susceptance as shown in the next Smith chart of Figure 7.11. Then $d_2$ is

$$d_2 = (0.25 + 0.161)\lambda = 0.411\lambda$$

To summarize, the two solutions are as follows:

Case A
$$d_1 = 0.22\lambda$$
$$d_2 = 0.0885\lambda$$

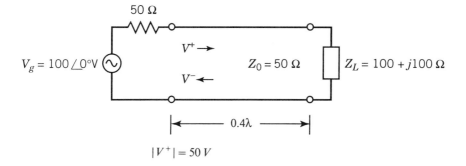

**FIGURE 7.6**

Case B
$$d_1 = 0.381\lambda$$
$$d_2 = 0.411\lambda$$

The case with the shorter lengths is preferred since losses will be smaller. Of course, one could add half-wavelengths to the above, but this serves no useful purpose.

Before the stub is put into place, the conditions on the line are as shown in Figure 7.6:

$$|V^+| = 50 \text{ V}$$

$$P_L = 15.385 \text{ W}$$

$$\text{VSWR} = S = 4.25 \quad \text{(read off the chart)}$$

At the stub plane, the problem is reduced to a lumped equivalent as follows (use case A). First, Thévenize the length $(0.4 - 0.22)\lambda$ and generator as shown in Figure 7.7 (Supplementary Example 5.9):

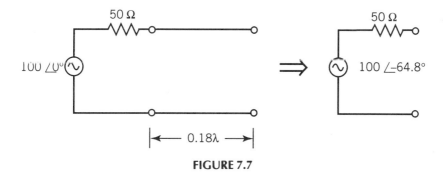

**FIGURE 7.7**

## 312 SINGLE FREQUENCY MATCHING

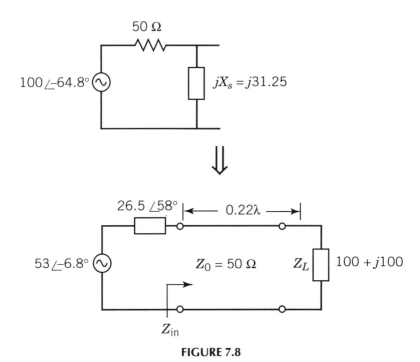

**FIGURE 7.8**

Now, using Figure 7.8, combine the stub's admittance and simplify:

$Y_s = (20 \, m\mho)(-j1.6)$

$$\therefore Z_s = \frac{1}{Y_s} = j31.25 \, \Omega$$

Calculate $Z_{in} = 26.5 \underline{/-57.6°}$ that is indeed $Z_G^*$. Now calculate $V^+$ (see Section 5.3):

$$|V^+| = 63.676 \, V$$

Notice that $|V^+|$ is greater when the stub is in place (it was 50 V before). Then

$$P_L = \frac{|V^+|}{2(50)}(1 - |\Gamma_L|^2) = 25 \, W$$

As a check, note $P_{avail}$:

$$P_{avail} = \frac{|100|^2}{8(50)} = 25 \, W \quad \text{(before stub)}$$

$$P_{avail} = \frac{|53|^2}{8(14.045)} = 25 \, W \quad \text{(after stub is in place)}$$

## IMPEDANCE OR ADMITTANCE COORDINATES

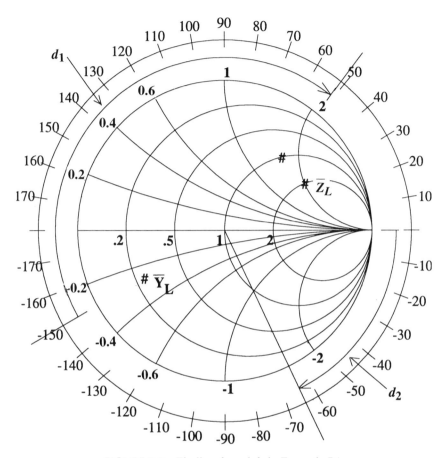

**FIGURE 7.9** Finding $d_1$ and $d_2$ in Example 7.1.

Thus, the generator is delivering its $P_{avail}$ and the load is absorbing it. The stub acts to boost $|V^+|$ (here from 50 to 63.676 V), so more "signal" impinges on the load. A simple way of looking at the situation is as follows. The wave on the stub arriving at the junction splits in such a way that the portion moving toward the generator helps to cancel with other wavelets moving in the same direction. The part that goes toward the load is in phase with that transmitted past the stub (from the generator), and this superposition increases $|V^+|$, which now carries more power toward the load. The load still reflects the same fraction of incident power, but the net absorbed is now $P_{avail}$. ▲

*Example 7.2* Discuss the matching procedure in the previous problem from a strictly lumped circuit approach.

## IMPEDANCE OR ADMITTANCE COORDINATES

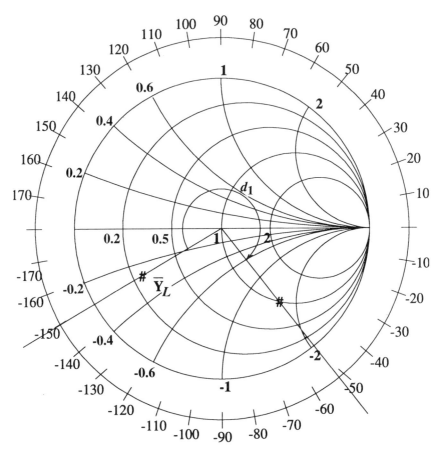

**FIGURE 7.10** The second solution for $d_1$ in Example 7.1.

The network is separated at plane $a-a$ (the plane of the stub) as shown in Figures 7.12 and 7.13

Convert the source side to a current source as shown in Figure 7.14:

Then the final lumped network becomes, see Figure 7.15,

The voltage across the circuit is

$$V = \frac{2\underline{/-64.8°}\,\text{A}}{(40+j32)\,\text{m}\mho} = 39.04\underline{/-103.46°}\,\text{V}$$

## IMPEDANCE OR ADMITTANCE COORDINATES

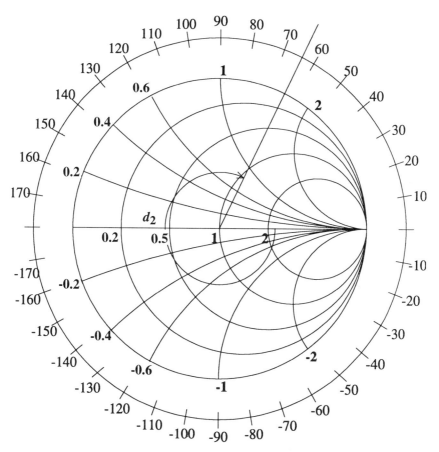

**FIGURE 7.11** Determination of $d_2$ (second solution) in Example 7.1.

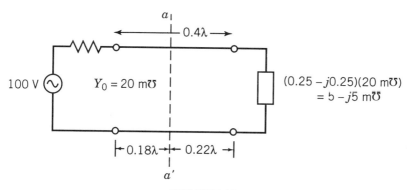

**FIGURE 7.12**

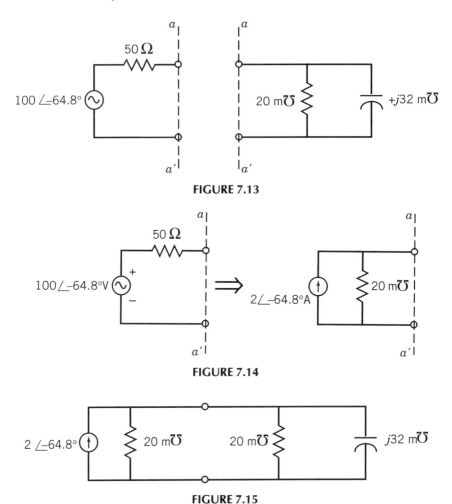

**FIGURE 7.13**

**FIGURE 7.14**

**FIGURE 7.15**

The currents in the capacitor and conductance are

$$I_C = (32\underline{/90°})(39.04\underline{/-103.46°}) = 1.249\underline{/-13.45°}$$
$$I_L = (20)(39.04\underline{/-103.46°}) = 0.781\underline{/-103.46°} \text{ A}$$

The power absorbed in the load is

$$P_L = \tfrac{1}{2}|0.781|^2 50 = 15.24 \text{ W}$$

Now when the stub is present, it looks inductive and has value $-j32$ m℧, which resonates out the capacitive component. Thus, Figure 7.16 reads,

BASICS, STUBS  317

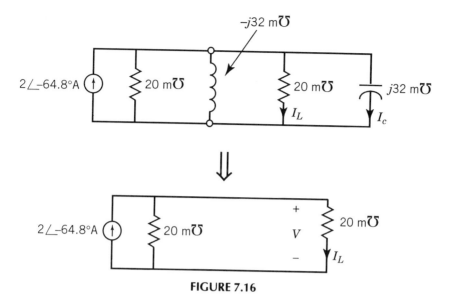

**FIGURE 7.16**

Now the currents and voltages are

$$I_L = 1\underline{/-64.8°} \text{ A}$$

$$V = \frac{1\underline{/-64.8°}}{20 \text{ m}\mho} = 50\underline{/-64.8°} \text{ V}$$

$$I_c = (32\underline{/90°})(50\underline{/-64.8°}) = 1.6\underline{/25.2°} \text{ A}$$

and the power absorbed is

$$P_L = \tfrac{1}{2}|1|^2(50) = 25 \text{ W} \qquad \blacktriangle$$

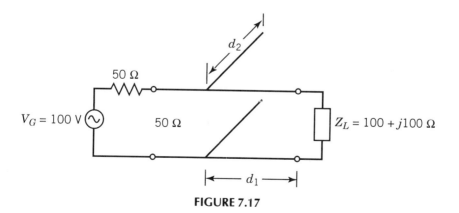

**FIGURE 7.17**

**Example 7.3** Match the load with an open 50-Ω stub. Then let the stub have $Z_0 = 75\,\Omega$ with a short as shown in Figure 7.17

Since the load is $100 + j100\,\Omega$, we know that the length $d_1$ is $0.22\lambda$ (choose the shorter length). Again, the stub must present $-j1.6$ normalized susceptance. For an open, we have $\bar{Y} = 0$ at the plane of the open. To arrive at $-j1.6$, we need to move $0.25 + 0.0885$ as shown on the Smith chart. Thus,

$$d_2 = 0.3385\lambda$$

For the case as in Figure 7.18, when the stub characteristic impedance is $75\,\Omega$.

When the stub and main line have different characteristic impedance values, it is best to work analytically to avoid errors. Recall that the same impedance will "jump" to a new point when the normalization of a chart is changed. The length $d_1$ stays the same in any case, so again choose $d_1 = 0.22\lambda$. The stub must present $(-j1.6)(20\,\text{m}\mho) = -j32\,\text{m}\mho$ susceptance. Here, it is easier to calculate $d_2$ rather than use a new Smith chart. The equation for a shorted line is

$$Y_{in} = -jY_0 \cot \beta d$$

Here, $Y_{in} = -j32\,\text{m}\mho$ and $Y_0 = 1/75\,\Omega = 13.33\,\text{m}\mho$. Then

$$\cot \beta d = \frac{32}{13.33} = 2.4$$

$$\therefore \beta d = 0.0628\lambda \qquad \blacktriangle$$

**Example 7.4** Summarize the procedure for a single shorted stub match.

1. Enter the normalized admittance $\bar{Y}_L$ in the admittance Smith chart as shown in Figure 7.19.

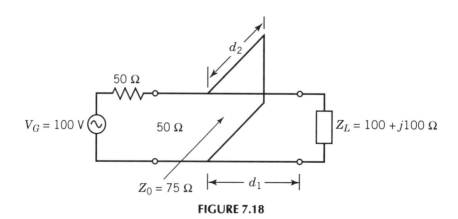

FIGURE 7.18

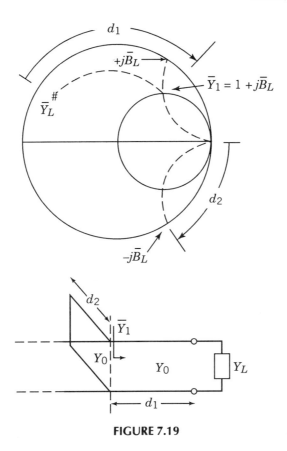

**FIGURE 7.19**

2. Move along a constant radius until the $g = 1$ circle is intersected. (There are always two such intersections.)
3. The distance moved is the length $d_1$, the distance from the load plane to the plane of the stub.
4. At intersection of the $g = 1$ circle, the normalized admittance is $\bar{Y}_1 = 1 + j\bar{B}_L$. Now find the value $-j\bar{B}_L$ on the periphery of a Smith chart and draw an arc starting from the right-most point on the chart. The length of this arc is the length $d_2$ (the stub's length). ▲

**Example 7.5** Match the given load with a series stub as shown in Figure 7.20. Notice that such a construction can be completed with a coaxial system as indicated.

The normalized impedance and admittance are $\bar{Z}_L = 0.5 + j1$, $\bar{Y}_L = 0.4 - j0.8$:

$$\bar{Y}_L \quad \text{at } 0.1151 \text{ WTL scale}$$
$$\therefore d_1 = (0.1151 + 0.178)\lambda = 0.2931\lambda$$

# SINGLE FREQUENCY MATCHING

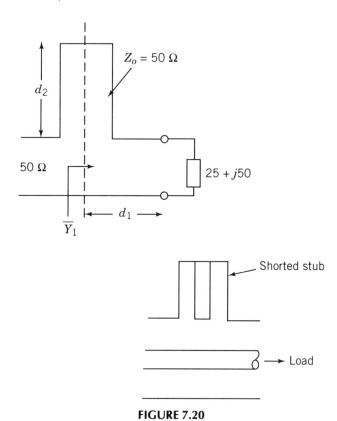

**FIGURE 7.20**

Thus,

$$\bar{Y}_1 = 1 + j1.575$$

Since we are using a series stub, use the normalized impedance:

$$\bar{Z}_1 = \frac{1}{\bar{Y}_1} = 0.2873 - j0.4525$$

as shown in Figure 7.21.

Thus, the stub adds $+j0.4525$. Then the length is determined by the formula:

$$Z_{in} = jZ_0 \tan \beta d_2$$
$$\therefore \beta d_2 = 0.0676\lambda$$

▲

## IMPEDANCE OR ADMITTANCE COORDINATES

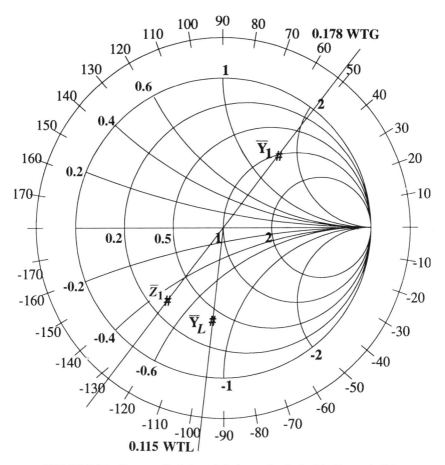

**FIGURE 7.21** Steps to find $d_1$ and $d_2$ (second solution) in Example 7.5.

## 7.2 DOUBLE-STUB TUNER

The single-stub tuner discussed in the previous section has a major drawback when it is implemented in a laboratory setting. Only two positions from the load are available, and at each position, a precisely defined length of stub is required. The double-stub tuner (shown in Fig. 7.22) reduces the above constraints, and with adjustable-length stubs, a large range of loads may be matched for a fixed separation between stubs. The distance $d$ is arbitrary, so we replace the actual load $\mathbf{Z}'_L$ and line of length $d$ by its equivalent $\mathbf{Z}_L$ at the plane of stub 1. The separation between stubs is fixed, and for convenience we set it to $\lambda/8$. The principle of this matching scheme is to choose $l_1$ such that its susceptance,

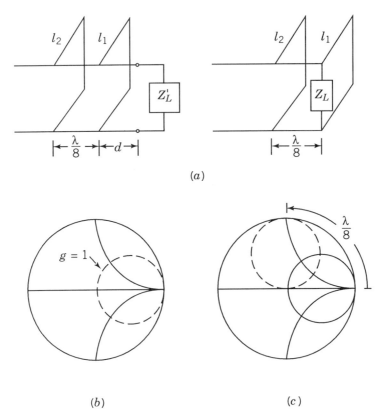

**FIGURE 7.22** The double-stub tuner is shown in (a). It allows a wide range of loads to be matched by changing the lengths $l_1$ and $l_2$. Basically, the two lengths $d_1$ and $d_2$ in the single-stub arrangement are replaced by the two above. In (b) the $g = 1$ circle is highlighted to indicate the needed locus presented to the right of stub 2. (c) Shows the $g = 1$ rotated $\lambda/8$ toward the load (WTL), which is the locus on which the net load at the plane of stub 1 must lie.

combined with that of the load $Y_L$, forms a particular admittance for the next step. The step is moving this combined admittance the distance between the stubs. If the admittance just to the right of stub 2 is on the $g = 1$ circle, then stub 2 can act like that of a single stub, cancelling the susceptance. To find the admittance that stub 1 must present, we have to "look ahead." With reference to Figure 7.22b, we know that the admittance presented to stub 2 must lie (we force this condition) somewhere on the $g = 1$ circle. This locus, when transformed back to the plane of stub 1, must look as shown in Figure 7.22c. The dotted circle is just the entire $g = 1$ circle rotated toward the load. Any admittance on this dotted circle, when rotated toward the generator, falls on the $g = 1$ circle. This is the key step in understanding such a scheme. Since the equivalent load at the plane of

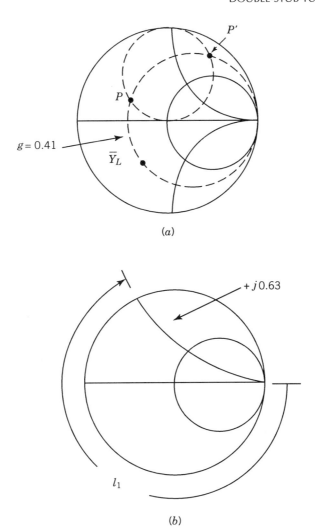

**FIGURE 7.23** An arbitrary load $\bar{Y}_L$ and its constant $g$ circle highlighted. (a) One intersection of this circle and the rotated $g = 1$ circle is denoted as point $P$. (b) Indicates the normal procedure of determining the length of a stub, given its required input admittance.

stub 1 is restricted to this rotated $y = 1$ circle, it determines the length of stub 1 as follows. With reference to Figure 7.23a, we show the rotated $g = 1$ circle and an arbitrary load $\bar{Y}_L$. Sketched is the constant $g$ circle for that load, and we denote one intersection of it and the rotated $g = 1$ circle as point $P$. Thus, we require stub 1 to combine with $\bar{Y}_L$ so as to arrive at this point. For example, let

$$\bar{Y}_L = 0.41 - j0.44$$

We recognize that the stub's admittance is purely imaginary, so the real part of the combined admittance always has the real part of $\bar{Y}_L$. Therefore, for all lengths $l_1$, the real part stays on the 0.41 normalized conductance circle. From the chart, we read the coordinates of $P$ as $0.41 + j0.19$. Then we require

$$\bar{Y}_{s_1} + \bar{Y}_L = 0.41 + j0.19 \tag{7.17}$$

which is the defining equation for the length of stub 1. We find

$$\bar{Y}_{s_1} = j0.19 + j0.44 = j0.63$$

Thus, the input admittance of stub 1 is determined. If the other point of intersection $P'$ had been chosen, its coordinates would have been the right side of Eq. (7.17). The length of stub 1 is found in the normal fashion (Fig. 7.23b). Here,

$$l_1 = (0.25 + 0.09)\lambda$$
$$= 0.34\lambda$$

Now rotate point $P$ to the $g = 1$ circle and read $1 + j0.975$. Thus, stub 2 must present $-j0.975$, and its length is found just as in the single-stub case:

$$l_2 = (0.3775 - 0.25)\lambda$$
$$= 0.1275\lambda$$

As a final note, observe that the distance between the stubs and the load $\mathbf{Z}_L$ cannot be chosen completely arbitrarily. In Figure 7.24, we show a situation where the separation between stubs is greater than $\lambda/8$, but less than $\lambda/4$. The constant $g$ circle tangent to this rotated $g = 1$ circle is indicated with asterisks and shaded. Notice that any load $\bar{Y}_L$ lying within this region is on a constant $g$ circle that cannot intersect the rotated $g = 1$ circle. Thus, no point $P$ exists, and no solution is possible. However, any load $\bar{Y}_L$ outside this region is on a $g$ circle that intersects the rotated $g = 1$ circle twice, and two solutions exist. If the load happens to lie on the constant $g$ circle (with asterisks), then it has just one solution; that is, the point of tangency is the point $P$. For a given separation, one can vary $d$ to push $\mathbf{Z}_L$ into a favorable region.

**Example 7.6** Use a double-stub tuner with separation $\lambda/8$ to match $30 - j30\,\Omega$ to $50\,\Omega$. What loads cannot be matched by this tuner? See Figure 7.25.

Plot
$$\bar{Y}_L = 0.833 + j0.833$$

Move along the $g = 0.833$ circle and intersect the rotated $g = 1$ circle at

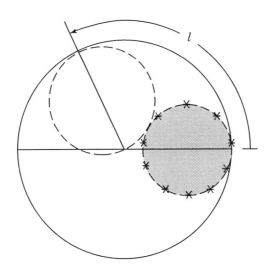

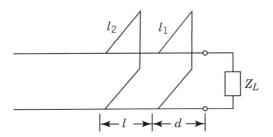

**FIGURE 7.24** A situation where certain loads $\bar{Y}_L$ cannot be matched by a specific separation between the two stubs of the tuner. For the separation shown (between $\lambda/8$ and $\lambda/4$), those loads $\bar{Y}_L$ within the shaded region have $g$ circles that do not intersect the rotated $g = 1$ circle, so no solution exists. In such a case, either the separation between stubs is changed, or $d$ is changed to push $\bar{Y}_L$ out of the shaded area.

$0.833 + j2.0$ as done in Figure 7.26. Then $\bar{Y}_{s_1}$ is determined as

$$\bar{Y}_{s_1} + \bar{Y}_L = 0.833 + j2.0$$
$$\bar{Y}_{s_1} + 0.833 + j0.833 = 0.833 + j2.0$$
$$\bar{Y}_{s_1} = j1.167$$

Thus,

$$l_1 = (0.25 + 0.137)\lambda = 0.387\lambda$$

Rotate point $P$ to the $g = 1$ circle; read $1.0 - j2.13$, so stub 2 must present $+j2.13$.

**326** SINGLE FREQUENCY MATCHING

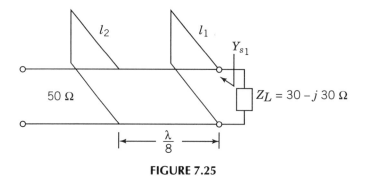

**FIGURE 7.25**

## IMPEDANCE OR ADMITTANCE COORDINATES

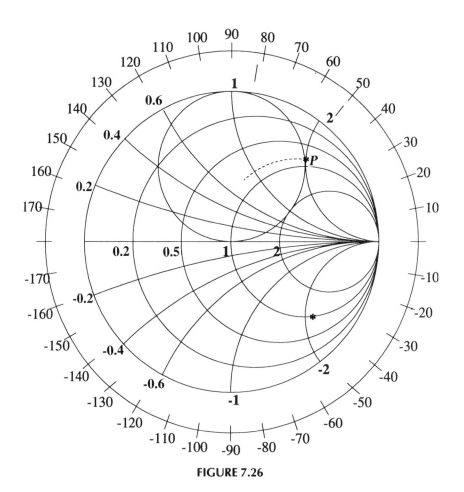

**FIGURE 7.26**

This gives

$$l_2 = (0.25 + 0.1807)\lambda$$
$$= 0.4307\lambda$$

The loads that cannot be matched by this separation have normalized real parts within the $g = 2$ circle. ▲

## 7.3 QUARTER WAVE TRANSFORMERS

The properties of quarter wave transformers are deduced by first examining the general input impedance formula:

$$\mathbf{Z}_{in} = \mathbf{Z}_0 \frac{\mathbf{Z}_L + j\mathbf{Z}_0 \tan \beta l}{\mathbf{Z}_0 + j\mathbf{Z}_L \tan \beta l}$$

Evaluating this for the special case $\beta l = \pi/2$ ($l = \lambda/4$) gives

$$\mathbf{Z}_{in} = \frac{\mathbf{Z}_0^2}{\mathbf{Z}_L} \tag{7.18}$$

Thus, the section inverts the load impedance and scales by the factor $\mathbf{Z}_0^2$. This relationship says the section of line can "match" the load to a generator if $\mathbf{Z}_{in} = \mathbf{Z}_G^*$.

Then

$$\mathbf{Z}_0 = \sqrt{\mathbf{Z}_G^* \mathbf{Z}_L} \tag{7.19}$$

However, if $\mathbf{Z}_G$ and $\mathbf{Z}_L$ are complex, then $\mathbf{Z}_0$ is in general complex, which is an undesirable situation. Recall that complex $\mathbf{Z}_0$ implies a lossy line. Thus, the load and generator must both be purely real to use a realizable line between them. Therefore,

$$\mathbf{Z}_0 = \mathbf{Z}_T = \sqrt{R_G R_L} \tag{7.20}$$

This equation gives a unique value for the intervening line, and since a single section can perform a match, the next logical step is a cascade of $N$ transformers. The two-section ($N = 2$) transformer shown in Figure 7.27b is defined by

$$R_G = \left(\frac{\mathbf{Z}_{01}}{\mathbf{Z}_{02}}\right)^2 R_L \tag{7.21}$$

which we see does not uniquely define $\mathbf{Z}_{01}$ and $\mathbf{Z}_{02}$; just their ratio is restricted by

# 328 SINGLE FREQUENCY MATCHING

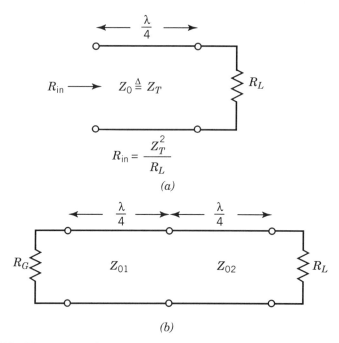

**FIGURE 7.27** (a) Single $\lambda/4$ transformer. (b) Two-section transformer.

$R_G$ and $R_L$. The choice of values for $Z_{01}$ and $Z_{02}$ for the widest matching bandwidth, or inband ripple, and/or ease of calculation has been an area of considerable study and will be dealt with next. Before this area is investigated, however, the next example discusses the use of a stub and transformer together as a matching scheme. The stub cancels the load reactance; then the resulting real part is transformed by the quarter wave section.

**Example 7.7** Match the load in Figure 7.28 with a shorted 50-$\Omega$ stub and quarter wave transformer.

The normalized load is $\bar{Z}_L = (100 + j100)/50 = 2 + j2$. Then

$$\bar{Y}_L = 0.25 - j0.25$$

We must add $+j0.25$, which the stub produces in the normal fashion (Fig. 7.29).

$$d = (0.25 + 0.039)\lambda$$
$$= 0.289\lambda$$
$$\therefore \bar{Y}_{\text{tot}} = 0.25 + j0$$

## QUARTER WAVE TRANSFORMERS 329

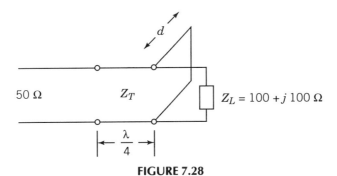

**FIGURE 7.28**

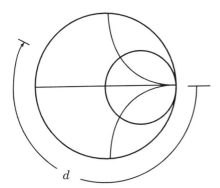

**FIGURE 7.29**

Then the normalized impedance seen by the transformer section is

$$\bar{Z}_{tot} = \frac{1}{\bar{Y}_{tot}} = 4 \qquad Z_{tot} = 4(50) = 200\,\Omega$$

Using the formula for the transformer gives the result

$$Z_T = \sqrt{(200)(50)}$$
$$= 100\,\Omega \qquad \blacktriangle$$

Multisection transformers have a long history (Ref. 1), and design formulas are available in many sources, Refs. 2 and 3. The geometry and nomenclature for such transformers are given in Figure 7.30. For the special case of four transforming sections, the match condition states

$$R_G = \left(\frac{R_1}{R_2}\right)^2 \left(\frac{R_3}{R_4}\right)^2 R_L \qquad (7.22)$$

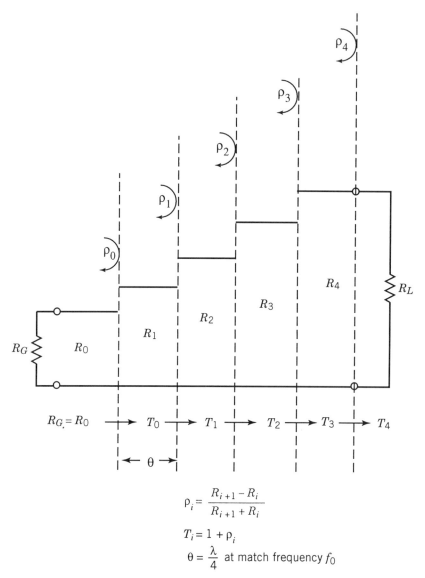

**FIGURE 7.30** A four-section transformer indicating the reflection and transmission coefficients at each junction.

The choices for the transformer values $R_1, R_2, R_3, R_4$ follow from the theory of small reflections (Ref. 4). If $R_G \simeq R_L$, then the reflections at each step are small, and one proceeds to develop an approximate theory for the structure. Recall that at the junction between two lines (see Fig. 7.31)

$$V_{\text{tot}} = V^+ + V^- = V^+ + \Gamma V^+ \triangleq TV^+ \tag{7.23}$$

where the equations accompanying the figure are:

$$\rho_i = \frac{R_{i+1} - R_i}{R_{i+1} + R_i}$$

$$T_i = 1 + \rho_i$$

$$\theta = \frac{\lambda}{4} \text{ at match frequency } f_0$$

QUARTER WAVE TRANSFORMERS   331

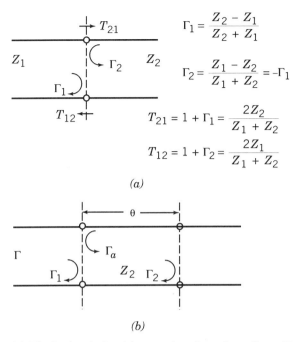

**FIGURE 7.31** (a) The basic relationships at a junction of two lines. (b) The details of a single section of line of length $\theta$.

which defines the transmission coefficient $T$. In Figure 7.31b, we analyze the reflections at two steps separated distance $\theta$. Assume that an initial wave of unit amplitude arrives from the left. Then

$$\Gamma = \Gamma_1 + T_{21}\Gamma_2 T_{12} e^{-j2\theta} + T_{21}\Gamma_2 \Gamma_a \Gamma_2 T_{12} e^{-j4\theta} + T_{21}\Gamma_2 \Gamma_a \Gamma_2 \Gamma_a \Gamma_2 T_{12} e^{-j6\theta} + \cdots$$

$$= \Gamma_1 + T_{12}T_{21}\Gamma_2 e^{-j2\theta} + T_{12}T_{21}\Gamma_2 (\Gamma_a \Gamma_2) e^{-j4\theta} + T_{12}T_{21}\Gamma_2 (\Gamma_a \Gamma_2)^2 e^{-j6\theta} + \cdots$$

$$= \Gamma_1 + T_{12}T_{21}\Gamma_2 e^{-j2\theta} \sum_{n=0}^{\infty} \Gamma_a^n \Gamma_2^n e^{-j2n\theta}$$

$$= \Gamma_1 + \frac{T_{12}T_{21}\Gamma_2 e^{-j2\theta}}{1 - \Gamma_a \Gamma_2 e^{-j2\theta}} \tag{7.24}$$

Now

$$T_{12}T_{21} = (1 + \Gamma_2)(1 + \Gamma_1) = 1 + \Gamma_1 + \Gamma_2 + \Gamma_1 \Gamma_2 \simeq 1$$

where we have used $\Gamma_2 = -\Gamma_1$. Therefore,

$$\Gamma \doteq \Gamma_1 + \Gamma_2 e^{-j2\theta} \tag{7.25}$$

**332** SINGLE FREQUENCY MATCHING

Carrying this further, when the first reflection is subscripted 0, one finds

$$\Gamma = \Gamma_0 + \Gamma_1 e^{-j2\theta} + \Gamma_2 e^{-j4\theta} + \cdots + \Gamma_N e^{-j2N\theta} \quad (7.26)$$

for an $N$ section transformer. With reference to Fig. 7.30, the reflection coefficient at the plane of $R_G$ is

$$\Gamma = \rho_0 + \rho_1 e^{-j2\theta} + \rho_2 e^{-j4\theta} + \cdots + \rho_N e^{-j2N\theta} \quad (7.27)$$

where the last reflection is

$$\rho_N = \frac{R_L - R_N}{R_L + R_N} \quad (7.28)$$

Here, we assume $R_L > R_G$, but if $R_L < R_G$, just replace each $\rho_i$ by $-\rho_i$ in any equation. Assume the transformer is symmetrical:

$$\rho_0 = \rho_N$$
$$\rho_1 = \rho_{N-1}$$
$$\rho_2 = \rho_{N-2}$$
etc.

Then Eq. (7.27) becomes

$$\Gamma = e^{-jN\theta}[\rho_0(e^{jN\theta} + e^{-jN\theta}) + \rho_1(e^{j(N-2)\theta} + e^{-j(N-2)\theta}) + \cdots]$$

The last term is

$$\rho_{(N-1)/2}(e^{j\theta} + e^{-j\theta}), \quad N \text{ odd}$$
$$\rho_{N/2}, \quad N \text{ even}$$

Then

$$\Gamma = 2e^{-jN\theta}\{\rho_0 \cos N\theta + \rho_1 \cos(N-2)\theta + \cdots + \rho_N \cos(N-2n)\theta + \cdots\} \quad (7.29)$$

The last term is

$$\rho_{(N-1)/2} \cos \theta, \quad N \text{ odd}$$
$$\tfrac{1}{2}\rho_{N/2}, \quad N \text{ even}$$

The next step is to choose the behavior for $|\Gamma|$ and, from that choice, determine the transformer impedance values. The first case is the "binomial" or maximally

## QUARTER WAVE TRANSFORMERS

flat (when $N = 2$) case. Choose

$$\Gamma = A(1 + e^{-j2\theta})^N \tag{7.30}$$

$$\rho = |\Gamma| = |A2^N(\cos\theta)^N| \tag{7.30a}$$

We determine $A$ by using $\theta = 0$ or $\pi$, that is, the transformer becomes transparent,

$$\rho = A2^N = \frac{R_L - R_G}{R_L + R_G} \tag{7.31}$$

$$\therefore A = 2^{-N} \frac{R_L - R_G}{R_L + R_G} \tag{7.31a}$$

This allows us to express $\Gamma$ as

$$\Gamma = 2^{-N} \frac{R_L - R_G}{R_L + R_G}(1 + e^{-j2\theta})^N = 2^{-N} \frac{R_L - R_G}{R_L + R_G} \sum_{n=0}^{N} C_n^N e^{-j2n\theta} \tag{7.32}$$

where

$$C_n^N = \frac{N!}{(N-n)!n!} \tag{7.33}$$

is the binomial coefficient. Comparing Eqs. (7.27) and (7.32) shows

$$\rho_n = \rho_{N-n} = 2^{-N} \frac{R_L - R_G}{R_L + R_G} C_n^N \tag{7.34}$$

since

$$C_n^N = C_{N-n}^N$$

With the above formulas, one calculates the transformer values; a useful approximation for finding them is as follows. Observe that

$$\ln x = 2\left[\frac{x-1}{x+1} + \frac{1}{3}\left(\frac{x-1}{x+1}\right)^3 + \frac{1}{5}\left(\frac{x-1}{x+1}\right)^5 + \cdots\right], \quad x > 0$$

Then

$$\ln \frac{R_{n+1}}{R_n} \doteq 2 \frac{R_{n+1} - R_n}{R_{n+1} + R_n} = 2\rho_n, \quad \rho_n < 0.35 \tag{7.35}$$

## SINGLE FREQUENCY MATCHING

where only the first term has been kept. Note also

$$2\frac{R_L - R_G}{R_L + R_G} \doteq \ln\frac{R_L}{R_G}$$

Then

$$\ln\frac{R_{n+1}}{R_n} = 2\rho_n = 2^{-N} C_n^N \ln\frac{R_L}{R_G} \quad (7.36)$$

where we assume $0.5R_G < R_L < 2R_G$ for the approximation to hold. Another formula for multisection transformers is (see Ref. 5)

$$\ln\frac{R_{n+1}}{R_n} = \frac{C_n^N \ln(R_L/R_G)}{\sum_{n=0}^{N} C_n^N} \quad (7.37)$$

Notice that the above equations would be the same if

$$2^N = \sum_{n=0}^{N} C_n^N \quad (7.38)$$

which is an identity. Thus, the same formula may have several appearances in the literature (see Ref. 6). For $N = 2$, the response is maximally flat, and for larger $N$, the behavior is acceptable. Using Eq. (7.36), we observe

$$\rho_1 + \rho_2 + \rho_3 + \cdots + \rho_N = \tfrac{1}{2}\ln\frac{R_L}{R_G} \quad (7.39)$$

As an example of the use of this form, let $R_L = 1000\,\Omega$ and $R_G = 100\,\Omega$, and use a five-section transformer. Since $\rho_0 = \rho_5$, $\rho_1 = \rho_4$, $\rho_2 = \rho_3$,

$$2(\rho_0 + \rho_1 + \rho_2) = \tfrac{1}{2}\ln\frac{1000}{100} = 1.15 \quad (7.40)$$

Using Eq. (7.34), we obtain

$$\rho_0 = 2^{-5}\frac{900}{1100}\, C_0^5 = 0.025568$$

and

$$2\rho_0 = \ln\frac{R_1}{R_0}, \text{ that is } (R_0 = R_G) \rightarrow R_1 = (1.05247)R_G = 105\,\Omega$$

Using Eq. (7.36):

$$\rho_1 = \frac{1}{2^6} C_2^5(1.15) = 0.17968$$

Then

$$2\rho_1 = \ln \frac{R_2}{R_1} \rightarrow R_2 = 150.4 \, \Omega$$

Solving Eq. (7.40) gives

$$\rho_2 = 0.36975$$

Thus,

$$2\rho_2 = \ln \frac{R_3}{R_2} \rightarrow R_3 = 315 \, \Omega$$

and

$$2\rho_2 = 2\rho_3 = \ln \frac{R_4}{R_3} \rightarrow R_4 = 660 \, \Omega$$

$$2\rho_1 = 2\rho_4 = \ln \frac{R_5}{R_4} \rightarrow R_5 = 945 \, \Omega$$

$$2\rho_0 = 2\rho_5 = \ln \frac{R_6}{R_5} \rightarrow R_6 = 995 \, \Omega$$

(should be $1000\,\Omega$, so close enough!). See Figure 7.32 for the structure, and

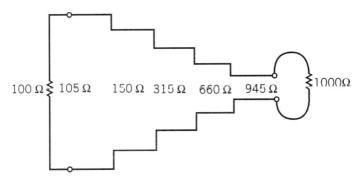

**FIGURE 7.32** A five-section transformer for matching 1000 to $100\,\Omega$.

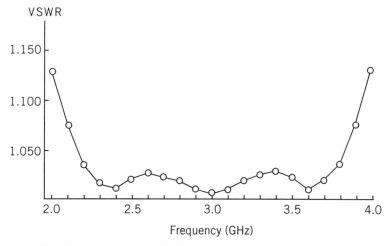

**FIGURE 7.33** VSWR for the five-section transformer in Figure 7.32.

Figure 7.33 for the frequency response. For the case $N = 2$, analysis shows

$$Z_1 = R_1 = R_L^{1/4} R_G^{3/4} \tag{7.41a}$$
$$R_2 = R_L^{3/4} R_G^{1/4} \tag{7.41b}$$

then for $R_L = 400$, $R_G = 100$,

$$Z_{01} = (4.472)(31.62) = 141.4 \, \Omega$$
$$Z_{02} = (89.44)(3.16) = 282.8 \, \Omega$$

**Example 7.8** Find the bandwidth of the binomial transformer.
From Eqs. (7.30a) and (7.31b), we have

$$\rho = \left| \frac{R_L - R_G}{R_L + R_G} (\cos \theta)^N \right|$$

Let $\rho_m$ be the largest value at the input (which defines the bandwidth). Then

$$|\rho_m| = \left| \frac{R_L - R_G}{R_L + R_G} (\cos \theta_m)^N \right|$$

or

$$\cos \theta_m = \left\{ \left| \frac{R_L + R_G}{R_L - R_G} \right| |\rho_m| \right\}^{1/N} \quad \text{Choose value} < \frac{\pi}{2} \tag{1}$$

## QUARTER WAVE TRANSFORMERS

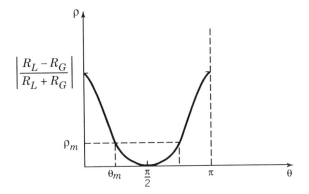

**FIGURE 7.34** Reflection coefficient of a binomial transformer.

Write

$$\theta = \left(\frac{\pi}{2}\right)\frac{f}{f_0} \qquad (2)$$

where $f = f_0$ is the frequency at which each section is exactly a quarter wavelength. Then define (see Fig. 7.34)

$$\frac{\Delta f}{f_0} = \frac{2(f_0 - f_m)}{f_0}$$

Thus,

$$\theta_m = \frac{\pi}{2}\frac{f_m}{f_0}$$

so

$$\frac{\Delta f}{f_0} = 2 - \frac{4}{\pi}\theta_m \qquad (3)$$

and this is greater than a single transformer. The BW for a single transformer is determined by noting

$$\mathbf{Z}_{in} = \mathbf{Z}_0 \frac{R_L + j\mathbf{Z}_0 \tan \beta l}{\mathbf{Z}_0 + jR_L \tan \beta l}$$

Thus after some algebra and using $\mathbf{Z}_0 = \sqrt{R_G R_L}$, we get

$$\Gamma = \frac{\mathbf{Z}_{in} - R_G}{\mathbf{Z}_{in} + R_G} = \frac{R_L - R_G}{R_L + R_G + j2\sqrt{R_L R_G}\tan \beta l}$$

# 338 SINGLE FREQUENCY MATCHING

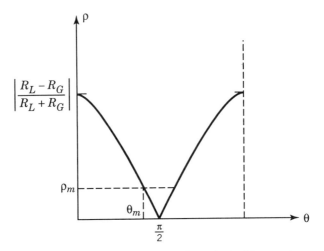

**FIGURE 7.35** Reflection coefficient for a single $\lambda/4$ transformer.

Then

$$\rho = |\Gamma| = \frac{1}{\left[1 + \left(\frac{2\sqrt{R_G R_L}}{R_L - R_0} \sec \theta\right)^2\right]^{1/2}}$$

which is sketched in Figure 7.35.

at $\theta = \theta_m$, $\rho = \rho_m$ and thus

$$\theta_m = \cos^{-1}\left[\frac{2\rho_m \sqrt{R_G R_L}}{(R_L - R_1)\sqrt{1 - \rho_m^2}}\right] < \frac{\pi}{2}$$

Choose a solution $< \pi/2$. Then

$$\frac{\Delta f}{f_0} = 2 - \frac{4}{\pi}\theta_m \qquad \blacktriangle$$

**Example 7.9** Match $R_L = 400\,\Omega$ to $R_G = 100\,\Omega$ using transformers of one to five sections.

The single case gives

$$Z_T = \sqrt{R_L R_G} = \sqrt{40{,}000} = 200\,\Omega \tag{1}$$

The binomial two-section has been given in Eqs. (7.41a) and (7.41b):

$$Z_1 = R_1 = 141.4\,\Omega \tag{2a}$$

$$R_2 = 282.8\,\Omega \tag{2b}$$

The three-section proceeds as

$$\ln \frac{R_1}{R_0} = 2^{-3} C_0^3 \ln 4, \qquad C_0^3 = 1$$

$$\ln \frac{R_2}{R_1} = 2^{-3} C_1^3 \ln 4, \qquad C_1^3 = 3$$

$$\ln \frac{R_3}{R_2} = 2^{-3} C_2^3 \ln 4, \qquad C_2^3 = 3$$

$$\ln \frac{R_L}{R_3} = 2^{-3} C_3^3 \ln 4, \qquad C_3^3 = 1$$

equating $2^{-3} \ln 4$ yields

$$\ln \frac{R_1}{R_0} = \tfrac{1}{3} \ln \frac{R_2}{R_1} = \tfrac{1}{3} \ln \frac{R_3}{R_2} = \ln \frac{R_L}{R_3}$$

or

$$3 \ln \frac{R_1}{R_0} = \ln \frac{R_2}{R_1} = \ln \frac{R_3}{R_2} = 3 \ln \frac{R_L}{R_3} \tag{3}$$

which shows that the ln of the ratios moving from one end of the transformer to the other goes as the binomial coefficient sequence 1:3:3:1. From Eq. (3), we obtain

$$R_2^2 = R_1 R_3$$
$$R_1 R_3 = R_L R_G = 4 \times 10^4$$
$$\therefore R_2^2 = 4 \times 10^4$$

and

$$R_2 = 200\,\Omega$$
$$R_1 = 119\,\Omega$$
$$R_3 = 336\,\Omega$$

Notice that when $N$ is odd, the middle section is $\sqrt{R_G R_L}$. If we return to the

## SINGLE FREQUENCY MATCHING

two-section case, we can use this principle. Find an intermediate value:

$$R_{int} = \sqrt{R_L R_G} = 200 \, \Omega$$

Then find

$$R_1 = \sqrt{R_G R_{int}} = \sqrt{(100)(200)} = 141 \, \Omega$$
$$R_2 = \sqrt{R_{int} R_L} = \sqrt{(200)(400)} = 282 \, \Omega$$

The four-section has immediately (the binomial sequence is 1:4:6:4:1)

$$\ln \frac{R_1}{R_0} = \tfrac{1}{4} \ln \frac{R_2}{R_1} = \tfrac{1}{6} \ln \frac{R_3}{R_2} = \tfrac{1}{4} \ln \frac{R_4}{R_3} = \ln \frac{R_L}{R_4} \tag{4}$$

Considerable algebra yields $R_3$ as

$$R_3^{3.2} = (R_0 R_L) R_L^{1.2} \tag{5}$$

and the final values are

$$R_1 \doteq 109 \, \Omega$$
$$R_2 \doteq 154 \, \Omega$$
$$R_3 \doteq 259 \, \Omega$$
$$R_4 \doteq 367 \, \Omega$$

The five-section starts out as

$$\ln \frac{R_1}{R_0} = 2^{-5} C_0^5 \ln(4) = \left(\frac{1}{32}\right)(1) \ln 4$$

or

$$R_1 = R_0 4^{1/32} = 104.43 \, \Omega$$

$$\ln \frac{R_2}{R_1} = 2^{-5} C_1^5 \ln 4 = \left(\frac{1}{32}\right)(5)$$

or

$$R_2 = R_1 4^{5/32} = 129.7 \, \Omega$$

$$\ln \frac{R_3}{R_2} = 2^{-5} C_2^5 \ln 4$$

or

$$R_3 = R_2 4^{10/32} = 200 \, \Omega$$

$$\ln \frac{R_4}{R_3} = \frac{1}{32} C_3^5 \ln 4$$

$$R_4 = R_3 4^{10/32} = 308.4 \, \Omega$$

QUARTER WAVE TRANSFORMERS    341

use

$$\frac{R_1}{R_G} = \frac{R_L}{R_5}$$

$$R_5 = \frac{R_L R_G}{R_1} = 383.0\,\Omega$$

or

$$R_5 = R_4 4^{5/32} = 383.0\,\Omega$$

The plot in Figure 7.36 shows the bandwidths available for the above matching networks. The center frequency was chosen to be 3.0 GHz, and the input

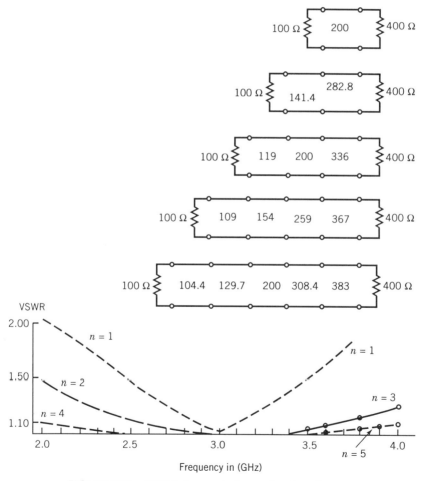

**FIGURE 7.36**   VSWR for one to five-section transformers.

VSWR is shown over the band from 2.0–4.0 GHz. The single transformer case ($n = 1$) response is shown over the entire band, whereas the other responses are only shown over half of the band. This is for the purpose of clarity only, since the responses are symmetrical about the center frequency. ▲

**Example 7.10** Another prescription for a two-section transformer has been given (Ref. 7), and it is

$$\frac{R_1}{R_G} = \frac{R_L}{R_2} = \left(\frac{R_L}{R_G}\right)^{1/(2+A)}, \quad 1.5 \leq A \leq 2$$

$A = 2$ gives the binomial case. For $A = 1.5$ and $R_G = 100\,\Omega$, $R_L = 400\,\Omega$, we find

$$R_1 = 148.6\,\Omega$$
$$R_2 = 269.2\,\Omega$$

These are not too far from the binomial case, but they do not satisfy Eq. (7.22). ▲

**Example 7.11** An N-section exponential transformer has $R_i$ defined by (see Ref. 8)

$$R_i = (R_L^i R_G^{N-i+1})^{1/(N+1)}$$

For $R_G = 100\,\Omega$, $R_L = 400\,\Omega$, and $N = 4$, we find

$$R_1 = 131.95\,\Omega$$
$$R_2 = 174.1\,\Omega$$
$$R_3 = 229.7\,\Omega$$
$$R_4 = 303.1\,\Omega$$

which differ quite a bit from the binomial case. Notice that the solution does not satisfy Eq. (7.22). The following plot (Fig. 7.37) shows the VSWR over a 2.0-GHz band centered about 3.0 GHz. The response of the four-section binomial case of Example 7.9 is shown in Figure 7.37 for comparison. ▲

**Example 7.12** When $R_G \simeq R_L$, a "short-step" transformer is available (see Ref. 9). The lengths of the steps are equal and are defined by

$$\tan \beta l = \sqrt{\frac{R_L/R_G - 1}{(R_1/R_G)^2 - R_2/R_1}}$$

# QUARTER WAVE TRANSFORMERS 343

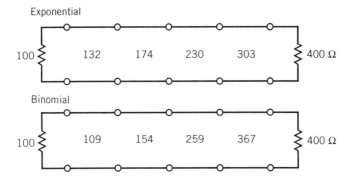

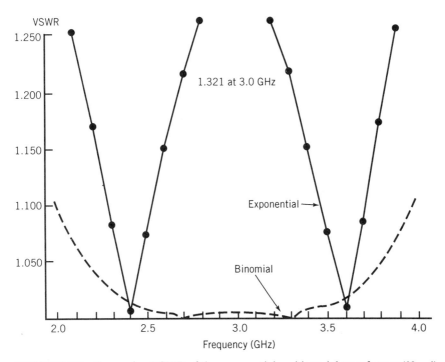

**FIGURE 7.37** Comparing VSWR of the exponential vs. binomial transformer ($N = 4$).

# SINGLE FREQUENCY MATCHING

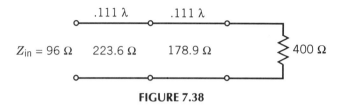

**FIGURE 7.38**

Here, $R_1 R_2 = R_G R_L$, but choose $R_2/R_1$ rather low to shorten the sections. For example, let

$$\frac{R_2}{R_1} = 0.8$$

$$R_G = 100\,\Omega$$

$$R_L = 400\,\Omega$$

$$\tan \beta l = 0.84515$$

$$\theta = \beta l = 40° = 0.111\lambda$$

which is depicted in Figure 7.38.

Figure 7.39 shows the VSWR for the 2.0 to 4.0 GHz band. Its performance for this transformation ratio (4:1) is about the same as that for a single-section quarter wave transformer; however, its length is only about 80°. ▲

The VSWR for the binomial transformer is approximately

$$\text{VSWR} = \frac{1 + |\Gamma|}{1 - |\Gamma|} \sim 1 + |\Gamma| = 1 + (\cos \theta)^N \ln\left(\frac{R_L}{R_G}\right) \tag{7.42}$$

which is useful for initial design purposes. This transformer lends itself to rather straightforward manipulation, but the Chebychev transformer gives a larger BW and is considered next. The choice for $\Gamma$ is now

$$\Gamma = e^{-jN\theta} \frac{R_L - R_G}{R_L + R_G} \frac{T_N(\sec \theta_m \cos \theta)}{T_N(\sec \theta_m)} \tag{7.43}$$

The maximum $\rho$ in the passband is

$$\rho_m = \frac{|R_L - R_G|}{(R_L - R_G)|T_N(\sec \theta_m)|} \tag{7.44}$$

The VSWR is

$$\text{VSWR} = 1 + \ln\left(\frac{R_L}{R_G}\right) \frac{T_N\left(\frac{\cos \theta}{\cos \theta_m}\right)}{T_N(\sec \theta_m)} \tag{7.45}$$

QUARTER WAVE TRANSFORMERS 345

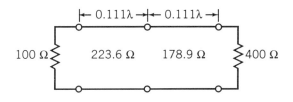

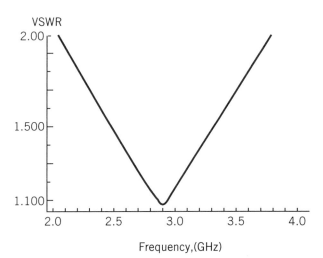

**FIGURE 7.39** VSWR of short step transformer.

and the passband characteristics are sketched in Figure 7.40. The calculations for this transformer are tedious and the design may be summarized as follows:

1. Given $R_L$, $R_G$, calculate

$$\rho_T = \frac{R_L - R_G}{R_L + R_G} \qquad (7.46)$$

2. Choose the maximum of $\rho_m$ in the band; then find

$$T_N(\sec \theta_m) = \frac{\rho_T}{\rho_m} \qquad (7.47)$$

3. Choose the number of $N$ sections (generally no more than four).
4. Write Eq. (7.29)

$$\Gamma = 2e^{-jN\theta}[\rho_0 \cos N\theta + \rho_1 \cos(N-2)\theta + \cdots + \rho_n \cos(N-2n)\theta + \cdots] \qquad (7.48)$$

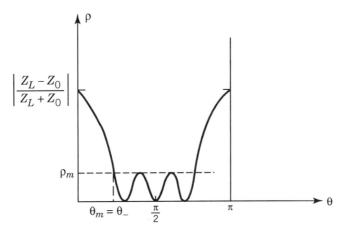

**FIGURE 7.40** The magnitude of the reflection coefficient for a Chebychev transformer.

and force this to match up with a Chebychev polynomial as

$$\Gamma = \rho_m T_N(\sec \theta_m) \tag{7.49}$$

Then depending on $N = 1, 2, 3,$ or $4$, expand

$$T_1(\sec \theta_m \cos \theta) = \sec \theta_m \cos \theta$$
$$T_2(\sec \theta_m \cos \theta) = 2(\sec \theta_m \cos \theta)^2 - 1$$
$$= \sec^2 \theta_m(1 + \cos 2\theta) - 1$$
$$T_3(\sec \theta_m \cos \theta) = \sec^3 \theta_m(\cos 3\theta + 3\cos \theta) - 3\sec \theta_m \cos \theta$$
$$T_4(\sec \theta_m \cos \theta) = \sec^4 \theta_m(\cos 4\theta + 4\cos 2\theta + 3)$$
$$- 4\sec^2 \theta_m(\cos 2\theta + 1) \tag{7.50}$$

Matching up Eqs. (7.48) and the appropriate form in Eq. (7.50) permits the calculation of the $\rho_n$, $n = 0, 1, 2, 3, 4$. Then the impedance values are obtained by

$$Z_1 = \frac{1 + \rho_0}{1 - \rho_0} R_G$$

$$Z_2 = \frac{1 + \rho_1}{1 - \rho_1} Z_1$$

etc.

## QUARTER WAVE TRANSFORMERS

**Example 7.13** Calculate the $R_i$ values for a four-section Chebychev transformer using the procedure outlined in Ref. 2.

The notation is

$$f_0 = \text{center frequency} \quad \left(\theta = \frac{\pi}{2}\right)$$

$$f_{+,-} = \text{upper and lower band edges}$$

$$\frac{f_+}{f_-} = \text{frequency ratio} \equiv \frac{1+F}{1-F}$$

$$F = \text{frequency coefficient} = \frac{f_+/f_- - 1}{f_+/f_- + 1}$$

$$C = \cos\left[(1-F)\frac{\pi}{2}\right]$$

| N | $\dfrac{\ln(R_L/R_G)}{\ln(R_1/R_G)}$ | $\dfrac{\ln(R_2/R_G)}{\ln(R_1/R_G)}$ |
|---|---|---|
| 2 | $4 - 2C^2$ | |
| 3 | $8 - 6C^2$ | |
| 4 | $16 - 16C^2 + 2C^4$ | $5 - 4C^3$ |

Let $R_G = 100\,\Omega$, $R_L = 400\,\Omega$ and choose the band at 8 to 12 GHz. Calculate

$$F = 0.2$$

$$C = \cos\left[(0.8)\frac{\pi}{2}\right] = 0.309$$

$$C^2 = 0.09549$$

$$C^3 = 0.02951$$

$$C^4 = 0.00912$$

$$16 - 16C^2 + 2C^4 = 14.49$$

$$5 - 4C^3 = 4.882$$

$$\ln\left(\frac{400}{100}\right) = 1.386$$

$$\therefore \ln\left(\frac{R_1}{R_0}\right) = \frac{1.386}{14.49} \Rightarrow R_1 = 110\,\Omega$$

$$\ln\left(\frac{R_2}{R_G}\right) = \frac{1.386}{14.49} \Rightarrow R_2 = 159.5\,\Omega$$

**348** SINGLE FREQUENCY MATCHING

Use

$$\rho_0 = \rho_4 = \frac{110 - 100}{110 + 100} = 0.04762$$

$$R_4 = 100\left(\frac{1-\rho_4}{1+\rho_4}\right) = 377.3\,\Omega$$

$$\rho_3 = \rho_1 = \frac{R_2 - R_1}{R_2 + R_1} = \frac{49.5}{267.5} = 0.1837$$

$$R_3 = R_4\frac{1-\rho_3}{1+\rho_3} = \frac{0.81633}{1.1837}(377.3) = 260.2\,\Omega$$

Comparing this with the binomial design in the table below

|       | Binomial | Chebyshev |
|-------|----------|-----------|
| $R_1$ | 109      | 110       |
| $R_2$ | 154      | 159.5     |
| $R_3$ | 259      | 260.2     |
| $R_4$ | 367      | 377.3     |

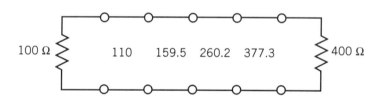

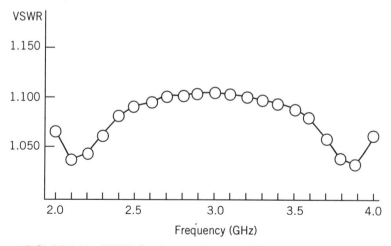

**FIGURE 7.41** VSWR for the transformer design described in Ref. 2.

so the values are not too much different. The VSWR over the band is given in Figure 7.41.

## 7.4 THE L-MATCH

The L-match uses two reactances configured as shown in Figure 7.42, which we recognize as either a low-pass or high-pass filter structure. The matching condition $\mathbf{Z}_s = \mathbf{Z}_p^*$ says (combine $jX_p$ and $R_p$)

$$R_s + jX_s = \frac{R_p X_p^2}{R_p^2 + X_p^2} - j\frac{R_p^2 X_p}{R_p^2 + X_p^2}$$

Thus,

$$R_s = \frac{R_p X_p^2}{R_p^2 + X_p^2} \tag{7.51}$$

$$X_s = -\frac{R_p^2 X_p}{R_p^2 + X_p^2} \tag{7.52}$$

From Eq. (7.51),

$$X_p = \pm\sqrt{\frac{R_s R_L}{R_L - R_s}}, \quad R_L > R_s \tag{7.53}$$

From Eq. (7.52),

$$X_s = -\frac{R_s R_p}{X_p} \tag{7.54}$$

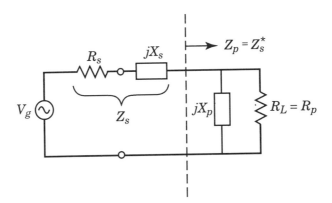

**FIGURE 7.42** The matching of two resistances using two reactances that form an inverted L (sometimes called a Γ match). The matching elements are combined with the resistances to form the "source and load" values.

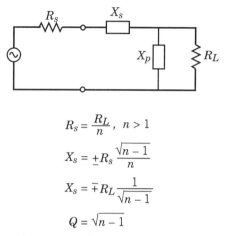

$$R_s = \frac{R_L}{n}, \quad n > 1$$

$$X_s = \pm R_s \frac{\sqrt{n-1}}{n}$$

$$X_s = \mp R_L \frac{1}{\sqrt{n-1}}$$

$$Q = \sqrt{n-1}$$

**FIGURE 7.43** Summary of the L-match.

Define a $Q$ for the $R_p$, $X_p$ elements (parallel):

$$Q_p = \frac{R_p}{|X_p|} = \sqrt{\frac{R_p}{R_s} - 1} \tag{7.55}$$

and for the series elements:

$$Q_s = \frac{|X_s|}{R_s} = \frac{R_p}{|X_p|} = Q_p \tag{7.56}$$

where we have used Eqs. (7.54) and (7.55). Observe from Eq. (7.54) that $X_s$ and $X_p$ have opposite signs. Equation (7.55) says that the $Q$ and thus the BW depend on the ratio of $R_p$ and $R_s$. The larger the ratio, the larger the $Q$, and consequently, the smaller the matching bandwidth. Figure 7.43 summarizes the L-match.

Quite often in analysis, it is convenient to convert between a series or parallel representation for a given complex impedance $\mathbf{Z} = R_s + jX_s$. This transformation is given in Figure 7.44 for future reference.

The L-match is quickly performed on a Smith chart, and with reference to Figure 7.45, we recognize that the shunt reactance puts the net load at the dotted plane somewhere on the $r = 1$ circle. (Here, the chart is normalized to $R_s$.) The series element then cancels the remaining reactance. In a manner similar to the double-stub tuner, we know that we must be on the $r = 1$ circle at the dotted plane. If we choose to use admittances at this plane, then all admittances are on a circle obtained by inverting the $r = 1$ circle through the chart center (the procedure for reciprocating). The "inverted $r = 1$" is sketched in Figure 7.46, and we see that the susceptance added in parallel with $R_L$ must place us on this admittance circle. We recognize that, as in the double-stub case, any $R_L < R_s$ will

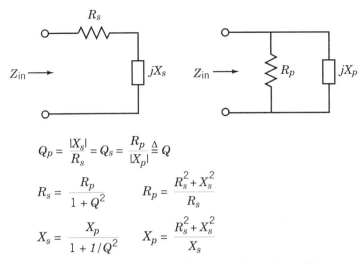

**FIGURE 7.44** Summary of the conversion between series and parallel representations for a given impedance.

have its admittance $1/R_L$ inside the $r = 1$ circle and be placed on the inverted $r = 1$ circle (for the sake of brevity, this circle will be called the target circle). This restriction was stated earlier in Eq. (7.53).

We may summarize the procedure on the Smith chart as follows:

1. Find

$$\bar{R}_p = \frac{R_p}{R_s}$$

2. Plot

$$\bar{Y}_p = \frac{1}{\bar{R}_p}$$

3. Move along a constant real part contour to intersect the target circle; we see that two solutions exist.
4. The arc used to reach the target circle defines the parallel element.
5. Convert the point $T$ on the target to its corresponding impedance value, shown in Figure 7.47.
6. The arc along the $r = 1$ circle defines the series reactance.

***Example 7.14*** Match $R_L = 1000\,\Omega$ to $R_s = 100\,\Omega$ as shown in Figure 7.48. Here, the load is

$$X_p = \pm \frac{R_p}{Q}$$

## 352  SINGLE FREQUENCY MATCHING

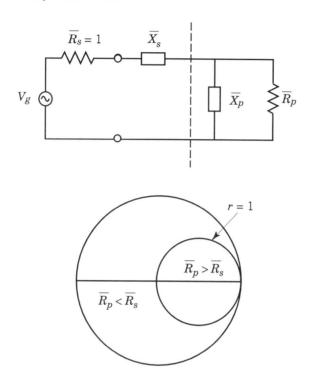

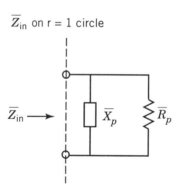

**FIGURE 7.45** The concept of the L-match on the Smith chart. The parallel element $\bar{X}_p$ places the net load $j\bar{X}_p \| \bar{R}_p$ somewhere on the $r = 1$ circle (normalization is with respect to $R_s$). Then the series element cancels the remaining reactance.

The $Q$'s are

$$Q_p = Q_s = \sqrt{10 - 1} = 3$$

Choose the low-pass form:

$$X_P = -\frac{R_p}{Q} = -333\,\Omega$$

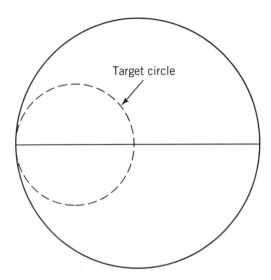

Normalized impedance plane

**FIGURE 7.46** The inverted $r = 1$ circle (called the target circle). Movement to the target circle along a constant real part contour determines the shunt matching element. After movement back to the $r = 1$ circle, one moves along this $r = 1$ contour to the chart center to determine the series matching element.

Then

$$X_s = QR_s = 3(100) = +300$$

as shown in Figure 7.49. ▲

*Example 7.15* In some instances, complex impedances can be matched by an L-match; the procedure is as follows. From the previous problem, we obtained Figure 7.50.

We may separate the $j300\,\Omega$ in series with the $100\,\Omega$ and ascribe $jX_1$ to the generator; then $Z_G = 100 + jX_1$. Similarly, one can interpret $-jX_4$ as part of the load $Z_L = 1000 \| (-jX_4)\Omega$. Thus, the $X_2, X_3$ form the L. The restrictions are that the series elements are both inductive and their sum remains $j300$, and the capacitors in shunt combine to $-j333\,\Omega$. ▲

*Example 7.16* Match $R_L = 1000\,\Omega$ to $R_s = 100\,\Omega$ using the L-match procedure on a Smith chart.

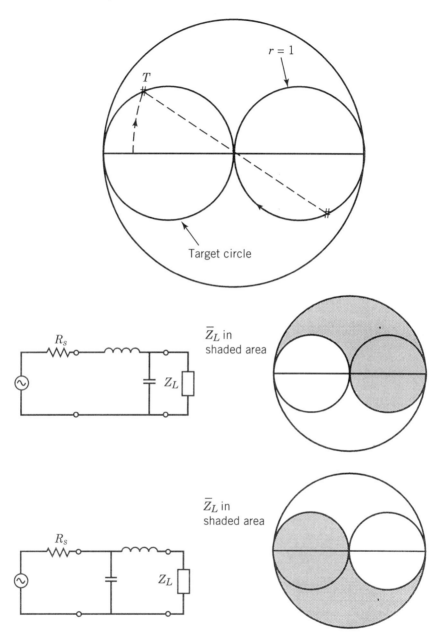

**FIGURE 7.47** Summary of the *L*-match on the Smith chart. Depending on the location of $\mathbf{Z}_L$, the "*L*" flips over. It turns out that the lower resistance side is always in series with the inductor.

THE L-MATCH    355

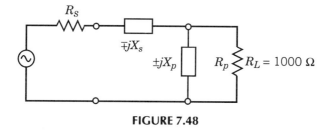

**FIGURE 7.48**

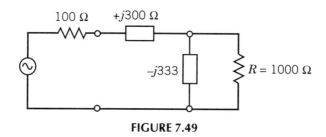

**FIGURE 7.49**

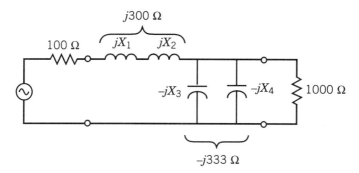

**FIGURE 7.50**

With reference to a Smith chart, the normalized load admittance is

$$\bar{R}_L = \frac{R_L}{R_s} = 10$$

$$\bar{Y}_L = \frac{1}{\bar{R}_L} = 0.1$$

To get to the target circle, we can add $\pm jB$, and we will choose $+jB$. Moving along the constant $g = 0.1$ circle, we intersect the target circle at $0.1 + j0.3$. Thus, the susceptance needed is $+j0.3$. The actual impedance is

$$\bar{X} = \frac{1}{B} = -j\frac{10}{3}$$

$$jX_p = -j\frac{10}{3}(100) = -j333\,\Omega$$

Now convert this to $\bar{Y}$ to $\bar{Z} = 1 - j3$, so the series element we add is $+j3$. Thus, the inductive reactance is

$$jX_s = j3(100) = +j300 \qquad \blacktriangle$$

**Example 7.17** Match $Z_L = 103.87 - j201.576$ to $50\,\Omega$ at $f = 300$ MHz.
The first step is

$$\bar{Z}_L = 2.077 - j4.03$$
$$\bar{Y}_L = 0.1 + j0.2$$

Then use a shunt capacitor.

$$\begin{array}{r} j0.293 \\ -j0.200 \\ \hline +j0.093 \end{array}$$

$$\therefore \quad j\omega\bar{C}_1 = +j0.093$$
$$\therefore \quad C_1 = 0.987\,\text{pF}$$

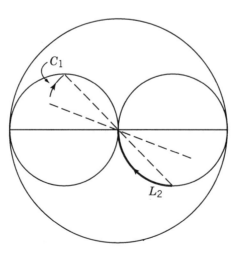

**FIGURE 7.51**

THE L-MATCH    357

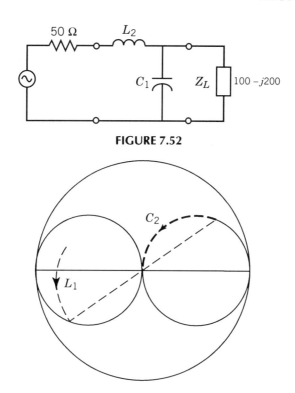

**FIGURE 7.52**

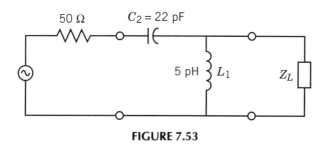

**FIGURE 7.53**

Reciprocate to find (See Fig. 7.51) $1 - j3.05$

$$\therefore \quad j\omega L_2 = +j3.05$$

Thus,

$$L_2 = \frac{(3.05)(50)}{2\pi(300) \times 10^6} = 80.9 \text{ nH}$$

The network is shown as Figure 7.52. The alternate configuration is Figure 7.53.

▲

## SUPPLEMENTARY EXAMPLES

**7.1** Verify the line lengths for the special case shown below in Figure 7.54.

$$\beta l_1 = \tan^{-1}\left(\frac{R_L}{R_0}\right)^{1/2}$$

$$\beta l_2 = \tan^{-1}\left(\frac{\sqrt{R_L R_0}}{R_L - R_0}\right)$$

From Eq. (7.12), we have

$$Y_s + Y_1 = Y_0 = \frac{1}{R_0} \tag{1}$$

where

$$Y_s = -j\left(\frac{1}{R_0}\right)\cot \beta l_2$$

$$Y_1 = Y_0 \frac{Y_L + jY_0 \tan \beta l_1}{Y_0 + jY_L \tan \beta l_1} = \frac{1}{R_0}\frac{R_0 + jR_L \tan \beta l_1}{R_L + jR_0 \tan \beta l_1}$$

Then Eq. (1) is

$$-j\cot \beta l_2 + \frac{R_0 + jR_L \tan \beta l_1}{R_L + jR_0 \tan \beta l_1} = 1$$

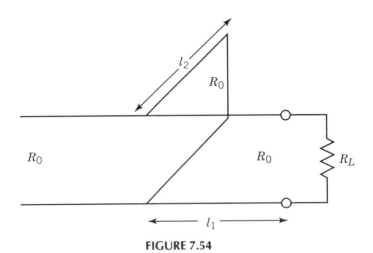

**FIGURE 7.54**

Let
$$X = \cot \beta l_2, \qquad Y = \tan \beta l_1$$
$$\therefore -jX + \frac{R_0 + jR_L Y}{R_L + jR_0 Y} = 1$$

Equate real parts:
$$R_0 XY + R_0 = R_L \tag{2}$$

Equate imaginary parts:
$$-XR_L + YR_L = R_0 Y \tag{3}$$
or
$$X = Y\frac{R_L - R_0}{R_L}$$

Then Eq. (3) into Eq. (2) gives
$$Y = \sqrt{\frac{R_L}{R_0}}$$
or
$$\tan \beta l_1 = \sqrt{\frac{R_L}{R_0}}$$
and
$$X = \cot \beta l_2 = \frac{R_L - R_0}{\sqrt{R_L R_0}}$$
$$\therefore \tan \beta l_2 = \frac{\sqrt{R_L R_0}}{R_L - R_0}$$

**7.2** Several programs have been published that calculate the lengths $\beta l_1$ and $\beta l_2$ for single- and double-stub matching problems (see Refs. 28 through 31). Here, we give the equations to be solved for the single-stub case. We treat the general case shown in Figure 7.55.

The stub may be either an open or shorted one. The starting point is Eq. (7.12):
$$Y_s + Y_1 = Y_{03} \tag{1}$$
where
$$Y_s = -jY_{02} \cot \beta l_2 \quad \text{(shorted stub)} \tag{2a}$$
$$= jY_{02} \tan \beta l_2 \quad \text{(open stub)} \tag{2b}$$

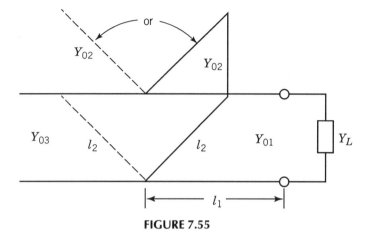

FIGURE 7.55

and

$$Y_1 = Y_{01}\frac{Y_L + jY_{01}\tan\beta l_1}{Y_{01} + jY_L\tan\beta l_1} = G_{in} + jB_{in} \quad (3)$$

is the load brought up the line a distance $l_1$.
Then Eq. (1) is

$$jB_s + G_{in} + jB_{in} = Y_{03}$$

and after equating real and imaginary parts, we obtain

$$G_{in} = Y_{03} \quad (4a)$$

$$B_s = -B_{in} \quad (4b)$$

From Eq. (3), we find

$$G_{in} = Y_{01}\frac{G_L Y_{01}(1 + \tan^2\beta l_1)}{(Y_{01} - B_L\tan\beta l_1)^2 + (G_L\tan\beta l_1)^2} \quad (5a)$$

$$B_{in} = Y_{01}\frac{(-\tan\beta l_1)|Y_L|^2 + B_L Y_{01} + Y_{01}^2\tan\beta l_1 - Y_{01}B_L\tan^2\beta l_1}{(Y_{01} - B_L\tan\beta l_1)^2 + (G_L\tan\beta l_1)^2} \quad (5b)$$

Using Eqs. (4a) and (5a) gives the length $\beta l_1$:

$$\beta l_1 = \tan^{-1}\left\{\frac{Y_{03}B_L \pm \{(Y_{03}B_L)^2 - Y_{01}(G_L - Y_{03})[G_L Y_{01} - (Y_{03}/Y_{01})|Y_L|^2]\}^{1/2}}{[(Y_{03}/Y_{01})|Y_L|^2 - G_L Y_{01}]}\right\} \quad (6)$$

Then the length $\beta l_2$ follows from Eq. (4b):

$$\beta l_2 = \cot^{-1}\left(\frac{B_{in}}{Y_{02}}\right) \quad \text{(shorted case)} \tag{7a}$$

$$\beta l_2 = \tan^{-1}\left(\frac{-B_{in}}{Y_{02}}\right) \quad \text{(open case)} \tag{7b}$$

When the radical in Eq. (6) is complex, no solution is possible; however, we know that if $Y_{01} = Y_{03}$, a pair of solutions always exists. In cases where the length $\beta l_1$ is negative, just add $\lambda/2$ to it.

**7.3** One technique that may be useful to match an arbitrary load $Z_L = R_L + jX_L$ is to start by placing a line with $Z_0 = \sqrt{R_L^2 + X_L^2}$ in front of it. For this value of line impedance, the $|\Gamma|$ of this load is a minimum. Choose the length $d$ such that $\bar{Z}_{in}$ is purely real (a length less than $\lambda/4$ is always possible). Then use a single or multisection transformer to complete the match. For example, in Figure 7.56

$$|\Gamma_L|^2 = \Gamma_L \Gamma_L^* = \frac{(R_L - Z_0)^2 + X_L^2}{(R_L + Z_0)^2 + X_L^2}$$

Then $\partial |\Gamma_L|^2 / \partial Z_0 = 0$ implies $Z_0 = \sqrt{R_L^2 + X_L^2}$, or $|\Gamma_L|$ is a minimum under this condition. For example, let $Z_L = 30 + j70\,\Omega$. Thus, $Z_0 = 76.2\,\Omega$ and $|\Gamma_L| = 0.659$, so

$$\bar{Z}_L = 0.394 + j0.919$$

We can rotate this 0.1255 wavelengths to arrive at the real axis. Use

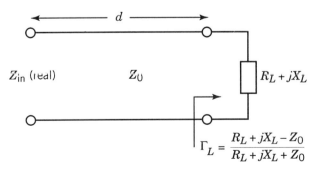

**FIGURE 7.56**

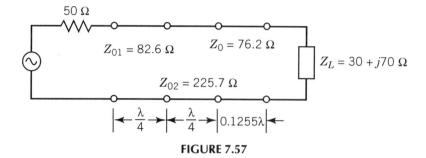

**FIGURE 7.57**

a two-section $\lambda/4$ transformer for convenience. Here, $R_L = 4.9(76.2) = 373\,\Omega$ and assume that $R_s = 50\,\Omega$.

Equation (7.41a) gives

$$Z_{01} = R_1 = R_L^{1/4} R_G^{3/4} = (4.3947)(18.803) = 82.6\,\Omega$$
$$Z_{02} = R_2 = R_L^{3/4} R_G^{1/4} = 225.7\,\Omega$$

and the final network appears as follows in Figure 7.57.

An alternative method is to move along the line a short distance to enable the placement of a shunt stub to cancel the susceptance. Then use a two-section transformer to complete the match. Here, we move a distance $0.0355\lambda$ and use a shorted stub of length $0.3355\lambda$ as shown in Figure 7.58. The real part at this plane is $262.76\,\Omega$, and the quarter wave transformers are found as before. The VSWRs for both methods are compared in Figure 7.59,

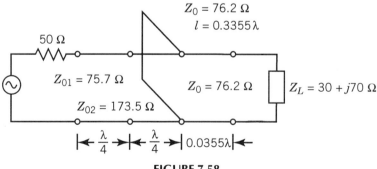

**FIGURE 7.58**

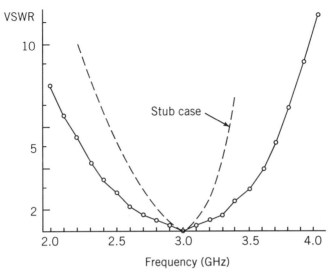

**FIGURE 7.59** VSWR responses for networks in Figures 7.57 and 7.58.

and the stub case is worse. The input impedance for the first method is also shown in Figure 7.60.

**7.4** An interesting matching method for a load that may be approximated as a conductance in parallel with a shorted stub has been given by Levy and Helszajn (Ref. 26). The method sets the insertion loss ratio (power loss ratio) $P_{LR}$ into the form:

$$P_{LR} \triangleq \frac{1}{1-|\Gamma|^2} = L = 1 + K^2 + \varepsilon^2 \left[ \frac{(1+\sin\theta_0)T_n(u) - (1-\sin\theta_0)T_{n-2}(u)}{2\sin\theta} \right]^2$$

$$u = \frac{\cos\theta}{\cos\theta_0}$$

Figure 7.61 shows a sketch of $L$ versus electrical length $\theta$.

Then equations are developed for the cases $n = 2, 3$ (one- and two-sections of quarter wave transformers). The networks appear (normalized admittances) in Figure 7.62.

Levy and Helszajn give analytic expressions for $Y_1$, $Y_2$, $Y_n$, and $G$ once the fractional bandwidth $w$ and the maximum and minimum VSWR in the band have been specified. For example, for the single-section case, the procedure is as follows.

## IMPEDANCE OR ADMITTANCE COORDINATES

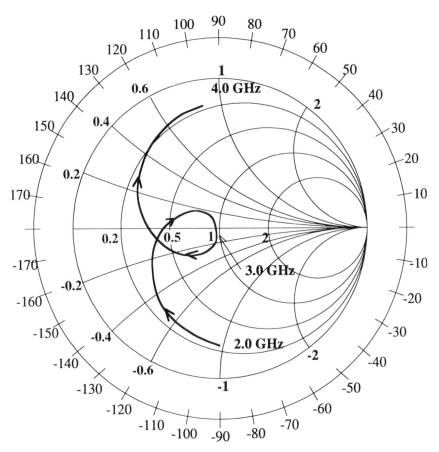

**FIGURE 7.60**  Variation of input impedance for network in Figure 7.57.

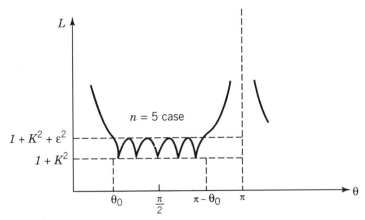

**FIGURE 7.61**  Power loss ratio versus electrical length. [Ref. 14] © 1982 IEEE.

Choose $w$, $S_{max}$, $S_{min}$. Then calculate

$$K^2 = \frac{(S_{min} - 1)^2}{4S_{min}}$$

$$\varepsilon^2 = \frac{(S_{max} - 1)^2}{4S_{max}}$$

$$\theta_0 = \frac{\pi}{4}(2 - w)$$

$$\beta = \tan^2 \theta_0 + \frac{\tan \theta_0}{\cos \theta_0}$$

$$a = K^2 + \varepsilon^2$$

$$b = 2\left(\beta\varepsilon^2 - \frac{K^2}{2}\right)$$

$$c = \beta^2 \varepsilon^2$$

$$d_0 = 2\sqrt{c}$$

$$n_1 = \sqrt{2\sqrt{(a+1)c - b + 1} - \sqrt{2\sqrt{ac - b}}}$$

$$n_2 = \sqrt{a+1} - \sqrt{a}$$

Thus,

$$Y_1 = \frac{n_1}{n_2}$$

$$Y_n = n_1 d_0$$

$$G = n_1^2$$

$$B' = \frac{\pi}{4} Y_n$$

$$Q = \frac{B'}{G}$$

Tables are provided for some cases at the end of the Levy and Helszajn article. The best bandwidth occurs when $S_{min} \simeq (1 + S_{max})/2$.

Since closed-form expressions are given, other tables may be generated quickly using a PC. Notice that one does not start with the load admittance, but rather calculates what $G$ and the characteristic admittance $Y_n$ must be to obtain such a match. The case for the single-section match given by $1 \leqslant S \leqslant 1.2$

| $w$ | $G$ | $B'$ | $Q$ | $Y_1$ | $Y_n$ |
|---|---|---|---|---|---|
| 0.667 | 2.077 | 1.336 | 0.643 | 1.579 | 1.70105 |

**366** SINGLE FREQUENCY MATCHING

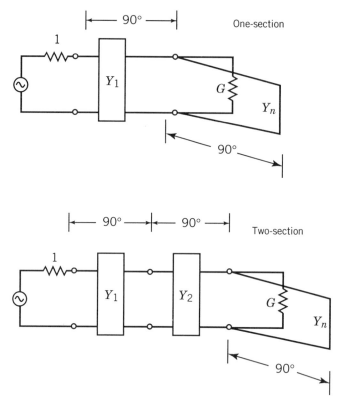

**FIGURE 7.62** Networks specified in Ref. 14. (© 1982 IEEE).

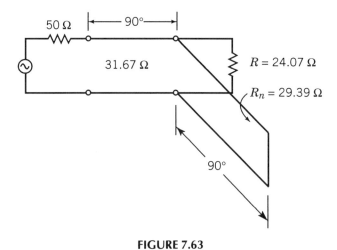

**FIGURE 7.63**

## ADMITTANCE COORDINATES
### Ex. 7.4

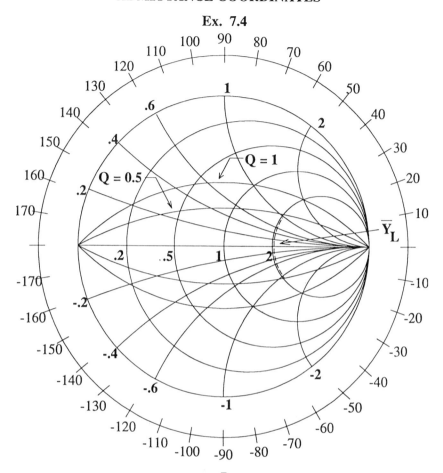

**FIGURE 7.64** Plot of $\bar{Y}_L$ along with Q-circles.

is shown in Figure 7.63 (converted to impedances). The input admittance for the load is sketched in Figure 7.64. The VSWR obtained at the input plane is shown in Figure 7.65. Notice that the $Q$ values do indeed correspond to the regions between the apropriate $Q$ circles, so quick bandwidth ranges are easily spotted.

**7.5** The method of matching given by Steinbrecher (Ref. 19) is interesting, and it is applied to a $Z = 30 + j70\,\Omega$ load shown below. Observe that the first line $Z_{01}$ is the one used to minimize the reflection coefficient as discussed in Problem 7.3. The response over the 2 to 4 GHz band is in Figure 7.67.

**368**   SINGLE FREQUENCY MATCHING

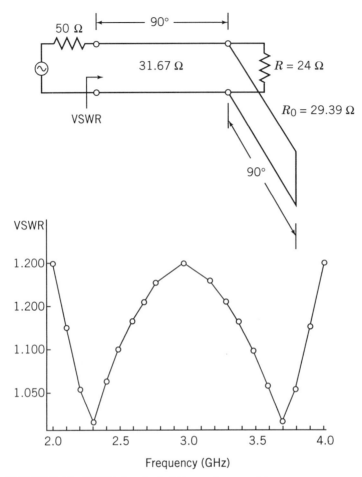

**FIGURE 7.65**   VSWR for single $\lambda/4$ transformer with appropriate stub.

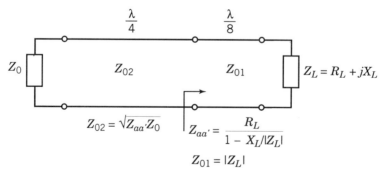

**FIGURE 7.66**   A matching scheme described in Ref. 15. (© 1967 IEEE).

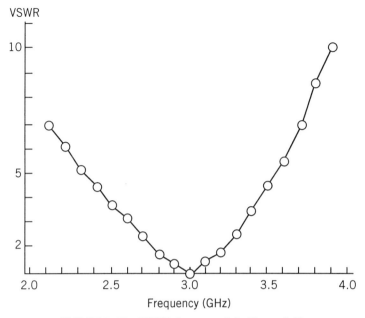

**FIGURE 7.67**  VSWR for network in Figure 7.66.

## PROBLEMS

**7.1** (a) Find both solutions to match the load in Figure P7.1 to the generator with a single shorted shunt stub:

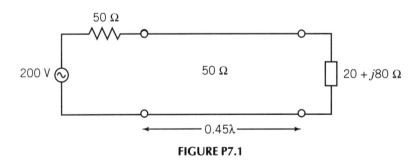

**FIGURE P7.1**

  (b) What is the percentage increase in power absorbed in the load after the match is completed?

**7.2** Match the above load to the source using a double-stub tuner. The separation between the stubs is fixed at $\lambda/8$. You may insert it anywhere on the line.

# SINGLE FREQUENCY MATCHING

**7.3** Match the load in Problem 7.1 above, using a single and double $\lambda/4$ transformer. The source and any line sections are $50\,\Omega$.

**7.4** Use a short-step transformer to match the load $Z_L = 35$ to $50\,\Omega$.

**7.5** Match the load $Z_L = 20 + j80\,\Omega$ to $50\,\Omega$ using an $L$-match. What is the 3-dB bandwidth. Let the center frequency be 2 GHz.

**7.6** Determine analytically the lengths $l_1$ and $l_2$ ($l_2$ is the stub length) for the stub match in Problem 7.1 above. Find both solutions.

**7.7** Assume a matching scheme as shown in Figure P7.7

If the stubs have a fixed length, $l = \lambda/16$, how are $Z_L$ and $d$ related?

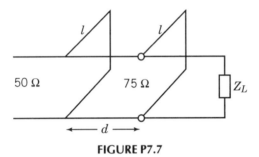

**FIGURE P7.7**

**7.8** Match $35\,\Omega$ to $50\,\Omega$ using a three-step Chebychev transformer. Match between 1 and 3 GHz.

**7.9** Summarize the procedure for designing a double-stub tuner.

**7.10** Sketch the standing wave envelopes on all TLs for the network in Figure P7.10. What is the RL at the source?

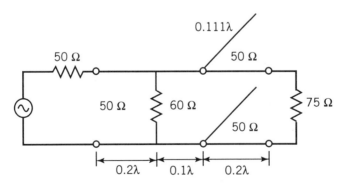

**FIGURE P7.10**

## REFERENCES

1. Young, L., "Tables for Cascaded Homogeneous Quarter-Wave Transformers," *IRE Trans. Microwave Theory Tech.*, *MTT-7*, April 1959; pp. 233–237, and *MTT-8*, 1960, pp. 243–244.
2. Bowman, D. F., *Antenna Engineering Handbook*, 2nd ed., R. Johnson and H. Jasik, eds., McGraw-Hill, New York, 1984, Chap. 43.
3. Saad, T., *Microwave Engineer's Handbook*, Artech House, Dedham, Mass. 1971, 17.
4. Collin, R. E., *Foundations for Microwave Engineering*, McGraw-Hill, New York, 1966.
5. Collin, R. E., "Theory and Design of Wide Band Multisection Quarter-Wave Transformers," *Proc. IRE*, *43*, Feb. 1955, pp. 179–185.
6. Stegun, R. J., "Simplified Determination of Impedances of Chebyshev Transformers," *IEEE Trans. Microwave Theory Tech.*, *MTT-14*, July 1966, p. 354.
7. Udelson, B. J., and McDonald, D. W., "The Design of Two-Section Transformers Having Variable Length," *Microwave J.*, Vol. 9, April 1966, pp. 68–70.
8. Hatcher, B. R., "Nomogram for Transmission Line Transformers," *Microwave J.*, Vol. 5, Nov. 1962, pp. 110–111.
9. Matthaei, G. L., "Short-Step Chebyshev Impedance Transformers," *IEEE Trans. Microwave Theory Tech.*, *MTT-14*, Aug. 1966, pp. 372–383.
10. Victor, A., "Calculator Aided Stub Matching," *RF Design*, Vol. 6, May/June 1983, pp. 30–42, corrections in same, July/Aug., p. 8, and Sept./Oct., Letters to Editor.
11. Remillard, W. J., "A Computer Program for Conjugate Matching," *Microwave J.*, Vol. 31, Dec. 1988, pp. 119–124.
12. Middelveen, R., "Double-Stub Tuning Made Easy by HP 41 Program," *Microwaves and RF*, Vol. 25, July 1986, pp. 113–116.
13. Remillard, W. J., "FORTRAN Program Solves Double-Stub Tuner Problem," *Microwaves*, Vol 17, Oct. 1978, pp. 74–76.
14. Levy, R. and Helszajn, J., "Specific Equations for One and Two Section Quarter-Wave Matching Networks for Stub-Resistor Loads." *IEEE Trans. Microwave Theory Tech.*, *MTT-30*, Jan. 1982, pp. 55–62.
15. Steinbrecher, D. H., "An Interesting Impedance Matching Network," *IEEE Trans. Microwave Theory Tech.*, *MTT-15*, June 1967, p. 382.

## BIBLIOGRAPHY

Ajayi, O. O., and Ajose, S. O., "Semi-Analytical Method for Computing Input Impedance of General Nonuniform Transmission Line," *IEE Proc. (London)*, 135 (4), Aug. 1988, pp. 269–271, Part H.

Becciolini, B., "Impedance Matching Networks Applied to RF Power Transistors," Motorola Application Note AN-721.

Bode, H., *Network Analysis and Feedback Amplifier Design*, Van Nostrand Reinhold, New York, 1945.

Boulouard, A., "Lumped Element Phase Shifting and Matching Networks," *RF Design*, Vol. 11, July 1988, pp. 58–60.

Bowick, C., *RF Circuit Design*, Howard Sams & Co., Carmel, Ind. 1982.

Burwasser, A. J., "TI-59 Program Computes Values for 14 Matching Networks," *RF Design*, Nov./Dec. 1983, pp. 12–27.

Burwasser, A. J. and Bossaller, E. F., "Simple Bandpass Filters, a Review of Fundamentals and a BASIC Program," *RF Design*, Vol. 9, March 1986, pp. 46–54. See errata in same, June 1985, p. 13, Oct. 1985, p. 14, and July 1986, pp. 12–13.

Chakmanian, J., "Control VSWR Bandwidth in T-Section Networks," *Microwaves*, Vol 20. July 1981, pp. 87–94.

Chen, W. K., "Explicit Formulas for the Synthesis of Optimum Broad-Band Impedance Matching Networks," *IEEE Trans. Circuits Systems*, CAS-24, April 1977, pp. 157–169.

Cheng, D. K. and Liang, C. H., "Computer Solution of Double-Stub Impedance-Matching Problems," *IEEE Trans. Education*, E-25, Nov. 1982, pp. 120–124.

Crystal, E. G., "Tables of Maximally Flat Impedance-Transforming Networks of Low-Pass-Filter Form," *IEEE Trans. Microwave Theory Tech.*, MTT-13, Sept. 1965, pp. 693–695.

Fano, R. M., "Theoretical Limitations on the Broadband Matching of Arbitrary Impedances," *J. Franklin Institute*, 249, Jan. 1950, pp. 57–83, and Feb. 1950, pp. 139–154.

Gonord, P., Kan, S., and Ruaud, J., "The Q Factor of Microstrip Matching Networks," *RF Design*, Vol. 9, Sept. 1986, pp. 45–48.

Gonzalez, G., *Microwave Transistor Amplifiers, Analysis and Design*, Prentice-Hall, Englewood Cliffs, NJ, 1984.

LaPenn, A. J., "Basic Program Computes Values for 14 Matching Networks," *RF Design*, Vol. 8, April 1985, pp. 44–47. See erraa in same, June 1985, p. 12, and Oct. 1985, p. 14.

Lefeuvre, S. and Matheau, J. C., "Impedance Transformation by Cascaded Lossless Line Sections," *IEEE Trans. Antennas Propagation*, AT-12, Sept. 1964, pp. 640–642.

Lev, J., "Calculator Program Finds Fano BW," *Microwaves and RF*, Vol. 24, Sept. 1985, pp. 153–155.

Levy, R., "Synthesis of Mixed Lumped and Distributed Impedance-Transforming Filters," *IEEE Trans. Microwave Theory Tech.*, MTT-20, March 1972, pp. 223–233.

Levy, R., "Explicit Formulas for Chebyshev Impedance-Matching Networks, Filters, and Interstages," *Proc. IEE (London)*, 111, June 1964, pp. 1099–1106.

Mathur, S. P. and Sinha, A. K., "Design of Microstrip Exponentially Tapered Lines to Match Helical Antenna to Standard Coaxial Transmission Lines," *IEE Proc. (London)*, 135(4), Aug. 1988, p. 272–274, Part H.

Matthaei, G. L., "Synthesis of Tchebycheff Impedance-Matching Networks, Filters, and Interstages," *IRE Trans. Circuit Theory*, CT-3, Sept. 1956, pp. 163–172.

Matthaei, G. L., Young, L., and Jones, E. M. T., *Microwave Filters and Impedance Matching Structures*, McGraw-Hill, New York, 1964.

Matthaei, G. L., "Tables of Chebyshev Impedance-transforming Networks of Low-Pass Filter Form," *Proc. IEEE*, Vol 52, Aug. 1964, pp. 939–963.

Novak, S., "CAD Amplifier Matching with Microstrip Lines," *RF Design*, Vol. 11, June 1988, pp. 43–53.

Novak, S., "CAD for Lumped Element Matching Circuits," *RF Design*, Vol. 12, Feb. 1989, pp. 102–108.

Ramachandran, V., "Design Charts of an Exponential Transmission Line for Impedance Matching," *IEEE Trans. Circuit Theory*, Dec. 1963, pp. 516–520.

Rosloniec, S., "Design of Transmission Line Matching Circuits," *RF Design*, Vol. 13, Feb. 1990, pp. 52–56.

Sifri, J., "Matching Techique Yields Optimum LNA Performance," *Microwaves and RF*, Vol. 25, Feb. 1986, pp. 87–90.

Solymar, L., "Some Notes on the Optimum Design of Stepped Transmission-Line Transformers," *IEEE Trans. Microwave Theory Tech., MTT-6*, Oct. 1958, pp. 374–378.

Termin, F. E., *Radio Engineers Handbook*, McGraw-Hill, New York, 1943, p. 212.

van der Walt, P. W., "Short-Step-Stub Chebyshev Impedance Transformers," *IEEE Trans. Microwave Theory Tech., MTT-34*, Aug. 1986, pp. 863–868.

# Index

ABCD parameters, 222
Absolute potential, 29
Admittance:
 per unit length, 204
 chart, 268
Attenuation, 210
Attenuation constant, 209
Available power, 205, 207, 231, 238, 305

Backward crosstalk constant, 140
Bandwidth, 336
Bandwidth of single transformer, 337
Barretter, 275
Battery of dc circuit theory, 16
Binomial transformer, 332, 344
Bounce diagram, 71, 75

Capacities in multiconductor systems, 154
Characteristic resistance, 14
Characteristic:
 complex impedance, 294
 impedance, 24, 209
Charged line, 90
Chebychev:
 polynomial, 346
 transformer, 344
Coaxial line, 10, 38
Coefficient:
 of electrostatic capacitance, 155
 of electrostatic induction, 155
 of potential matrix, 154
Complex:
 admittance plane, 265
 amplitude, 196
 impedance plane, 248
 number, 203

phasor-vector, 4
Poynting vector, 50
Reflection coefficient plane, 215, 248
Conductivity, 3
Conformal transformation, 252
Conservative field, 29
Constant:
 conductance circles, 272
 power circles, 244
 resistance circles, 272
Continuity equation, 1, 36
Conventional capacitance matrix, 158
Counter streaming waves, 187, 241
Contours of constant power, 238
Crosstalk, 132
Crystal diode, 275
Current:
 density, 1
 reflection coefficient, 68
 reflection coefficient plane, 265
Cyclotron frequency, 51

Damping, 215
Decibels (dB), 211
Denormalized, 254
Device under test (DUT), 164
Difference equation, 93
Diode, 122, 275
Direct capacities, 158
Displacement current, 13, 33
Distortionless condition, 209
Distributed:
 capacitance, 13
 circuit analysis, 6
 inductance, 13
Dividers, 256

375

# INDEX

Double-stub tuner, 321, 359
Down-converted, 278
Duals, 83

ECL circuit, 172
Electric field intensity, 1
Electric polarization, 1
Electrical length, 213
Electromotance, 29
Electromotive forces, 29
Emf:
  definition, 27
  producing field, 7
Energy of the fields, 4
Equivalent load planes, 275
Even mode, 149
Even mode propagation constant, 162

Faraday's Law, 7, 34
Ferrite, 50
Fifty ohm (50Ω) chart, 300
Forward crosstalk constant, 140

Geometric series, 199
Graphical methods, 175
Guided waves, 8

Hall voltage, 33
High-resistivity semiconductor, 37
Higher-order mode, 48

Ideal simple medium, 2
Immitance planes, 272
Impedance, 199
Input:
  admittance, 265
  impedance, 199, 248
Insertion loss ratio, 363
Internal inductance, 49
Intrinsic wave impedance, 19
Inverters, 181

Jump on Smith Chart, 318

Kirchoff's:
  current law, 7
  voltage law, 7

Ladder networks, 270
Laplace transform, 84
Laplace's equation, 17
Lenz's law, 33, 143
Line attenuation factor, 287
Load:
  mismatch loss, 288

reflection coefficient, 197
Lorentz:
  force equation, 2
  transformations, 260
Loss:
  (in dB) scales, 286
  magnetization, 4
  polarization, 4
  tangent, 5
Low loss line, 280
L-match, 349

Magnetic flux density, 1
Magnetization, 1, 3
Magnetization:
  frequency, 51
  loss, 5
Match, 17
Matching, 304
Matching bandwidth, 328
Matrix formalization, 133
Maximally flat transformer, 332, 333, 334
Maxwell capacitance matrix, 154
Maxwell's equations, 2
Mechanical dividers, 255
Microstrip, 170
Modified Maxwell capacitance matrix, 156
Motional emf, 31
Multisection transformers, 329
Mutual capacity, 132, 136

Nepers (Np), 210
Network analyzer, 248
Nonlinear, 175
Normalized impedance, 250

Odd mode, 150
Odd mode propagation constant, 162
Ohm's law, 3, 63
Ohmic losses, 284
One-dB (1-dB) step scale, 287
Open two-wire line, 10
Oscilloscope trace, 75

Peak values, 206
Periphery, 260
Permeability, 1
Permittivity, 1
Phase:
  angle, 10
  definition, 9
  velocity, 10, 12, 62
Phasor domain, 195
Polarization loss, 5
Potential difference, 27

Power, 202
Power:
 factor, 206
 factor angle, 206
 in the fields, 4
 reflection coefficient, 205
 waves, 209
Poynting vector, 5, 15, 18
Precharged line, 108
Programs (computer), 359
Propagation constant, 209
Pulse forming, 90

Quality factor Q, 49, 350
Q-circles, 294, 300, 367
Quarter wave transformers, 327

Radially scaled parameters, 286
Radiation, 7
Receiving end, 87
Reflection coefficient at any plane, 200, 249
Reflection loss, 283
Relative permeability, 3
Relative permittivity, 3
Return concept, 13
Return loss, 283
rms, 206

Saturation magnetization, 51
Scalar potential, 17, 28, 33
Self-resonance, 7
Semiconductors, 33
Short step transformer, 342
Single-stub tuner, 359
Sinusoidal steady state, 187
Skin effect, 37
Slotted line, 273
Smith chart, 247
Smith & Jones chart, 270
Smith, P. H. 263
Source reflection coefficient, 70

Speed of light, 219
Square law devices, 275
Standing wave:
 definition, 200
 loss coefficient, 286
 loss factor, 283
Stored energy, 5
Stub:
 series, 228, 319
 shorted, 318
 shunt, 394
Superposition, 64
Surface currents, 34, 51
Surge impedance, 15

Tandem lines, 99
Target circle, 351, 353
TE mode, 48
Thevenin source, 177
Three dB (3-dB) pad, 293
Time domain reflectometer, 164
Transmission coefficient, 99, 125, 126, 331
Transmission loss, 281, 286, 288
Transmission loss coefficient, 286
Transmission lines, 6, 10
Transverse electromagnetic wave (TEM), 9, 10, 17

Vector potential, 28, 33
Velocity field, 2
Voltage, 15, 27
Voltage reflection coefficient, 68
Voltage wave, 18
Voltage standing wave ratio (VSWR), 190, 201

Wave equation, 45
Wave machines, 190
Wavelength, 47
Wavelength scales, 260, 261
Weak coupling approximation, 135